Modification and Blending of Synthetic and Natural Macromolecules

NATO Science Series

A Series presenting the results of scientific meetings supported under the NATO Science Programme.

The Series is published by IOS Press, Amsterdam, and Kluwer Academic Publishers in conjunction with the NATO Scientific Affairs Division

Sub-Series

I. Life and Behavioural Sciences	IOS Press
II. Mathematics, Physics and Chemistry	Kluwer Academic Publishers
III. Computer and Systems Science	IOS Press
IV. Earth and Environmental Sciences	Kluwer Academic Publishers
V. Science and Technology Policy	IOS Press

The NATO Science Series continues the series of books published formerly as the NATO ASI Series.

The NATO Science Programme offers support for collaboration in civil science between scientists of countries of the Euro-Atlantic Partnership Council. The types of scientific meeting generally supported are "Advanced Study Institutes" and "Advanced Research Workshops", although other types of meeting are supported from time to time. The NATO Science Series collects together the results of these meetings. The meetings are co-organized bij scientists from NATO countries and scientists from NATO's Partner countries – countries of the CIS and Central and Eastern Europe.

Advanced Study Institutes are high-level tutorial courses offering in-depth study of latest advances in a field.
Advanced Research Workshops are expert meetings aimed at critical assessment of a field, and identification of directions for future action.

As a consequence of the restructuring of the NATO Science Programme in 1999, the NATO Science Series has been re-organised and there are currently Five Sub-series as noted above. Please consult the following web sites for information on previous volumes published in the Series, as well as details of earlier Sub-series.

http://www.nato.int/science
http://www.wkap.nl
http://www.iospress.nl
http://www.wtv-books.de/nato-pco.htm

Series II: Mathematics, Physics and Chemistry – Vol. 175

Modification and Blending of Synthetic and Natural Macromolecules

edited by

Francesco Ciardelli

Department of Chemistry and Industrial Chemistry,
University of Pisa,
Pisa, Italy

and

Stanislaw Penczek

Centre of Molecular and Macromolecular Studies,
Polish Academy of Sciences,
Lodz, Poland

Kluwer Academic Publishers

Dordrecht / Boston / London

Published in cooperation with NATO Scientific Affairs Division

Proceedings of the NATO Advanced Study Institute on
Modification and Blending of Synthetic and Natural Macromolecules for Preparing Multiphase Structure and Functional Materials
Pisa, Italy
6-16 October 2003

A C.I.P. Catalogue record for this book is available from the Library of Congress.

ISBN 1-4020-2734-6 (PB)
ISBN 1-4020-2733-8 (HB)
ISBN 1-4020-2735-4 (e-book)

Published by Kluwer Academic Publishers,
P.O. Box 17, 3300 AA Dordrecht, The Netherlands.

Sold and distributed in North, Central and South America
by Kluwer Academic Publishers,
101 Philip Drive, Norwell, MA 02061, U.S.A.

In all other countries, sold and distributed
by Kluwer Academic Publishers,
P.O. Box 322, 3300 AH Dordrecht, The Netherlands.

Printed on acid-free paper

Printed in the Netherlands.

Contents

Preface

This book presents in an extended form the text of lectures given by 15 invited expert speakers having international reputation in the topics of the NATO ASI "*Chemical Modification and Blending of Synthetic and Natural Macromolecules for Preparing Multiphase Structural and Functional Materials: Principles, Methods and Properties* " codirected by Francesco Ciardelli of the University of Pisa (Italy) and Stanislaw Penczek of the Academy of Science Poland in Lodz.The meeting took place in Tirrenia (Pisa), Italy at the Grand Hotel Golf during 6 - 16 October, 2003 with the participation of 65 scientists from 15 different countries.

The main objectives of this ASI were:

To provide a comprehensive presentation of the experimental approaches for modifying chemical structures of well established synthetic and natural macromolecules by means of sustainable one step processes. Particular attention has been given to the melt processing and reactions in water dispersion. Related process engineering, mechanism and product analysis has also been presented and discussed.

Reactive or interactive mixing of polymers with biopolymers was reviewed, including organic and inorganic products assisted by modified macromolecules as components or compatibilizers at the interface for the easy and safe preparation of materials, with micro and nanophase morphology. These materials have varied mechanical properties for structural applications or optical/electronic properties for application in advanced devices. Preparation of inorganic fillers with controlled size and surface structure has been described.

Basic and advanced knowledge in these subjects have been transfered to younger participants. For this purpose the above reviews and presentations have been prepared by internationality known experts with proven communication capacity. The active and productive attendance of young participants has been further encouraged by tutorial activity, also focused on the spin off of international cooperative research projects in the field.

Among the several potentially interested participants, an important number was not able to attend because of the presently busy schedule of many scientists and for financial problems :indeed with the available funds only 35 participants could be supported and others were not able to rise funds from their Country. Considering this aspect and on the other side the relevant scientific success of the ASI as confirmed by most attendants , we considered useful to prepare this book following the main line adopted during the Institute in Tirrenia. On these bases the present book follows strictly the scientific programme of the ASI which had the objective to provide the participants with a modern and fundamental presentation of the problems related to the experimental and theoretical aspects concerning the preparation, characterization and blending of complex macromolecular systems. This goal was, on the evaluation of all participants, fully accomplished and we hope they will attain the same feeling from the reading of the following chapters: At the same time we think that those not able to attend can find relevant information in the present volume.

In particular the order of the chapter maintains the same succession of the ASI with the chapters that follow, "*Poly(alkylene phosphate) chain and block copolymers thereof: from models of biomacromolecules to modification of mineral particles*" (Penczek), "*Combinations of transition metal catalysts for reactor blending*" (Bianchini et al), "*Free radical processes in water dispersion: new developments*" (Vairon et al), "*Functionalization of polyolefins in the melt*" (Ciardelli et al), and "*Hyperbranched architectures: synthesis-characterization-properties*" (Hadjichristidis) dealing with the preparation of complex macromolecules.

Then the successive six chapters : "*Effects of intramolecular composition and topology on intermolecular structure and bulk properties of complex polymer systems: experimental characterization and computer simulation*" *(Pakula)*, "*Preparation of nanocomposites with well-defined organic (CO)polymers*" *(Matyjasewski)*, "*Role of interfaces in multi-component polymer systems and biocomposites*" *(Marosi et al)*, "*Reactive processing of polymer blends using reactive compatibilisation and dynamic crosslinking: phase morphology control and microstructure-property relations*" (Groeninckx et al), "*Processing for ultimate properties*" (Lemstra), and " *Development and. processing. Of thermoplastic starch materials*" (Stepto) discuss the structure, preparation and properties of multiphase materials from both synthetic and natula macromolecules.

Theoretical aspects about reactivity and structure design of macromolecular systems are presented in the chapter "*Theoretical considerations on the reactions in polymer blends*" (Platé) and, "*Functional copolymer macromolecules: design, characterization and properties*" (Khokhlov) respectively.

Finally an example of methodology application is offered by the chapter "*High performance engineering thermoplastics via reactive compatibilization*" (Paul), while the last chapter "*Characterization of complex systems by fluorescence spectroscopy*" (Slomkowski) reports about a sophisticated approach to structural analysis

The Editors
Francesco Ciardelli and Stanislaw Penczek

Acknowledgements

The Editors wish to acknowledge the assistance of high quality provided by Professor Giacomo Ruggeri for the revision of the single chapters, their organization in this volume and for keeping the contacts with authors and publisher contact person Ms. Wil Bruins. Also they acknowledge the skilled and dedicate work by Mrs. Beatrice Pistoresi.
The sponsorship provided by the following Istitutions and Companies is also gratefully acknowledge: NATO Scientific Affair Division, University of Pisa (DCCI), Belotti Instrument.

Contributors

AGLIETTO MAURO, Dipartimento di Chimica e Chimica Industriale, Università di Pisa, Via Risorgimento 35, 56126 Pisa (Italy)

BERTALAN GYORGY, Department of Organic Chemical Technology, Budapest University of Technology and Economics 1111 Budapest, Megyetem rkp. 3, Hungary

BIANCHINI CLAUDIO, ICCOM-CNR, Area della Ricerca CNR ,via Madonna del Piano snc, I 50019 Sesto Fiorentino (Italy)

BURGUIERE CARINE, Unité de Chimie des Polymères – UMR 7610, Université P. et M. Curie, 4 Place Jussieu, 75005 Paris, France

CHARLEUX BERNADETTE, Unité de Chimie des Polymères – UMR 7610, Université P. et M. Curie, 4 Place Jussieu, 75005 Paris, France

CIARDELLI FRANCESCO, Dipartimento di Chimica e Chimica Industriale, Università di Pisa, Via Risorgimento 35, 56126 Pisa (Italy); ICCOM-CNR Pisa-Station c/o University of Pisa Via Risorgimento 35 I 50126 Pisa - Italy

COIAI SERENA, Scuola Normale Superiore, Piazza dei Cavalieri, Pisa, Italy

COLTELLI MARIA BEATRICE, Dipartimento di Chimica e Chimica Industriale, Università di Pisa, Via Risorgimento 35, 56126 Pisa (Italy)

FARCET C., Unité de Chimie des Polymères – UMR 7610, Université P. et M. Curie, 4 Place Jussieu, 75005 Paris, France

GROENINCKX GABRIEL, Catholic University of Leuven, Department of Chemistry, Division of Polymer Chemistry, Laboratory of Macromolecular Structural Chemistry, Celestijnenlaan, 200 F, B-3001 Heverlee-Leuven, Belgium

HADJICHRISTIDIS NIKOS, Chemistry Department, University of Athens, Panepistimiopolis Zografou, 15771, Athens (Greece)

HARRATS CHAREF, Catholic University of Leuven, Department of Chemistry, Division of Polymer Chemistry, Laboratory of Macromolecular Structural Chemistry, Celestijnenlaan, 200 F, B-3001 Heverlee-Leuven, Belgium

IATROU HERMIS, Chemistry Department, University of Athens, Panepistimiopolis Zografou, 15771, Athens (Greece)

KHOKHLOV ALEXEI R., Physics Department, Moscow State University, 119992 Moscow, Russia

KUDRYAVTSEV YAROSLAV V., A.V. Topchiev Institute of Petrochemical Synthesis, Russian Academy of Sciences Leninsky prosp. 29, Moscow B-71, 119991, Russia

LITMANOVICH ARKADY D., A.V. Topchiev Institute of Petrochemical Synthesis, Russian Academy of Sciences Leninsky prosp. 29, Moscow B-71, 119991, Russia

MAROSI GYORGY J., Department of Organic Chemical Technology, Budapest University of Technology and Economics 1111 Budapest, Megyetem rkp. 3, Hungary
MATYJASZEWSKI KRZYSZTOF, Center for Macromolecular Engineering, Department of Chemistry, Carnegie Mellon University, 4400 Fifth Avenue, Pittsburgh, PA 1521
MILLER HAMISH, ICCOM-CNR, Area della Ricerca CNR ,via Madonna del Piano snc, I 50019 Sesto Fiorentino (Italy)
PAKULA TADEUSZ, Max Planck Institute for Polymer Research, Postfach 3148, 55021 Mainz, Germany
PASSAGLIA ELISA, ICCOM-CNR Pisa-Station c/o University of Pisa Via Risorgimento 35 I 50126 Pisa - Italy
PAUL DONALD R., Department of Chemical Engineering and Texas Materials Institute, University of Texas at Austin, Austin, Texas 78712 US
PENCZEK STANISLAW, Center of Molecular and Macromolecular Studies Polish Academy of Sciences 90-363 Lodz, Sienkiewicza 112, Poland
PITSIKALIS MARINOS, Chemistry Department, University of Athens, Panepistimiopolis Zografou, 15771, Athens (Greece)
PLATE' NIKOLAY A., A.V. Topchiev Institute of Petrochemical Synthesis, Russian Academy of Sciences Leninsky prosp. 29, Moscow B-71, 119991, Russia
RUGGERI GIACOMO, Dipartimento di Chimica e Chimica Industriale, Università di Pisa, Via Risorgimento 35, 56126 Pisa (Italy)
SLOMKOWSKI STANISLAW, Center of Molecular and Macromolecular Studies, Polish Academy of Sciences, Sienkiewicza 112, 90-363 Lodz, Poland
STEPTO ROBERT F.T., Polymer Science and Technology Group, Manchester Materials Science Centre, UMIST and University of Manchester, Grosvenor Street, Manchester, M1 7HS, UK
VAIRON JEAN-PIERRE, Unité de Chimie des Polymères – UMR 7610, Université P. et M. Curie, 4 Place Jussieu, 75005 Paris, France

POLY(ALKYLENE PHOSPHATE) CHAINS AND BLOCK COPOLYMERS THEREOF

From Models of Biomacromolecules to Modification of Mineral Particles

STANISLAW PENCZEK

Center of Molecular and Macromolecular Studies Polish Academy of Sciences
90-363 Lodz, Sienkiewicza 112, Poland

Abstract: Poly(alkylene phosphate) backbones are at the basis of two important classes of biomacromolecules, nucleic and teichoic acids. Both are known to strongly bind metal cations; teichoic acids interact specifically with Ca^{2+} and Mg^{2+} cations transporting these cations in the biological milieu. This review paper describes the work of this laboratory directed towards synthesis of the backbones interaction with cations and some applications in nonbiological systems, although related to the ability to interact with cations. Thus, poly(alkylene phosphates) are described as liquid membranes and in the form of block copolymers as regulating crystal growth by interacting with cations rich surface. Moreover, poly(alkylene phosphates) function as strong acids: cationation of polyaniline (doping) leading to the intermolecular complex, in which poly(pentamethylene phosphate) specifically recognize the distance between the nitrogen atoms in polyaniline.

Key words: poly(alkylene phosphates), biomimicking polymers, liquid membranes, polyaniline, crystal growth modifiers.

1. INTRODUCTION

Polyesters based on phosphoric acid can now be prepared by using several methods elaborated in our laboratory and are well suited for modification of polymers able to interact with strong acidic groups.

Poly(alkylene phosphate) chains are related to several classes of biomacromolecules. Besides polynucleotides there are teichoic acids-either

F. Ciardelli and S. Penczek (eds.),
Modification and Blending of Synthetic and Natural Macromolecules, 1–14.

based on glycerol, namely 1,2- or 1,3-poly(glycerol phosphates) (PGP) or on cyclic sugars, like D-ribitol[1,2].

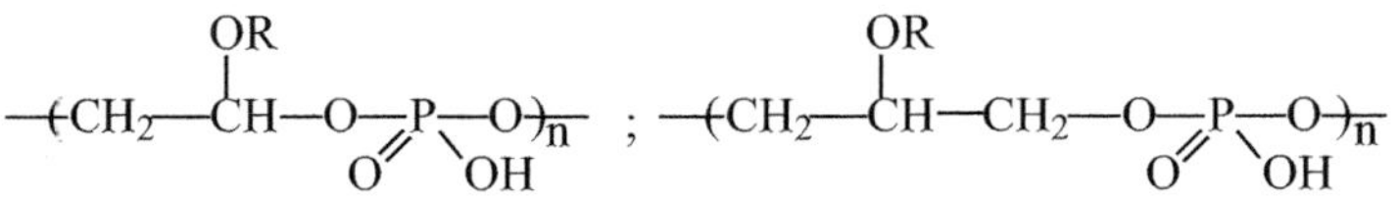

(glycerol-1,2- and 1,3-teichoic acids)

$DP_n = 25\text{-}40$

(R = alanyl, sugar)

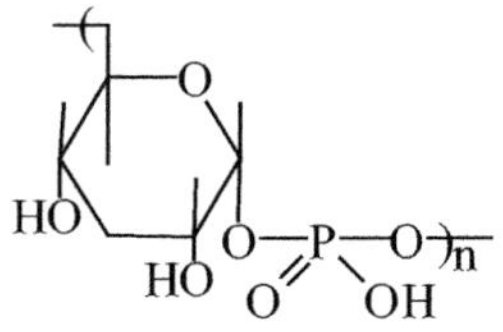

(D-ribitol-1,5-teichoic acid)

Teichoic acids (TA) are known to be responsible for the ion transport through the bacteria cell wall (wall teichoic acids) or binding Mg^{2+} on the external surface of the cytoplasmic membrane (lipoteichoic acids)[3,4]. The main physiological function of TA is to collect and preconcentrate the divalent cations, particularly magnesium ions, from the surrounding medium, and to facilitate their permeation towards the phospholipid bilayer, where their concentration is kept at the proper level. Both 1,2- and 1,3-glycerol TA could actively (i.e. against a gradient of concentration) transport Ca^{2+} or Mg^{2+} cations but it was not clear for a long time which one (i.e. 1,2- or 1,3-) is responsible for a given cation. Indeed, the TAs have mostly been isolated from bacterial sources and could not be standardized sufficiently. Magnesium ions are associated with the cell walls in two different ways[5].

The stronger bound cations are associated with two adjacent phosphate groups in TA chains as it is shown below in scheme 1.

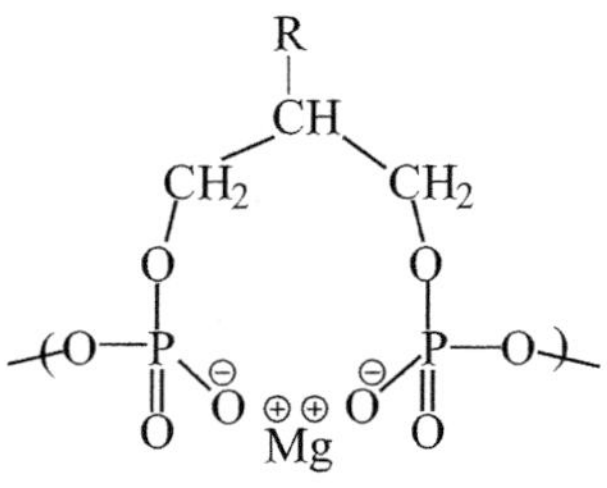

Scheme 1

At pH = 5, $K_a = 2.7{\cong}10^3$ (mol≅L^{-1}) for Mg^{2+} and glycerol TA whereas $K_a = 0.61{\cong}10^3$ (mol≅L^{-1}) for ribitol TA. This is because two adjacent units with rather structurally stiff sugar moiety are not allowing these units to rotate freely around the ester bond, as required for assuming conformation shown in scheme 1. Therefore it has become of interest to study the influence of the distance between the phosphate units on the binding efficiency in systems where ionic interactions are important.

For this review paper we selected a few systems studied in our laboratory. These are: interaction of poly(alkylene phosphates) with Na^P Ca^{2+} and Mg^{2+} cations, active transport in liquid membranes, interaction of polymers containing blocks of units with monoesters of phosphoric acid with inorganic crystals or particles, and finally interaction of poly(alkylene phosphates) with polyaniline.

2. SYNTHESIS OF MODEL MACROMOLECULES

Syntheses based on the ring-opening polymerization were described in our earlier papers[6,7,8,9].

Since our original interest was related to the biomacromolecules and their models, we first elaborated synthesis of macromolecules with repeating units with two or three carbon atoms between the phosphate groups. The corresponding polymers were thus prepared by ring-opening polymerization of the five- or six-membered cyclic esters of phosphoric acid or their derivatives. This strategy can be illustrated taking as an example synthesis of poly (1,2-glycerol phosphate). Five membered rings are strained (~5 kcal/mol) and polymerization proceed to completion. The ring strain was determined in our studies of the polymerization thermodynamics[10] and earlier by Westheimer[11] in his works on hydrolysis of the five membered esters:

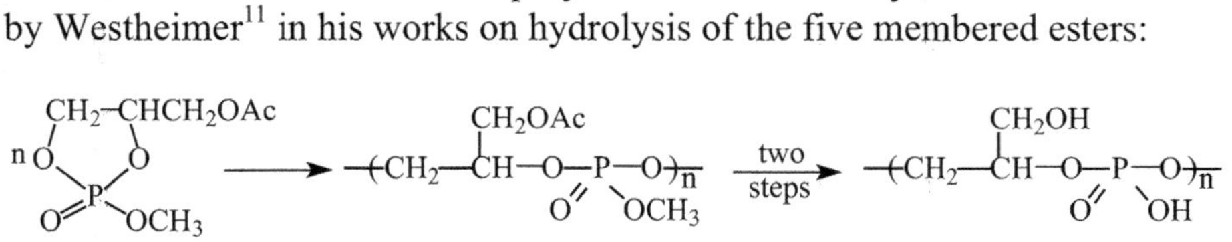

(1)

or

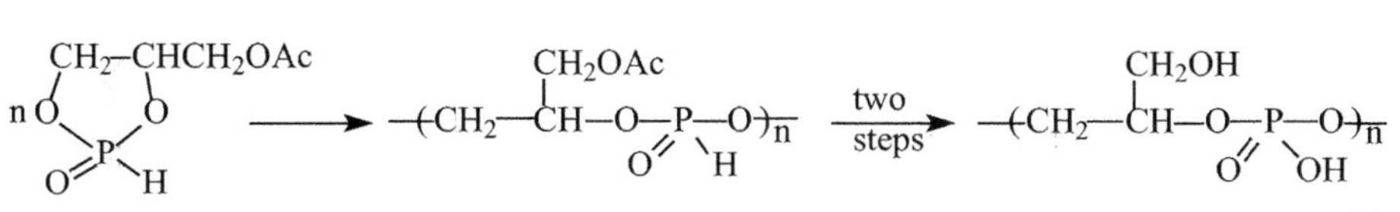

(2)

In both approaches, namely the "triesters route" and "H-phosphonate route" polymers of relatively high molar masses (>10^4) were prepared. The "H-phosphonate route" allowed also preparation of the high molar mass polymers with six atoms repeating units, like in the 1,3-glycerol phosphate.

Recently we elaborated a more general method leading to poly(alkylene phosphates), not restricted to structures related to cyclic monomers with strained rings. This method is based on the well known condensation of alcohols with H-phosphonates[12,13]:

$$\text{(excess) ROH} + (\text{R'O})_2\text{P(O)H} \rightleftharpoons (\text{RO})_2\text{P(O)H} + 2\text{R'OH} \tag{3}$$

When diol is used then polycondensation takes place and removal of an alcohol formed shifts the equilibrium to the side by the desired polymer:

$$n\,\text{HOROH} + n\,(\text{R'O})_2\text{P(O)H} \rightleftharpoons \text{HOR}\left[\text{OP(O)(H)OR}\right]_{n-1}\text{OP(O)(H)OR'} + \uparrow\text{R'OH} \tag{4}$$

This polycondensation proceeds in the presence of the catalytic amounts of sodium or potassium alcoholate. When R' is CH_3 or C_2H_5, as usually used in these reactions, high molar mass products could not be prepared, because dealkylation is accompanying the major reaction of the chain growth (e.g. for dimethyl H-phosphonate):

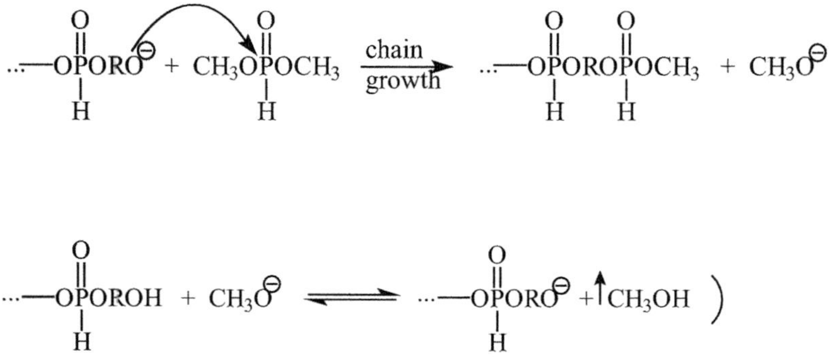

(5)

Chain termination involves "wrong direction of attack" of the growing anion chain end and involving carbon atom in place of the phosphorus atom, as it is shown for the chain growth:

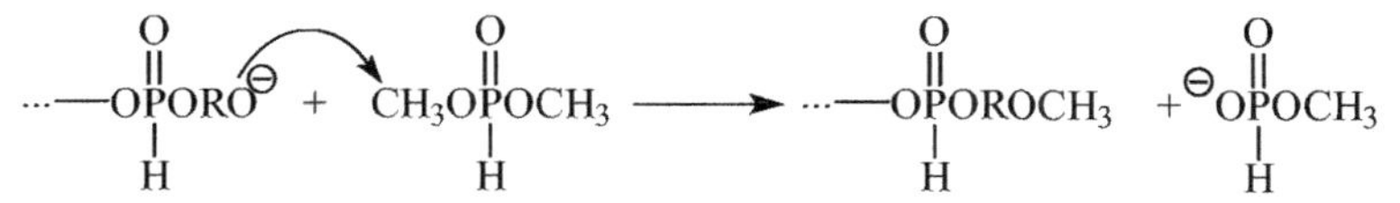

(6)

This dealkylation, well known in the phosphorus organic chemistry[14], was also described as a reason of molar masses limitation in the polycondensation of diols with dialkyl H-phosphonates[15]. Indeed, as shown in equ. (6) every dealkylation would stop the growth of the polymer molecule. We have found, however, that changing the polycondensation into the polytransesterfication, proceeding at higher temperature, allows preparing polymers with molar masses higher than $30 \cong 10^3$. Mechanism of this novel process has also been discussed in our papers[12,13].

Direct polytransesterfication was successfully used in preparation of polymers based on various glycols with exception of 1,3-propylene glycol and 1,4-butylene glycol; in these two instances cyclic compounds resulted. However, both ring opening polymerization and transesterfication gave high molar mass poly(alkylene phosphates), starting from two methylene groups in the repeating units up to twelve[16]. Besides, poly(alkylene phosphates) were prepared from oligomeric glycols, like poly(ethylene glycol) or poly(tetramethylene glycol)[17].

Dealkylation can be fully avoided if in polycondensation diphenyl H-phosphonate is used in place of the aliphatic dialkyl H-phosphonates[18]. This process is then proceeding smoothly, even at the room temperature, since the less stable aromatic ester is replaced by a more stable aliphatic one. The equilibrium is so much shifted to the side of macromolecules, that phenol formed in this process does not have to be removed during polycondensation:

$$n(C_6H_5O)_2P(O)H + nHOROH \rightleftharpoons HO[ROP(O)(H)O]_{n-1}ROP(O)(H) + C_6H_5OH \quad (7)$$

Poly(alkylene phosphates), depending on the number of atoms in repeating units, differ in their abilities to complex metal cations. This stems from different in distances between the phosphate groups along the chains, and the arrangements differing for the even and odd number of atoms in the repeating unit of the backbone.

If we assume the zig-zag conformation, then for an uneven number of atoms in repeating unit every phosphate group along the chain is exposed to the same side of the macromolecule. For an even number of repeating units only every second is directed to the same side. This is ilustrated below in scheme 2:

uneven number (e.g.5):

Scheme 2

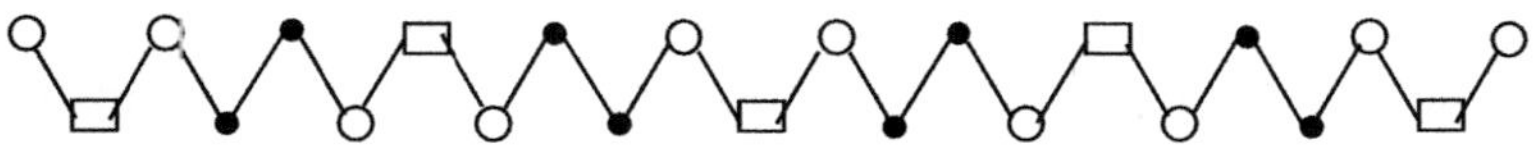

even number (e.g.6):

where □ is the P atom; the other elements of the chain are thus self-explanatory.

3. INTERACTION OF SYNTHETIC POLY(ALKYLENE PHOSPHATES) WITH METAL CATIONS

There are in principle two ways these novel macromolecules can interact with cations. Either by complexation with non-ionic units, like in poly(methylethylene phosphate), thus in a similar way as poly(ethylene oxide) or crown ethers are complexing metal cations, or by acidic functions and formation of the macromolecular salts or by using both ways.

Dipol-ion interaction

Complexation with neutral chains was studied with poly(methylethylene phosphate) and sodium cations[19]. ^{23}Na NMR was used, and the corresponding equilibrium constants were determined from the results of dynamic NMR data. Line broadening, expressed by the line width ($w_{1/2}$) allowed determination of the corresponding complexation constants K_1, K_2, K_3..., where subscripts denote the average number of repeating units responsible for complexing of one Na^{ρ} cation:

Scheme 3.

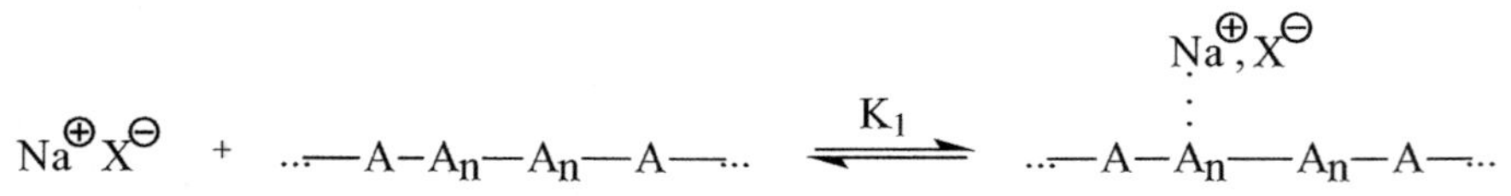

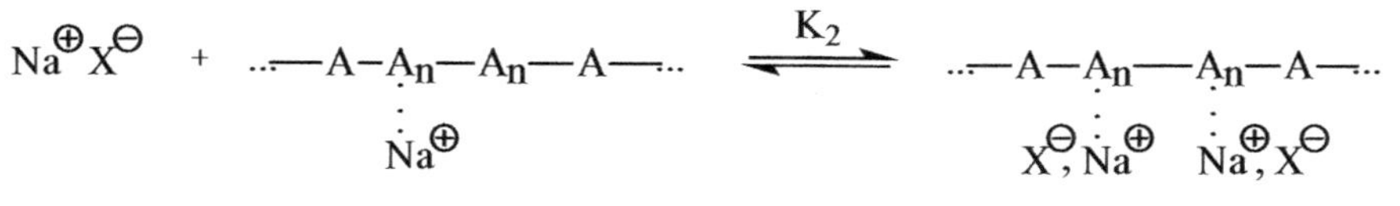

It was found that the equilibrium constants K_i increase with n, until n becomes equal to 3, thus until the structure resembles the following one:

Scheme 4

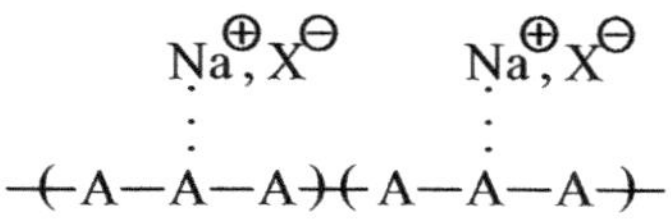

In the formulae above A = -(OP(O)(OCH$_3$)OCH$_2$CH$_2$)-

The final conclusion reached was, therefore that K_3>K_2>K_1. This is probably due to the formation of clusters, like pairs or triplets of ion-pairs.

K_3 is for Na^{ρ} close to the similar complexation constants determined for poly(ethylene oxide).

4. IONIC INTERACTIONS

As it has been indicated in the introduction, there was a long-lasting discussion related to the responsability of either 1,2- or 1,3- glycerol phosphate for active transport of Ca^{2+} or Mg^{2+} cations to the cell walls. The models prepared throughout our work allowed determination of the corresponding preferences of the 1,2- and 1,3- polymers towards magnesium and calcium cations[20]. The transport phenomena were studied in a simple apparatus with three compartments separated by solid membranes of Nafion[21]. In the first compartment (feeding phase) water solution of Ca^{2+} and Mg^{2+} of known concentrations was placed, in the middle one water solution of either polyphosphate of ethylene glycol or (in another experiment) of polyphosphate of 1,3-propylene glycol, and in the third one (stripping phase) water solution of HCl. In such a setup macromolecule of poly(alkylene phosphate) comes to the membrane, picks up mostly this cation for which it is specific, and goes to the other wall, where cations are discharged and exchanged with the membrane with H^{ρ} ions. Flame photometry is used to measure concentrations of Ca^{2+} and Mg^{2+} in the stripping phase. In this way it was shown that 1,3-polymer is more active in transporting Mg^{2+} than Ca^{2+}.

The ratio of fluxes and association constants are reflecting the corresponding preferences[20].

Table 1. Characteristics of the preference of the polyphosphates 1,2-PGP and 1,3-PGP towards magnesium ions in the competitive exchange diffusion processes.

Specification	Carrier	
	1,2-PGP	1,3-PGP
$R_1 = (S_s/S_f)_{max}$:		
Proton-coupled transport	1,65	3,70
Sodium-coupled transport	1,48	
Magnesium flux:		
$J_{Mg} \cdot 10^9/(mol \cdot cm^{-2} \cdot s^{-1})$	3,50	14,00
Calcium flux:		
$J_{Ca} \cdot 10^9/(mol \cdot cm^{-2} \cdot s^{-1})$	35,00	
Ratio of fluxes:		
$R_2 = J_{Mg}/J_{Ca}$	0,10	>>1,00

Therefore 1,3-PGP prefers Mg^{2+}, although Mg^{2+} cation is "smaller" than Ca^{2+} cation; Mg is 12th in the Periodic Table, whereas Ca is the 20th. This result agrees well with scheme 2, showing that phosphate groups in 1,3-PGP are in fact closer one to another than in 1,2-PGP. The results of these studies allowed asking the next question: can other ions be separated by these or similar structures. Of practical interest for industrial electrochemists is separation of Ni^{2+} and Co^{2+} [20]. After screening of various structures the most efficient one that was finally found is poly[poly(ethylene glycol) phosphate], where the poly(ethylene glycol) (PEG) has an average polymerization degree n = 45. We cannot offer the exact structure of the ionic complexes formed, but apparently both dipol-ionic and ionic-ionic interactions are involved, providing this remarkable specificity in interactions, (remembering, that Co is 27th and Ni is 28th in the Periodic Table). It should be added, that neither PEG alone nor the low molar mass esters of phosphoric acid show any efficiency in separation of Ni^{2+} and Co^{2+}. In Fig. 1 the kinetics of separation of Ni^{2+} and Co^{2+} by poly[poly(ethylene glycol) phosphate] is illustrated[20].

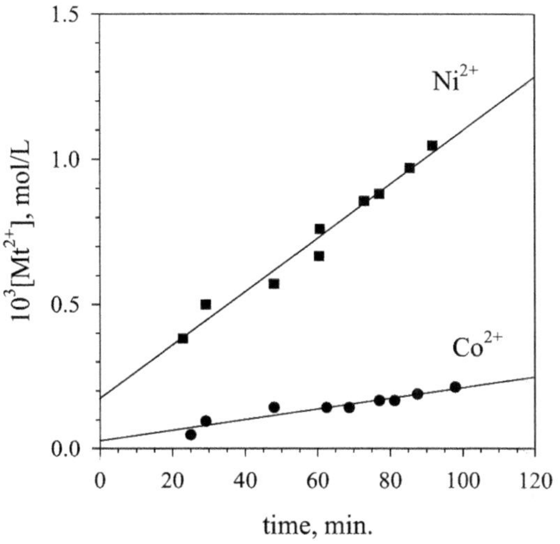

Figure 1. Competitive transport of Ni^{2+} and Co^{2+} within a liquid membrane with poly[(PEG 1000) phosphate] as a carrier.

5. IONIC-NONIONIC DIHYDROPHILIC BLOCK COPOLYMERS (WITH PHOSPHATE UNITS) AND THEIR INTERACTION WITH INORGANIC PARTICLES

Strong interaction of mono- and diesters of phosphoric acid with various cations, and particularly with Ca^{2+} and Mg^{2+} cations prompted us to explore the interaction of related polymers with inorganic particles and their influence on the inorganic crystal growth.

It has already been shown, that dihydrophilic block copolymers with ionic and nonionic blocks are successfully used in modifying of the surface of inorganic particles[23]. This modification prevents aggregation of existing particles and/or influences the crystallization process, providing crystals of the shape and size both depending on the structure and amount of the polymer used[24].

The ionic blocks used till now have mostly been based on carboxylic acids; e.g. copolymers of acrylic or methacrylic acid[24,25].

Block copolymers with ionic block with esters of phosphoric acid have been prepared for the first time in our laboratory[26]. One of the approaches used is illustrated below:

$$CH_3(OCH_2CH_2)_nO^{\ominus}Mt^{\oplus} + {}_m\,CH_2\!-\!CHCH_2OB\ (\text{epoxide, } \backslash O/) \longrightarrow CH_3(OCH_2CH_2)_n(OCH_2\underset{\displaystyle CH_2OB}{CH})_m\!- \longrightarrow$$

$$\longrightarrow CH_3(OCH_2CH_2)_n(OCH_2\underset{\displaystyle CH_2OH}{CH})_m\!- \longrightarrow CH_3(OCH_2CH_2)_n(OCH_2\underset{\displaystyle CH_2OP(O)(OH)_2}{CH})_m\!-$$

$(B = CH(CH_3)OCH_3)$

(8)

Block copolymers were prepared with various block lengths and various degrees of phosphorylation. The efficiency of these copolymers in influencing crystallization of $CaCO_3$ was then compared with efficiencies of similar copolymers, but bearing carboxy and sulphonyl groups. In figures below, taken from our paper[26], the influence of these three different kinds of copolymers on the crystallization of $CaCO_3$ is shown. It is clearly seen, that size of crystal and their structures change dramatically, when crystallization proceeds in the presence of block copolymers. The most efficient, as it follows from comparison of the figures, are block copolymer with phosphate esters.

Figure 2. SEM micrographs of $CaCO_3$ crystallized: (A) - in absence of polymeric additives and in the presence of: (B) - copolymer with sulphonyl, (C) - carboxymethyl, and (D) – phosphoryl side groups. Copolymer structures explained in the text. Extend of substitution of -CH_2OH groups the same for (B), (C), and (D), and equal to 50%.

The mechanism of interaction of inorganic crystals and/or particles and dihydrophilic block copolymers can be presented in the following way: on the surface of inorganic salts there are positively charged sites, due to the presence of metal cations; e.g. Ca^{2+}. Block copolymer having a block consisting of anions interacts with these sites and forms chemically adsorbed layer with the second, and highly hydrophilic block sticking out from the surface. The actual volume of every hydrophilic block, including the solvating molecules of water, can extend several times the actual volume of macromolecule. This is the sterically hindering layer that prevents two particles to aggregate. The major reason, as it is belived today, is such that any close enough approach would require elimination of water molecules bound already by hydrogen bonds to a given hydrophilic block, sitting on the surface of a crystal.

6. INTERACTION OF POLY(ALKYLENE PHOSPHATES) WITH POLYANILINE

Synthetic poly(alkylene phosphates) (PAP) described in the previous sections are unique among the synthetic polyanions, because the ionic groups can be located along the chain at the chosen at will distances indicated below by X_n:

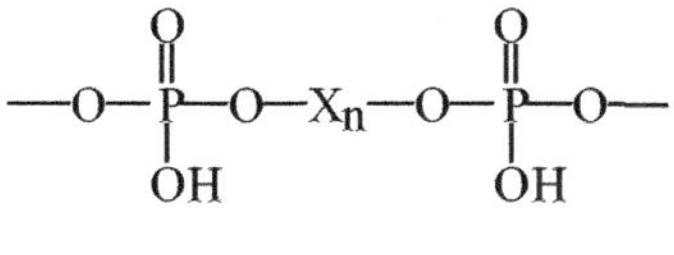

This feature makes poly(alkylene phosphates) particularly interesting for studying interaction with other polymers bearing centers rich in electrons in the chains (e.g. amino groups) at the fixed distances.

Since it is known, that polyaniline (PANI) has to be doped (e.g. protonated) in order to show increased electrical conductivity[27,28], we studied the influence of the distances between the phosphate groups in poly[(polymethylene) phosphates] ($X = CH_2$) on the conductivity of the corresponding complexes with polyaniline in order to establish to most efficient molecular fit, judged by the electrical conductivity of the complexes.

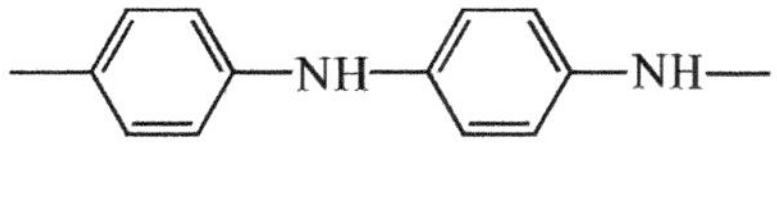

According to the UV-VIS spectra of the mixture of PANI with poly(alkylene phosphates) the corresponding complex is indeed formed as it may be judged from two new transitions (ca. 425 and 900 nm) appearing both characteristic for the protonated state of PANI[29,30]. Moreover, the presence of the isobestic points at 460 and 770 nm proves that only two optically different phases coexist (protonated and non-protonated). The extent of protonation depends on the number of methylene units (spacer) between the phosphate groups:

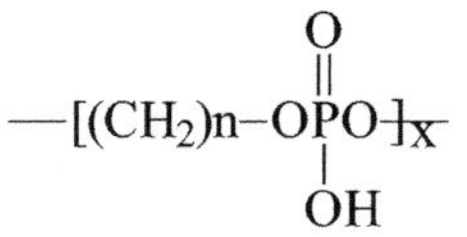

Only for n = 5 the complete protonation could be achieved.
Inspection of the molecular models have shown, that when n = 5 there is a perfect fit of the phosphate groups in PAP and amino bridges in PANI[31]. This is show below, in Fig. 3.

Figure 3. Computer model of polyaniline-poly(pentamethylene phosphate) complex. Atoms marked with following colours: H – white, C – navy blue, N – green, O – red, and P – yellow.

This observation is consistent with the result of conductivity equal to $1.5 \cdot 10^{-1}$ $S \cdot cm^{-1}$ was observed for n = 5, whereas e.g. only $6 \cdot 10^{-4}$ $S \cdot cm^{-1}$ was measured for n = 12.

A thin film of PANI doped with 1 $mol \cdot dm^{-3}$ HCL solution deprotonates almost completely after its immersion into distilled water for 0.5 h, whereas PANI/PAP films has features typical of the protonated state even after 20 h of washing with water.

ACKNOWLEDGMENTS

This review-lecture is based on several review papers and book chapters published previously with G. Lapienis, P. Klosinski, K. Kaluzynski and J. Pretula.

Partial support from KBN grant 7 T09 A 15621 is highly appreciated.

REFERENCES

1. Munson R.S. and Glaser L., Biology of Carbohydrates, Ed. V. Ginsburg, Wiley, New York 1981, vol. 1, p. 91.
2. Naumowa I.B., *Usp. Sovrem. Biol.*, **75**, 357 (1973).
3. Heptinstall S., Archibald A.R. and Baddiley J., *Nature*, **225**, 519 (1970).
4. Hughes A.H., Hancock I.C. and Baddiley J., *Biochem. J.*, **132**, 83 (1973).
5. Baddiley J., Hancock J.C. and Sherwood P.M.A., *Nature*, **243**, 43 (1973).
6. Kaluzynski K., Libiszowski J. and Penczek S., *Macromol.* **9**, 365 (1976).
7. Libiszowski J., Kaluzynski K. and Penczek S., *J. Polym. Sci. Polym. Chem. Ed.* **16**, 1275 (1978).
8. Klosinski P. and Penczek S., *Macromol.* **16**, 316 (1983).
9. Penczek S. and Klosinski P., Models of Biopolymers by Ring-Opening Polymerization, Ed. S. Penczek, CRC Press, Inc., 1990, chapter 4, pp 291-378.
10. Sosnowski S., Libiszowski J., Slomkowski S. and Penczek S., *Makromol. Chem. Rapid Commun.*, **5**, 239 (1984).
11. Westheimer H., *Acc. Chem. Res.* **70**, 1 (1968).
12. Pretula J. and Penczek S., *Makromol. Chem., Rapid Commun.*, **9**, 731 (1988).
13. Pretula J. Kaluzynski K., Szymanski R. and Penczek S., *J. Polym. Sci. Part A: Polym. Chem.*, **37**, 1365 (1999).
14. Nifanteev E.E., Chimia Gidrofosforilnych Soedin., Nauka : Moskva, 1983.
15. Vogt W. and Balasubramanian S.,*Makromol. Chem.*, **163**, 111 (1973).
16. Penczek S. and Pretula J., *Macromol.*, **26**, 2228 (1993).
17. Pretula J. and Penczek S., *Makromol. Chem.*, **191**, 671 (1990).
18. Pretula J., Kaluzynski K., Szymanski R. and Penczek S., *Macromol.*, **30**, 8172 (1997).
19. Szymanski R. and Penczek S., *Makromol. Chem.*, **194**, 1645 (1993).
20. Wodzki R. and Klosinski P., *Makromol. Chem.*, **191**, 921 (1990).
21. Wodzki R. and Kaluzynski K., *Makromol. Chem.*, **190**, 107 (1989).
22. Narebska A., Wodzki R. and Wyszynska A., *Makromol. Chem.*, **190**, 1501 (1989).
23. Ramachandran V.S., Malhotra V.M., Jolicoeur C. and Spiratos N., Superplasticizers: Properties and Applications in Concrete, Ed. Minister of Public Works and Government Services, Canada, 1998, chapter 7, pp 193-221.
24. Sedlak M., Antonietti M. and Cölfen H., *Makromol. Chem. Phys.*, **199**, 247 (1998).
25. Marentette J.M., Norwig J., Stoeckelmann E., Meyer W.H. and Wegner G., *Adv. Mater.*, **8**, 9 (1997).
26. Kaluzynski K., Pretula J., Lapienis G., Basko M., Bartczak Z., Dworak A. and Penczek S., *J. Polym. Sci.*, submitted.
27. Ginder J.M. and Epstein A.J., *Phys. Rev. Lett.*, **64**, 1184 (1990).
28. Ginder J.M. and Epstein A.J., *Phys. Rev. Lett.*, **B41**, 674 (1990).
29. Kulszewicz-Bajer I., Pretula J. and Pron A., *J. Chem. Soc., Chem. Commun.*, 642 (1994).
30. Kulszewicz-Bajer I., Sobczak J., Hasik M. and Pretula J., *Polymer*, **37**, 25 (1996).

31. Pretula J., Doctoral Thesis, Center of Molecular and Macromolecular Studies, Polish Academy of Sciences, Lodz, 1999.

COMBINATIONS OF TRANSITION METAL CATALYSTS FOR REACTOR BLENDING

CLAUDIO BIANCHINI[1], HAMISH MILLER[1], and FRANCESCO CIARDELLI[2]

[1] ICCOM-CNR, Area della Ricerca CNR ,via Madonna del Piano snc, I 50019 Sesto Fiorentino (FI) – Italy; [2] ICCOM-CNR Pisa-Station c/o University of Pisa Via Risorgimento 35 I 50126 Pisa - Italy

Abstract: The formation of complex polyolefins based materials is described based on the combination of different transition metal catalysts in a single reactor (Reactor Blending). The paper describes how polymer properties can be controlled by selecting different homogeneous and heterogeneous catalyst and combining them in different ratio while at the same time varying polymerization conditions. For clarity sake and rational organization of the many contribution reported in the recent literature, the use of homogeneous and heterogeneous catalysts is discussed in two separate sections. In each of them the processes starting from ethylene and aimed to modulate MWD and chain branching, or propylene aimed to type and distribution of stereoregularity, are also reported in distinct sections.

Key words: Reactor blending, transition metal catalysts, ethylene polymerization, propylene polymerization, stereoblock polypropylene, chain branching, molecular weight distribution.

1. INTRODUCTION

This abstract summarizes recent advances in the combination of different transition metal catalysts in a single reactor for the formation of new polyolefinic materials. The current search for new polyolefinic materials is guided by the need of better performance materials with numerous industrial applications. In polymer processing and applications, molecular weight (MW), molecular weight distribution (MWD) and branching represent basic

F. Ciardelli and S. Penczek (eds.),
Modification and Blending of Synthetic and Natural Macromolecules, 15–38.

characteristics, which serve as major determinants of polymer properties. MW largely relates to the mechanical properties, while MWD is responsible for rheological properties. Although high-MW polyethylene has superior physical properties it is difficult to process. On the other hand, an increase in MWD tends to improve the flow behavior at high shear rate, which is important for blow molding and extrusion techniques.

Traditional methods for controlling MW and MWD include:

i. Polymers of different MWs can be physically blended. However, this solution faces problems of energy consumption, operational costs and miscibility limitations.
ii. A series of reactors (multi-stage reactors) can be used each one run under different polymerization reactions conditions. This method can be expensive, cumbersome and time-consuming.
iii. The third method utilizes the variation of operation conditions, such as temperature, comonomer concentration and hydrogen pressure (non-steady-state polymerization), in a single reactor during polymerization.

A fourth method, known as multi-component polymerization, involves combining two or more types of catalysts to produce polymers with different MW and MWD in a single reactor. The advantages of this method, which is capable of producing more easily polymers with good properties by using just a single polymerization process, have achieved considerable attention by industrial laboratories as can be seen by the number of patents issued in recent years[1].

Like MW and MWD, side branches in the polyolefin influence remarkably material properties such as melting temperature and crystallinity. For example, linear low density polyethylene (LLDPE) is more easily processed than high-density polyethylene (HDPE) for the presence of short chain branches that also impart a combination of stiffness and stress-crack resistance to the material. LLDPE, commonly produced by the copolymerization of ethylene with an α-olefin, can be now obtained from ethylene alone using two or more catalysts in a single reactor[2]. A generic example of such processes, known as "tandem catalysis", is shown in Scheme 1.

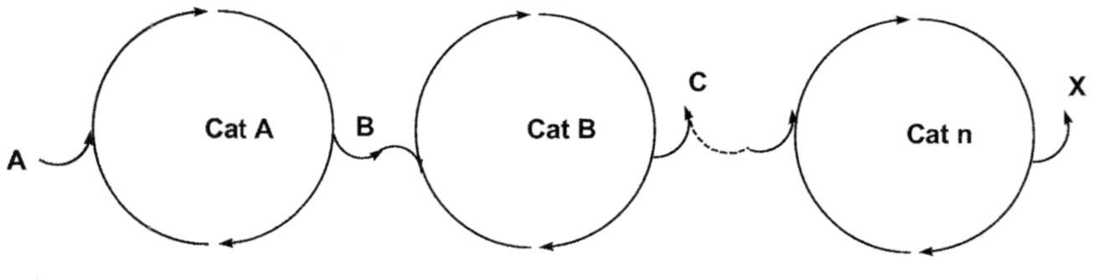

Scheme 1. Generic example of tandem catalysis[2].

Three main factors must be considered for a successful tandem or multi-component process:

i. The catalysts must be chemically compatible under the polymerization conditions, which means no or controlled interference between the active sites.
ii. The catalysts must show comparable tolerance to the activators (when these are required).
iii. The single catalysts must show comparable activity towards the corresponding substrate in order to maintain an appropriate concentration of all substrates all over the process.

The objective of this short abstract is to describe the way in which polymer properties can be controlled by using combinations of different types of homogeneous and heterogeneous single-site catalysts under different polymerization conditions. To this end specific examples will be highlighted along with a comprehensive list of references related to the subject. Some examples of the use of heterogeneous catalysts in this respect will also be described.

2. HOMOGENEOUS SYSTEMS

2.1 Ethylene Polymerization

Recent developments in ethylene polymerization by catalyst reactor blending have been concerned with controlling the final polymer properties in two main processes which are shown in Scheme 2.

In process 1, the catalyst precursors polymerize ethylene independently generating different polyethylenes during the polymerization reaction in a single reactor. In some cases the catalysts do not act independently but form a hybrid system to produce novel materials different from that obtained by mixing the polymers produced by each individual catalyst system. In process 2, one or more catalysts produce α-olefins *in situ* and another catalyst polymerizes ethylene and incorporates the α-olefins into the growing chain. Each process aims to control the final polymer properties through either process 1, obtaining a broad and/or bimodal MWD, or process 2 controlling the level of branching in the polyethylene macromolecular structure.

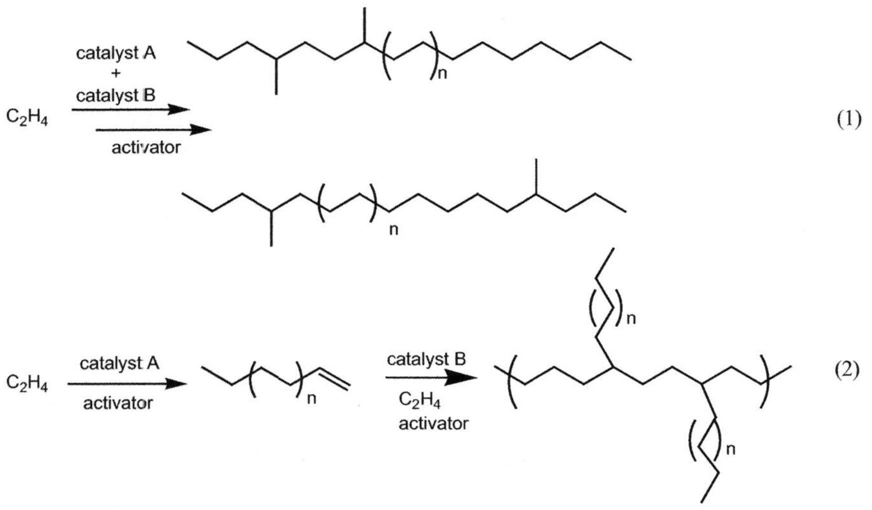

Scheme 2. General processes for ethylene polymerization using a binary system.

2.1.1 Bimodal or broad molecular weight distribution

Early reports[3] by Kaminski *et al.* relate to the production of polyolefins having broad MWDs using homogeneous systems consisting of mixtures of various metallocene catalysts including $Et(Ind)_2ZrCl_2$ (**1**), Cp_2ZrCl_2 (**2**), Ind_2ZrCl_2 (**3**) and Cp_2HfCl_2 (**4**) in the presence of MAO. Broad monomodal or bimodal MWDs can be achieved depending on the combination used. For example the **2**/**4**/MAO combination produced a PE with a modest MWD of 2.5. Alternatively, **1**/**3**/MAO and **1**/**4**/MAO produced bimodal PEs with MWDs in the range 5.5-10. In addition to variation of the mole fraction of each catalyst, control of the polymer produced could also be obtained by variation of reaction conditions such as temperature. For example combination of **1** with *rac*-$Et(Ind)_2HfCl_2$ (**5**) afforded PEs with bimodal MWDs, each catalyst acting independently of each other. At high temperature catalyst **1** was more active, and at lower temperature catalyst **5** produced most of the polymer with the consequent effect upon the MWDs of the polymerization products.

Similar studies by Soares *et al.* combined soluble metallocenes **1**, **2**, **4** and Cp_2TiCl_2 (**6**) and MAO in ethylene polymerization experiments[4]. These authors showed that the individual catalysts produced ethylene homopolymer just as they would independently, suggesting the chemical nature of the active sites is not effected by interactions between the individual catalyst components.

The polymerization of ethylene by late transition metal complexes including cationic nickel(II) and palladium(II) complexes with bidentate α-diimine ligands reported by Brookhart *et al.* has been the subject of numerous recent publications as these complexes have been found to be highly active in the production of branched high molecular weight polymers[5]. At the same time iron complexes with related tridentate nitrogen ligands have been reported which are highly active for the linear polymerization of ethylene to high density materials (HDPE)[6]. As a natural consequence of the development of this new class of polymerization catalysts their exploitation in combinations with traditional Ziegler-Natta and metallocene catalysts in reactor blending is now being investigated and has been the subject of recent patent publications[7].

Mecking has reported the reactor blending of different combinations of early and late metal polymerization catalysts to obtain blends of linear and branched polyethylenes using ethylene as the sole monomer[8]. The nickel complexes **7** or **8** in combination with the zirconium catalysts **9** or **10** functioned in the reactor independently producing blends with careful control of the temperature and pressure. Analysis of the branching by ^{13}C NMR spectroscopy showed the structure of the polymer produced by the nickel catalyst had methyl and long chain branching consistent with previous reports by Brookhart *et al.*[9].

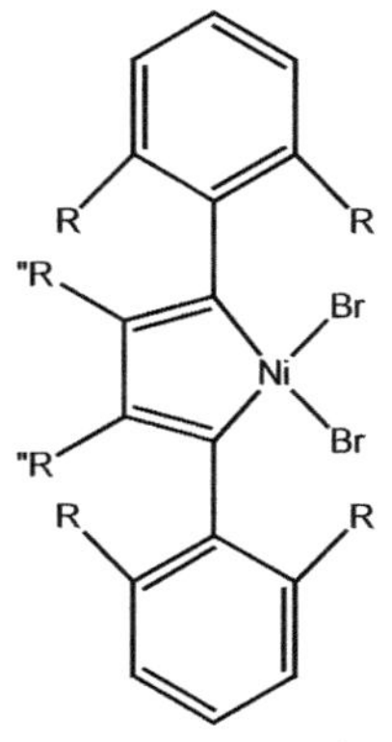

7: R = Me, R" = Me
8: R = ^{i}Pr, R" = H

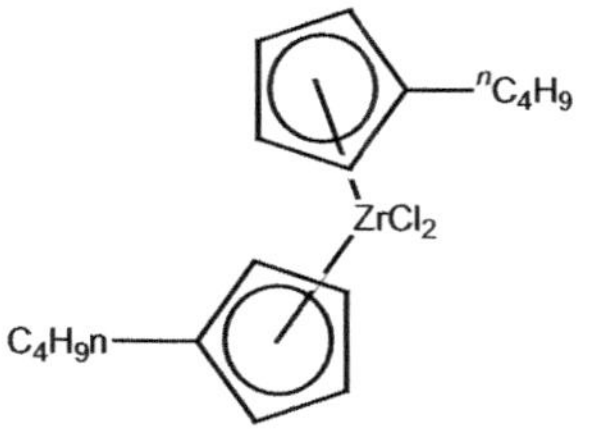

9

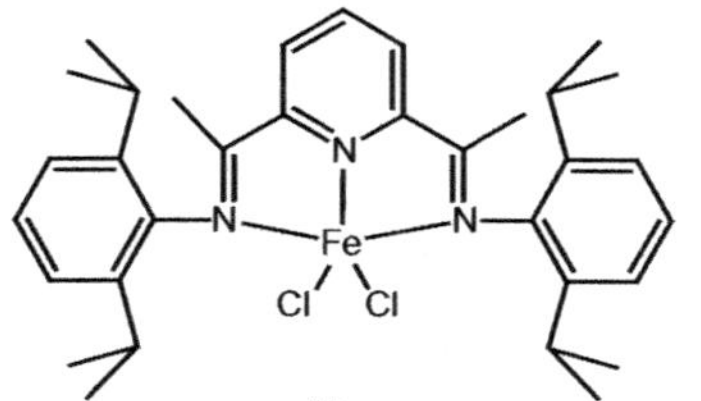

11

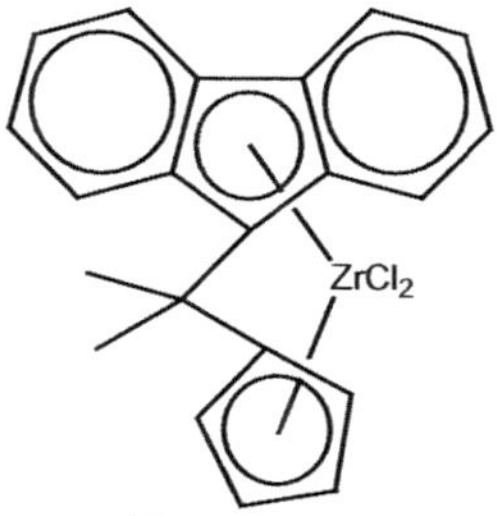

10

The average number of branches in the blends were consistent with the ratio of early/late metal catalyst employed so no direct interaction between the early and late metal centers was apparent.

The addition of hydrogen to the system resulted in a strong decrease in the molecular weight of the HDPE produced by the early metal catalyst with a related increase in the melt flowability with no effect upon the polymer produced by the nickel catalyst.

Reactor blends were also prepared with heterogeneous catalysts obtained by immobilization of the precursors on silica supported MAO (MAO-S). A net reduction in the amount of branching was observed as was seen with polymers obtained with solely late metal catalysts immobilized in this way.

Finally, Mecking combined two late metal catalysts in the reactor [8]. The addition of the iron catalyst **11**, which is known to polymerize ethylene in a strictly linear fashion to the nickel catalysts **7** or **8** which produce branched polyethylene gave blends of branched and linear polymers consistent with both catalysts being active.

PEs with broader MWDs and higher MWs have been obtained by Qian *et al.* by activation of a family of hetero-bimetallic metallocene complexes with MAO[10]. The nature of the alkyl substituent on one Cp ring bonded to Ti resulted of utmost importance for controlling either MWD or MW.

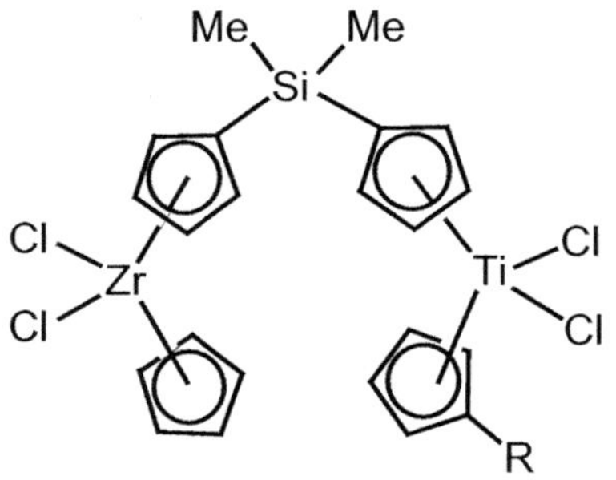

R = alkyl **12**

For R = Me, the PE product had a larger MWD (3.01) than those of the single components in comparable conditions (1.7). For R = $CH(CH_3)_2$ or C_6H_{11}, the MW was five times higher than that obtained with Cp_2ZrCl_2. It was proposed that the two metals influence each other by reducing the electron density at adjacent Cp rings.

Indirect evidence of a synergistic cooperation between the two catalysts was observed by Casagrande *et al.* for the production of highly branched polyethylene using the α-diimine complex **13**, an active catalyst for branched polyethylene in conjunction with MAO [9], and the tris-pyrazolyl titanium trichloride **14**, a catalyst for linear ultra high molecular weight polyethylene[11].

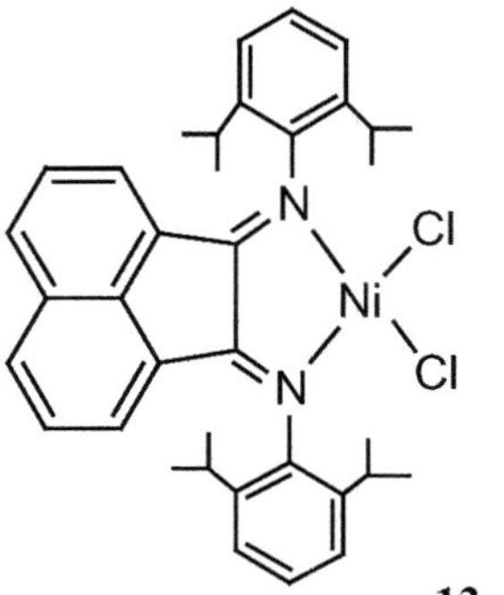

13

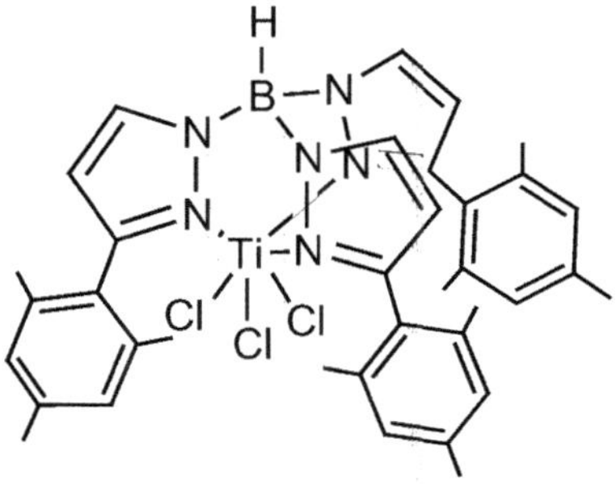

14

The properties of the polyolefin obtained were remarkably influenced by the molar fraction of nickel (χ_{Ni}). In particular, lower melting temperatures and higher branching degrees were observed as χ_{Ni} increased in the medium at a constant amount of MAO. For $\chi_{Ni} = 0.75$, a maximum in the melt flow index and a minimum in the intrinsic viscosity were obtained. All the polymer properties (productivity, intrinsic viscosity and melt flow index) showed a non-linear correlation with respect to χ_{Ni}, which suggested the formation of a new catalytic system based on both Ni and Ti.

Combinations of **13** with *rac*-ethylenebis ($IndH_4$)$ZrCl_2$ were studied with variation of polymerization conditions such as temperature[12]. A linear correlation was found between productivity and mole fraction of Zr in polymerization runs carried out at 0 and 50 °C suggesting that the catalyst precursors work independently of each other. However, the polymerization results at 30 °C showed a non linear relationship between the Zr mole fraction and the productivity with a peak at a Zr mole fraction of 0.67, which has been interpreted in terms a synergistic effect between the nickel and zirconium compounds. Under these reactor blending conditions, the PE mixtures showed monomodal MWD behavior in all cases. As the polymerization temperature was increased, phase separation increased also. The PEs produced at 0 °C showed very small particles of branched PE dispersed uniformly in the linear PE medium while at 50 °C the two polymers separated spontaneously.

2.1.2 Control of branching

Besides using specific catalysts for the copolymerization of ethylene with α-olefins[3] or catalysts capable of polymerizing ethylene with mechanisms involving effective chain-walking[9], branched polyolefins can be produced from ethylene alone by combining in the same reactor a selective α-olefin oligomerization catalyst with a catalyst capable of copolymerizing the produced α-olefins with ethylene. As previously said, these processes are commonly referred to as tandem catalysis, and generally do not involve a synergistic cooperation of the two catalysts that work independently of each other, yet they must have comparable activity in order to maintain appropriate concentrations of the substrates all over the reaction. In a few cases, however, the occurrence of synergic effects has been suggested, for example when non-linear correlations between productivity and catalysts ratio are observed.

The first example of tandem catalysis for the production of low density polyethylene (LDPE) was reported by Barnhart and Bazan who employed MAO-activated $[C_5H_5BOEt_3)]_2ZrCl_2$ (**15**) to form α-olefins and MAO-activated $[\eta^5\text{-}C_5Me_4)SiMe_2(\eta^1\text{-}NCMe_3)]TiCl_2$ (**16**) as copolymerization catalyst (Scheme 3)[13].

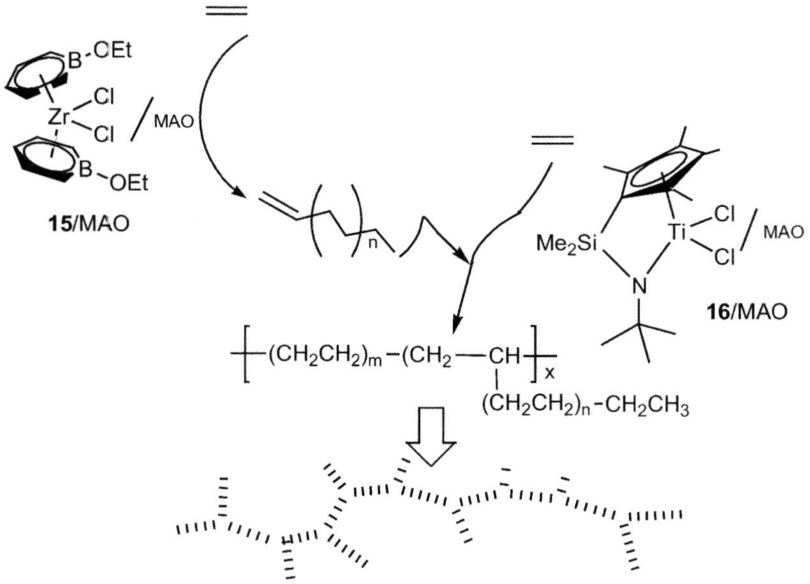

Scheme 3. Tandem catalysis scheme for the production of branched polyolefin from ethylene using the two independent catalysts 15/MAO and 16/MAO.

The highest productivity was obtained using an excess of Ti catalyst (Ti:Zr = 10) as well as a large excess of MAO (5000-10000 equiv.). In all cases, however, a poor control of the polymer structure was observed due to the fact that **15**/MAO produces a distribution of α-olefins, which are less reactive with increasing molecular weight. For analogous reasons, branched polyolefins with various degrees of branching were obtained using the Brookhart catalyst $\{[2\text{-}C_6H_4(Et)N{=}CMe)_2C_5H_3N]FeCl_2\}$/MAO (**17**), yielding Schulz-Flory distributions of α-olefins, in conjunction with the copolymerization catalysts $Me_2SiInd_2ZrCl_2$ (**18**)/MAO or $EtIndZrCl_2$

(**19**)/MAO[14]. Higher activity but less branching was obtained with the latter system, which gave a more homogeneous LDPE. It has been proposed that **19**/MAO polymerizes ethylene faster than **18**/MAO and competes with the iron catalyst for ethylene. By doing so, less α-olefins are produced and complete incorporation can be achieved.

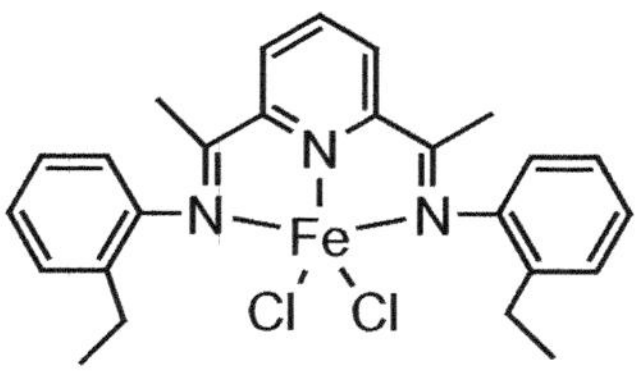

17

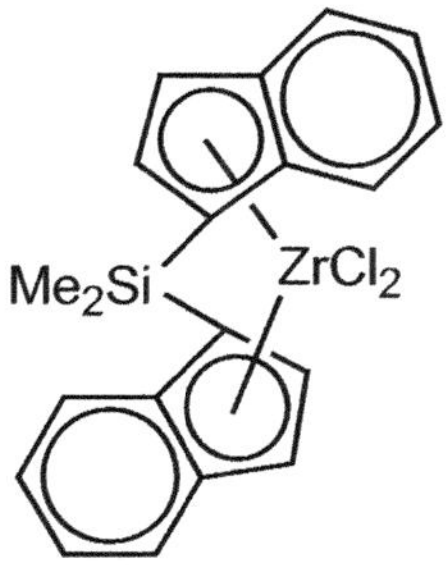

18

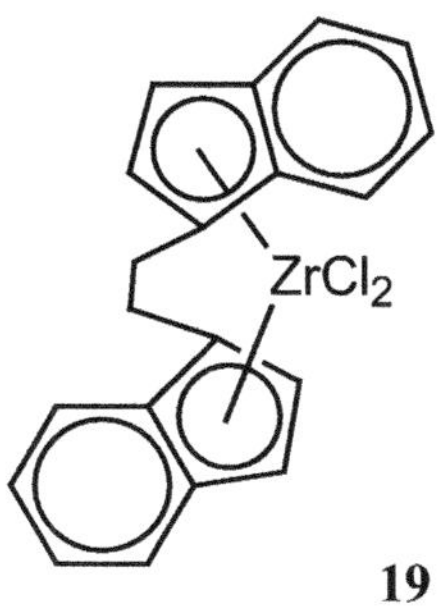

19

Effective control of branching in LLDPE was achieved by Bazan *et al.*. with the use of a tandem catalysis scheme involving a selective oligomerization catalyst in conjuction with a constrained geometry catalyst (CGC) for the copolymerization process. Both catalysts were activated with stoichiometric amounts of $B(C_6F_5)_3$[15]. To this purpose was designed the nickel(II) methallyl precursor $[(C_6H_5)_2PC_6H_4C(O)O\text{-}\kappa^2 P,O]Ni(\eta^3\text{-}CH_2CMeCH_2)$ (**20**) that was activated by $B(C_6F_5)_3$ yielding a selective catalyst for the selective transformation of ethylene into 1-butene and 1-hexene irrespective of the pressure (Scheme 4). In contrast, the oligomerization activity increased with the ethylene pressure.

The combination of **20**-$B(C_6F_5)_3$ with the copolymerization catalyst $\{[\eta^5\text{-}C_5Me_4)SiMe_2(\eta^1\text{-}NCMe_3)]\{MeB(C_6F_5)_3\}$ (**21**-$B(C_6F_5)_3$) gave LLDPE with both ethyl and butyl branches. Appropriate reaction conditions were chosen for both oligomerization by **20** and polymerization by **21** so as to have an ethylene consumption rate within the same order of magnitude for the individual catalysts and consequently almost total incorporation of the α-olefins. Remarkably, at low **20**/**21** ratios (< 0.89), only ethyl branches were formed, due the selective formation of 1-butene by **20** (poly[ethylene-*co*-(1-butene)] production), while increasing the concentration of **20** led also to incorporation of 1-hexene with consequent formation of poly[ethylene-*co*-(1-butene)-*co*-(1-hexene)] with small amounts of butyl branches. GPC analysis of the copolymer obtained at low nickel concentration showed a bimodal molecular weight distribution due to the presence of linear polyethylene formed by the titanium catalyst in the early stages of the reaction when the α-olefin concentration was low. Consistently, increasing the concentration of **20** gave a monomodal trace.

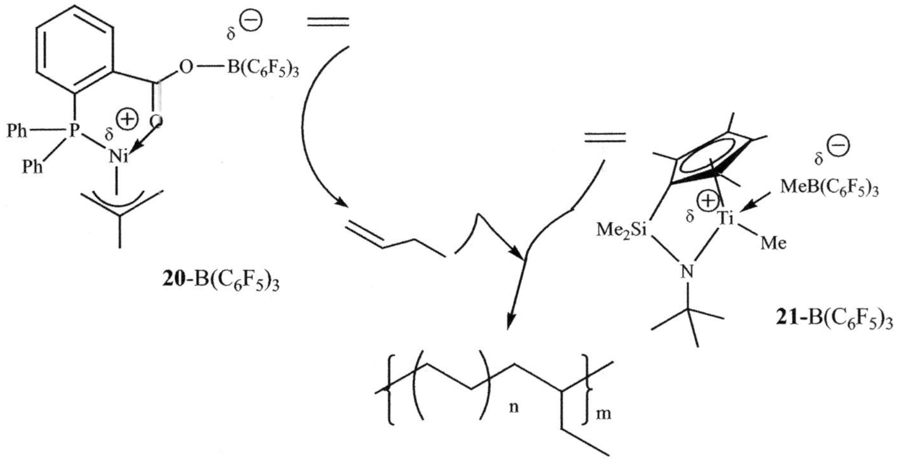

Scheme 4. LLDPE by tandem action of **20**-$B(C_6F_5)_3$ and **21**-$B(C_6F_5)_3$.

An NMR study of ethylene oligomerization by **20** showed the rate of initiation to the catalytically active species to be slow a compared to both propagation and chain-transfer. Therefore, in order to accelerate the initiation rate, maintaining the same catalytically active species, a new nickel catalyst was designed in which the methallyl group was replaced by the weaker η^3-benzyl ligand. The new catalyst, $[(C_6H_5)_2PC_6H_4C(O\text{-}B(C_6F_5)_3O\text{-}\kappa^3P,O\text{-}]Ni(\eta^3\text{-}CH_2C_6H_5)$ (**22**) was straightforwardly obtained by reacting a binuclear precursor with $B(C_6F_5)_3$ (Scheme 5)[16].

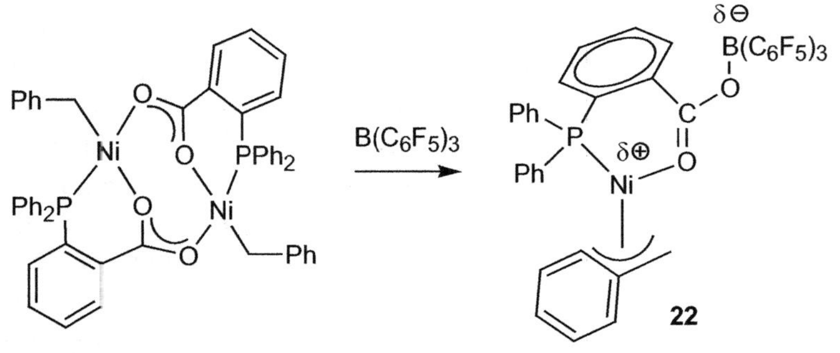

Scheme 5. Synthesis of the oligomerization catalyst **22**.

Tandem catalysis of the CGC **21** with the benzyl complex **22** gave LLDPE with a total branching of 9.2% even for a Ti/Ni ratio of 100 for which no branching at all was observed with the catalyst combination **20**/**21**.

A remarkable improvement of the tandem catalysis technique with stoichiometric activators has been recently achieved by Bazan *et al.* who were able to coordinate three catalysts in the same reactor and produce branched polyethylenes exhibiting a range of properties unattainable with two catalysts only[17].

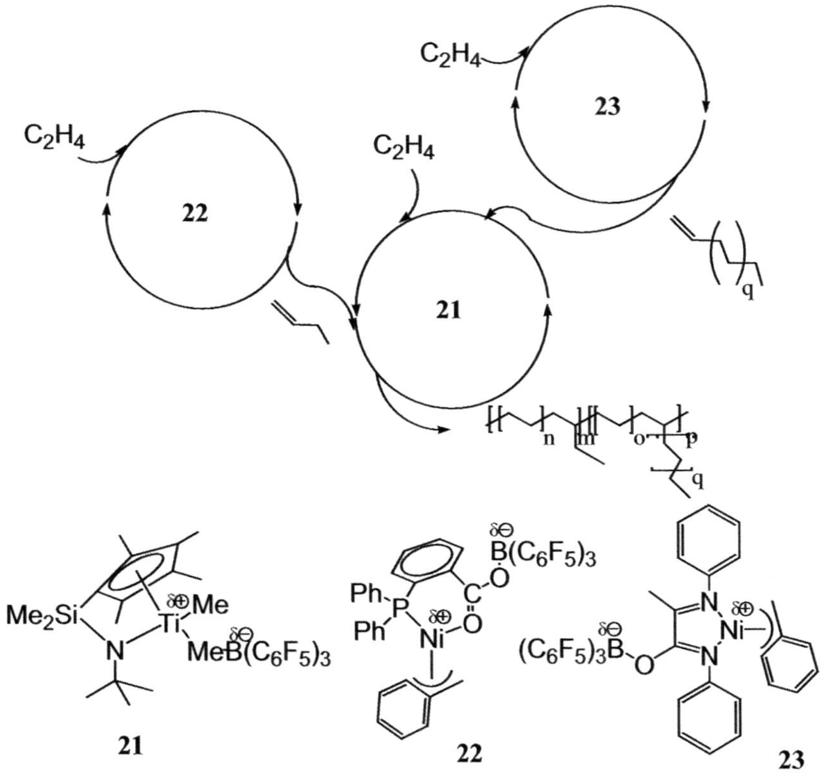

Scheme 6. Triple tandem catalysis for the production of branched polyethylenes.

The three catalysts employed in this study are reported in Scheme 6 that also illustrates a three-component tandem scheme. The choice of the α-diimine complex **23** as the third component was motivated by the fact that this catalyst is able to oligomerize ethylene yielding Schulz-Flory distributions of α-olefins with an activity comparable to those of **21** and **22** in the corresponding processes. Therefore, the 1-butene generated by **22** and the distribution of α-olefins from **23** could be incorporated by the titanium single-site catalysts **21** into a polyethylene backbone providing polymers with varying ratios of ethyl branches and longer branches.

In order to determine the experimental conditions that control the ethylene consumption rate of the individual catalytic systems, the three-component polymerizations were first designed on the basis of linear combinations of the two-components. Only ethyl branches were observed by

22/**23**/C_2H_4 combinations, consistent with a polyethylene-*co*-1-butene structure. In contrast, ethyl, butyl and longer chain branches were obtained with **21**/**23**/C_2H_4 combinations, the amount of each branch increasing with the loading of **23** relative to **21**. A high throughput screening technique was employed to find out the best catalyst ratio in the three-component polymerizations for the production of monomodal polymers. Notably, the structures of the PEs obtainable with the triple tandem catalytic procedure contain C_2:C_4:C_n ratios that cannot be obtained by either two-components combination.

Innovative approaches to the production of LDPE involve the design of binuclear catalysts containing both the oligomerization and copolymerization sites in the same complex structure. An elegant example is provided by the heteronuclear CGC (μ-CH_2CH_2-3,3'){(η^5-indenyl)[1-$Me_2Si(^tBuN)_2$]($TiMe_2$)($ZrMe_2$) (**24**) recently described by Marks and Jiaxi in which the Zr site is competent for ethylene oligomerization and the Ti site is competent for producing high molecular weight ethylene-α-olefin copolymers[18].

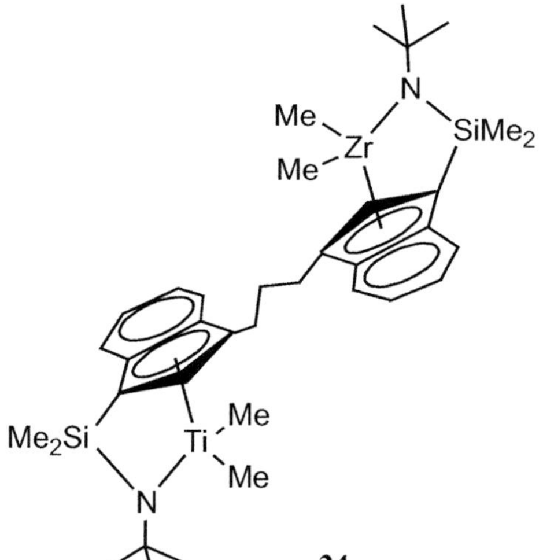

24

A homogeneous PE product with long chain branching was obtained upon activation with $[Ph_3C]^+B(C_6F_5)_4]^-$, whereas no branching was observed in control experiments with the corresponding mononuclear components.

A similar approach to branching control in PE had been previously developed by the same authors through the design of the homobimetallic Zr precursor **25**, in conjunction with a bifunctional (**A₂**) activator[19].

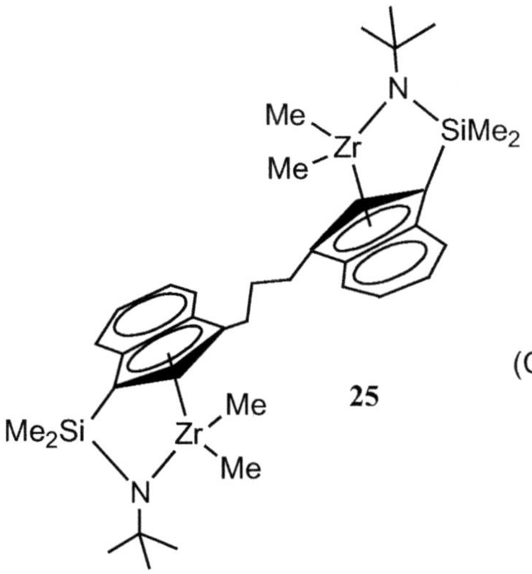

25

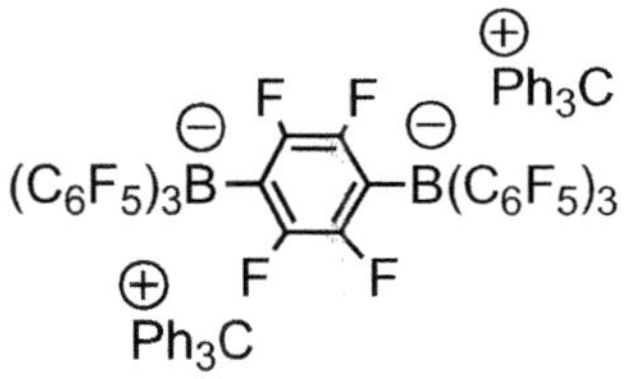

A_2

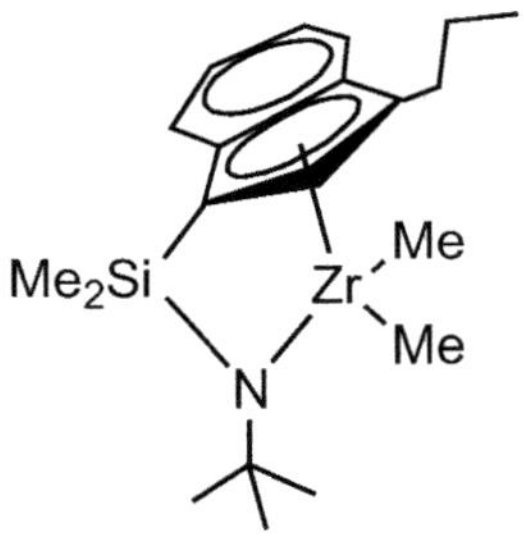

26

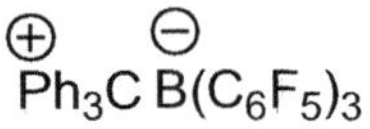

A_1

The bimetallic constrained geometry complex **25** in conjunction with the bifunctional activator **A_2** produced 11 times more ethyl branches in ethylene homopolymerization as compared to the monometallic **26**/**A_1** system. Evidence was provided supporting an intradimer process like that illustrated in Scheme 7.

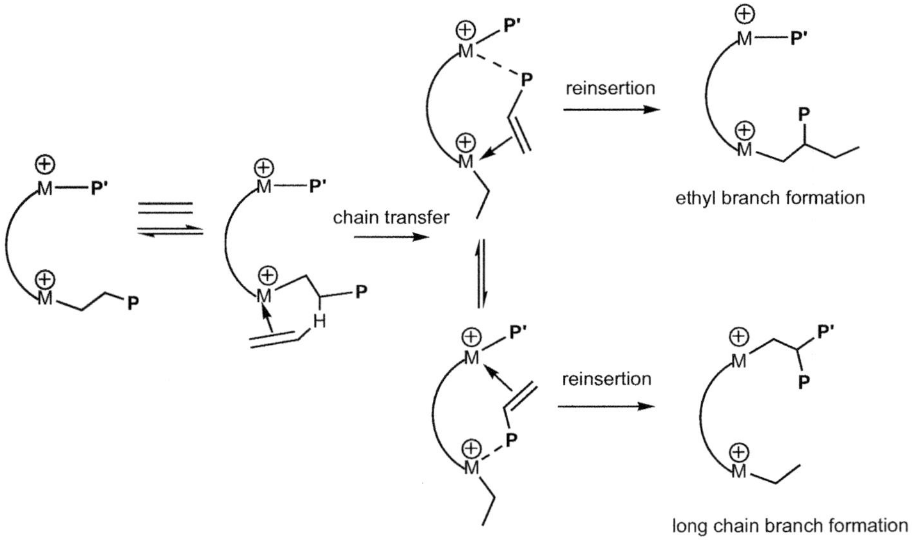

Scheme 7. Proposed pathways for ethyl branch formation in ethylene homopolymerization mediated by **26**.

Polymerization activities and product MWs did not vary between binuclear and mononuclear systems. Of crucial importance to cause the observed synergistic effect in increasing branching has been proposed to be the close spacial proximity and proper mutual orientations of the active sites, which would allow the olefin-terminated segment to have enhanced probability of being captured/enchained before diffusing away. In this respect, a crucial role seems to be plaayed by the bifunctional activator that favours the formation of a bimetallic pocket with a special affinity to small α-olefins.

2.2 Propylene Polymerization

Multi-component catalytic systems have been exploited also for the polymerization of propylene, particularly mixtures of metallocenes and of Ziegler-Natta/metallocene combinations. In these cases the search has been largely directed to control the molecular weight distribution of the resulting polymer blends, however interesting results have been also obtained in the production of stereoblocks polymers. The narrow MWD of metallocene-based polypropylenes is an advantage in processes of fiber spinning, production of biaxially oriented films and thin-wall injection molding. On the other hand metallocene-based polymers with narrow MWD show low shear sensitivity of the melts causing some processing problems. To improve processing, polymers with different molecular weights can be melt blended with the resultant problems and cost as described for ethylene polymerization above. Reactor blending of catalyst mixtures for propylene polymerization has been exploited to obtain polymer blends with control of the resultant MWD without these concurrent problems associated with melt blending.

It has been established that the use of mixtures of metallocene catalysts with different stereoselectivities can in principle generate stereoblock polymers if chain transfer between propagating species of different stereoselectivities takes place (Scheme 8)[20].

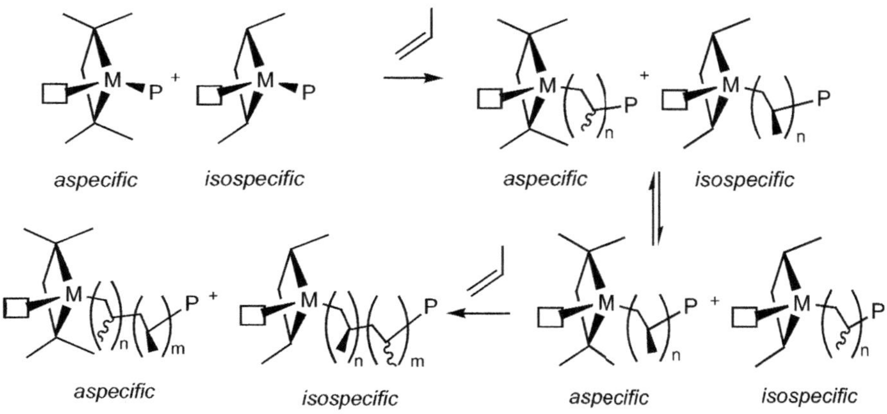

Scheme 8. Chain-transfer mechanism for the synthesis of steroblocks polypropylenes [20].

These new stereoblock materials have promise as thermoplastic elastomers or as heterophase compatibilizers. This has been demonstrated by Brintzinger *et al.*[21] who studied propylene polymerization using mixtures of the four zirconocene complexes **27-30** in conjunction with either MAO, triisobutylaluminium or triphenylcarbenium tetrakis(perflourophenyl borate) as activating agents.

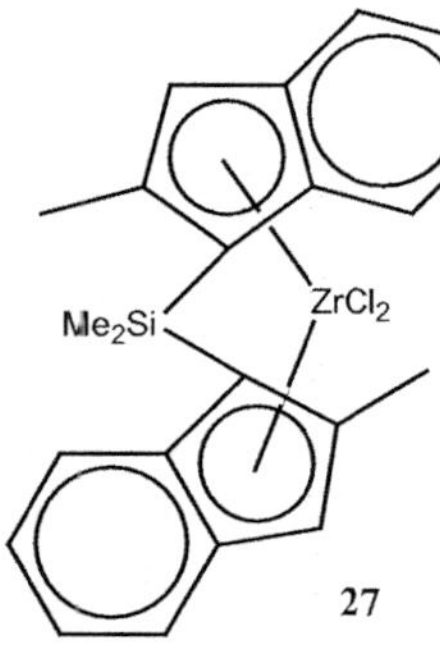

27

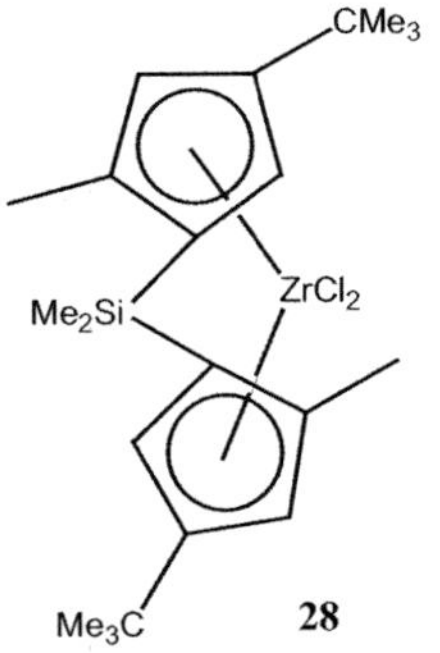

28

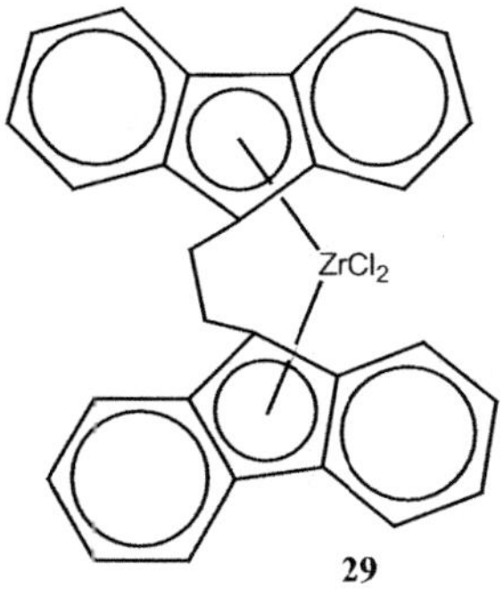

29

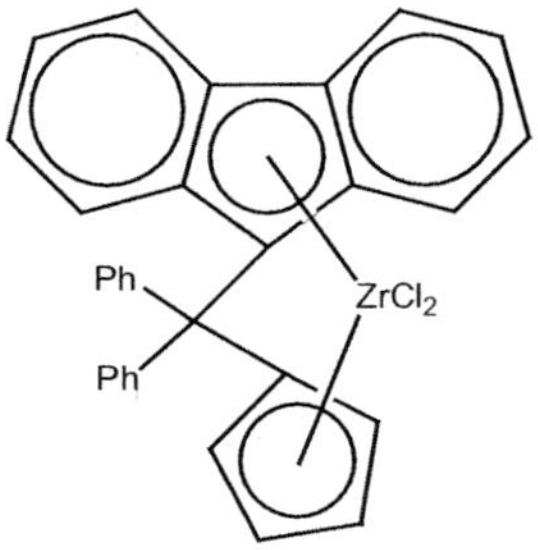

30

Propylene polymerization with MAO-activated mixtures of **27** and **29** gave completely separable mixtures of isotactic and atactic polymers characteristic for each of the individual catalysts. MAO-activated mixtures of **28** and **29**, however, gave polypropylene mixtures containing, besides isotactic and atactic polymers, polymer fractions in which isotactic and atactic polypropylene chain segments are inseparably linked. The formation of stereoblock polymers was concluded to be contingent upon an efficient polymeryl exchange between the Zr catalyst centers and the Al centers of the co-catalyst (Scheme 9).

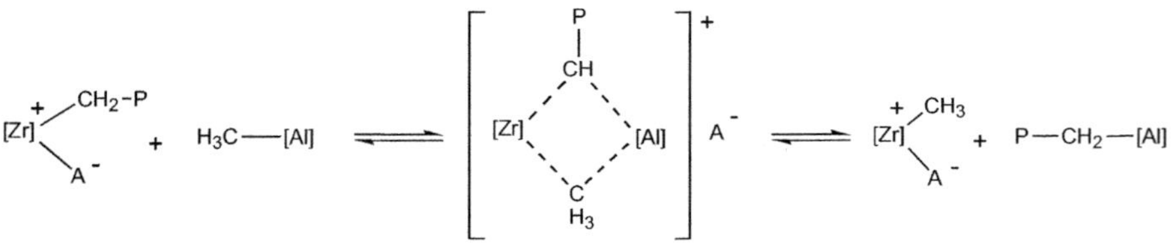

Scheme 9. Synthesis of steroblocks polypropylenes by chain transfer to aluminium.

This process is favoured by the highly substituted Zr catalysts in combination with MAO as the rather *free* Al centers favor the exchange of the bulky polymer chain from Zr in exchange for less sterically demanding CH_3 groups.

3. HETEROGENEOUS SYSTEMS

3.1 Ethylene Polymerization

The first reported examples of tandem catalysis for the production of LLDPE from ethylene alone were heterogeneous and involved Ziegler-Natta catalysts or combinations of Ziegler-Natta and late transition metal single-site catalysts as the oligomerization component. Problems associated with catalyst compatibility have increasingly oriented research towards the design of homogeneous tandem systems, some of which have been described in previous sections. Indeed, a more efficient control of the catalytic process, hence of the polyolefin structure, is generally achieved by single-site catalysis whose reactivity can be modulated by varying the electronic and steric properties of the metal-ligand assemblies.

Mixed heterogeneous catalysts have been obtained using two different transition metals derivatives supported on $MgCl_2$ [22] or mixtures of solid supported catalysts on $MgCl_2$ and SiO_2 for example[23]. The former catalyst system was obtained by co-supporting Hf and Ti tetrachlorides on $MgCl_2$ followed by activation with trialkylaluminium; the addition of various Lewis bases or acids allowed a fine tuning of MW and MWD of the obtained PE thanks two the differentiated reactivity of Hf or Ti containing sites towards protic and basic compounds. The other catalysts system was prepared by heterogenization of $TiCl_2$ and/or Ind_2ZrCl_2 on $MgCl_2$ and SiO_2 supports and required activation with MAO. Polymodal PE was produced with a MWD as high as 42 with the PEs arising from each catalyst independently.

Typical polymerization-copolymerization catalysts employed in tandem catalysis are ∂–$TiCl_3 \cdot 0.33\ AlCl_3$, $TiCl_4/MgCl_2$ supported on polyethylene, $TiCl_4/MgCl_2$ supported on ethyl anisate and $VOCl_3$ [24]. These have been employed to produce LLDPE in conjunction with oligomerization catalysts, yielding selectively 1-butene, such as $Ti(O^iPr)_4/AlCl_3$ or sulfonated nickel ylide complexes. Excellent results in terms of 1-butene incorporation and productivity have been obtained with $Ti(O^iPr)_4/AlCl_3$ in conjunction with either $TiCl_4/MgCl_2/PE$. With the latter system the content of ethyl branches incorporated into the polymer backbone was found to increase with the concentration of $Ti(O^iPr)_4/AlCl_3$ yielding up to 45 methyl groups every 1000 carbon atoms. However, at high concentration of oligomerization catalyst the productivity decreased due to poisoning of the polymerization catalyst.

In order to get high branching (56 Me/1000 C), ∂–$TiCl_3 \cdot 0.33\ AlCl_3$ and the nickel(II) ylide complex were added in sequential manner. The oligomerization catalyst was added first, and when the 1-butene concentration was high, the Ti catalyst was introduced into the reactor. High level of branching were similarly obtained by introducing $VOCl_3$ after the nickel catalyst.

A special class of homometal tandem catalysis has been developed by Phillips Petroleum[25]. Typically, a chromium compound, for example CrO_3, was supported on silica and reduced to act as polymerization/copolymerization catalyst upon activation with trialkylaluminium or alkylaluminium chloride. The oligomerization component was generated *in situ* by adding sub-stoichiometric amounts of pyrrole or pyrrole derivatives that, interacting with the Cr surface, modified some chromium sites. By properly adjusting the pyrrole to chromium ratio, polymers with a branching degreee as high as 6 Me/1000 C could be obtained.

A similar strategy was developed by Benham *et al.* to produce branched PE[26]. A Cr^{VI} oxide supported on silica was partially reduced by $Cr_4(CH_2SiMe_3)_8$ to give a new tandem catalyst containing both oligomerization sites (reduced chromium) and copolymerization sites (unreacted chromium oxide).

A chromium(II)/silica catalyst has been also employed in conjunction with nickel ylide complexes stabilized by different types of *P,O*-chelating ligand (**31**-**32**) as ethylene oligomerization catalysts[27].

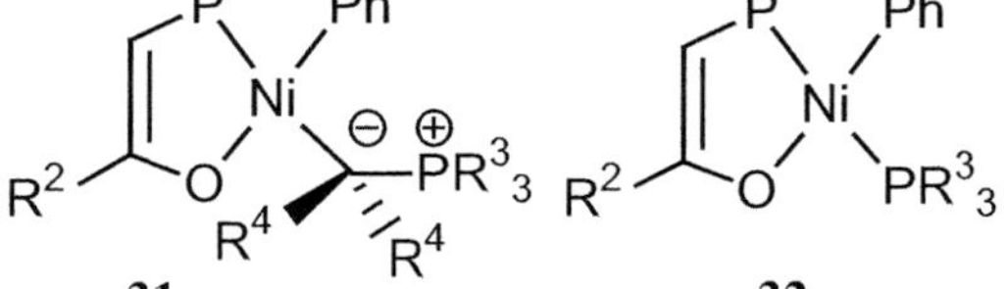

31 **32**

R = Ph,alkyl
R^2 = Ph, alkyl
R^3= Ph, alkyl, alkoxy
R^4 = Ph, alkyl, H

R = Ph,alkyl
R^2 = Ph, alkyl
R^3= Ph, alkyl, alkoxy

The nickel catalysts produced high molecular weight α-olefins and were used in lower concentration than the chromium catalyst to obtain PE with long chain branches. The degree of branching increased with the nickel concentration, which however had to be carefully scontrolled to avoid poisoning of the copolymerization catalyst. In general, these tandem systems are difficult to tune and the material produced is not perfectly reproducible.

Finally, a novel approach to branched polyethylenes by ethylene polymerization has been developed[28] by Okuda *et al.* using a tandem catalyst system consisting of a combination of silica supported linked cyclopentadienylamido titanium compounds and a homogeneous dibromo nickel catalyst having a pyridylimine ligand. The MAO activated titanium catalysts (**33**) were supported on pyridyl ethylsilane-modified silica and, in the presence of MAO-activated nickel complexes (**34**), were investigated for ethylene polymerization. The nickel complexes on their own produced a mixture of 1- and 2-olefin oligomers with methyl branches. When combined with the titanium catalysts polyethylene with long chain branches was produced suggesting incorporation of the Ni complex produced oligomers in the growing PE chain as described in Scheme 2. The MWDs obtained were broad and bimodal GPC curves were observed. A trend in decreasing melting temperature and increasing branching was associated with increasing Ni/Ti ratio. These systems were also compared to mixtures of the nickel and titanium homogeneous catalysts. The major component of the heterogeneously produced PE was the lower molecular weight fraction whereas the homogeneous system produced mainly the higher molecular weight fraction.

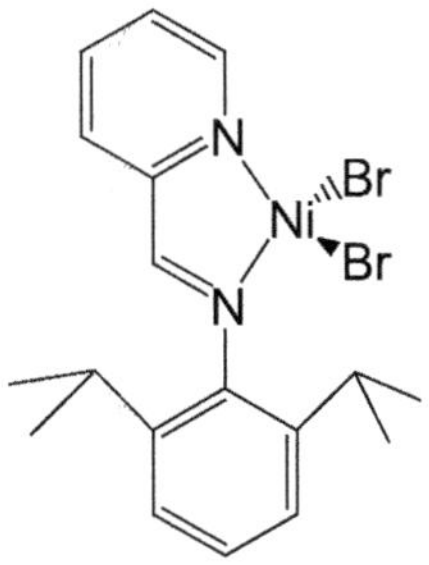

34

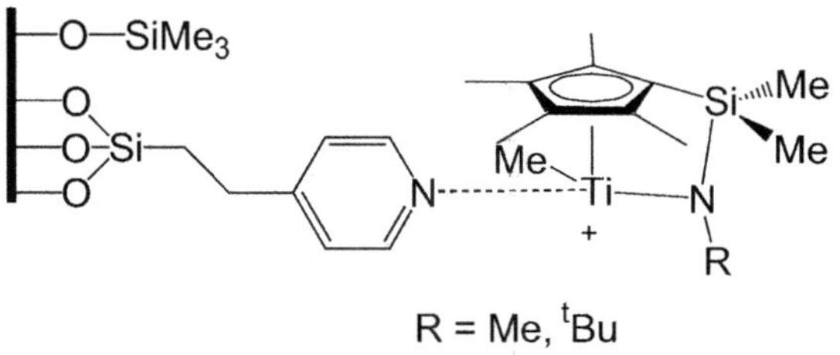

33

3.2 Propylene Polymerization

Mixtures of Ziegler-Natta and metallocene complexes are being widely employed to produce polypropylenes with a range of different properties. This approach was first introduced by Montell[29] by developing the "Catalloy" process which allowed the inclusion of variable amounts of EPM rubber in polypropylene in a reactor generated impact copolymer. The rubber content and the blend properties were strongly dependent on the catalyst and process used.

Ahn *et al.* used sequential addition of Ziegler-Natta catalysts and metallocene complexes to control the MWD of polypropylene[30]. First MAO and rac-ethylenebis(indenyl)zirconium dichloride, followed by triethylaluminium and $MgCl_2$-supported titanium tetrachloride produced polypropylene with a wide MWD curve and a shoulder peak. The MWD was much wider than that of the polypropylene produced by the individual catalysts. The peak that appeared at the lower MW region originated from the metallocene/MAO catalyst system and the peak at the higher MW region was due to the polypropylene from the $MgCl_2$-Ti catalyst. The height of each peak in the GPC could be controlled by variation of both the amount of each catalyst introduced into the system and also the polymerization time for each catalyst.

Rather than sequential addition of catalysts to the reactor, a number of different research groups have mixed both metallocene and Ziegler-Natta catalysts together at the same time during propylene polymerization producing individual polymers with different melting and crystallization temperatures. This approach has been exemplified by Lisovskii *et al*[31] who mixed commercial Ziegler-Natta catalysts with the zirconium metallocene complexes $C_2H_4(Ind)_2ZrCl_2$ and $Me_2Si(Ind)_2ZrCl_2$ to obtain polypropylene with a large MWD and improved processing. Polypropylenes were also prepared separately using each catalyst and melt blended as a comparison to the reactor blends.

The reactor blends obtained were beads containing intimately mixed metallocene and Ziegler-Natta derived semicrystalline polypropylene fractions. The properties of the polypropylene produced by the mixed-catalytic system were influenced by changing the composition of each catalytic system or by changing their relative ratio in the mixed system. The crystalline phases of each polymer showed high compatibility having a single melting temperature. The reactor blend was also shown to have a better stiffness/impact balance than the melt blended polymer as well as being more shear sensitive and consequently having an advantageous processing.

4. CONCLUSIONS

Reactor blending applying combinations of different metal catalysts is an innovative technique with great potential for the production of polyolefins with tailored properties. Particularly, variation of the parameters MW, MWD and level of branching can be achieved altering the physical properties of the polyolefins produced. Indeed, the large variety of olefin polymerization and oligomerization catalysts available nowadays provides an almost infinite combination of catalytic sites, which may lead to a very wide range of new polymeric materials by simply adjusting the composition of the catalytic mixture and the polymerization parameters.

Tandem catalysis can be applied in both homogeneous and heterogeneous processes. However, a much better control of the polyolefin structure is generally achieved by molecular catalysts as compared to heterogeneous catalysts due to the superior control of the number and type of catalytic sites. In both cases careful tuning is required to form a compatible catalyst blend which performs under the polymerization conditions. The tunability of these systems is ideal in that it enables the consequential design of the resultant polyolefinic material to suit a particular industrial application.

ACKNOWLEDGMENT

The European Commission (contract POLYCAT no. HPRN-CT-2000-00010) and COST Action D17 are thanked for financial support.

REFERENCES

1. WO9748735 (1997) Exxon, invs Canich, J. A. M., Vaughan, G. A., Matsunaga, P. T., Gindelberger, D. E., Shaffer, T. D., Squire, K. R.; WO02053603 (2001) Univation Tech LCC, invs Murray, R. E., Mawson, S., Szul, J. F., Erickson, K. A., Kwack, T. H., Karol, F. J., Schreck, D. J.; US2002107341 (2002) Univation Tech LCC, invs Murray, Rex E.; WO0037511 (2000) Univation Tech LCC, invs Murray, R. E., Mawson, S., Williams, C. C., Schreck, D. J.; US6136743 (1997) Mitsui Chemicals Inc, invs Sugimura, K., Yorozu, K., Suzuki, Y., Hayashi, T., Matsunaga, S.; US2002058584 (2002) Du Pont, invs Bennett, A. M. A; Bennett, Alison Margaret Anne; Coughlin, Edward Bryan; Citron, Joel David; Wang, Lin.
2. de Souza R. F. and Casagrande Jr, O. L. (2000), Recent advances in olefin polymerization using binary catalyst systems, *Macromolecular Rapid Communications*, 22, 1293-1301; Komon, Z. J. A. and Bazan, G. C. (2001), Synthesis of branched polyethylenes by tandem catalysis, *Macromolecular Rapid Communications*, 22, 467-478.
3. Ahlers, A. and Kaminsky, W. (1988), Variation of molecular weight distribution of polyethylenes obtained with homogeneous Z-N catalysts, *Macromolecular Chemistry Rapid Communications*, 9, 457-465.
4. D'agnillo, L., Soares, J. B. P. and Penlidis, A. (1998), Controlling molecular weight distributions of polyethylene by combining soluble metallocene/MAO catalysts, *Journal*

of Polymer Science, Part A: Polymer Chemistry, 36, 831-840; Beigzadeh, D., Soares, J. B. P. and Duever, T. A. (1999), Combined metallocene catalysts: an efficient technique to manipulate long-chain branching frequency of polyethylene, *Macromolecular Rapid Communications*, 20, 541-545.

5. Johnson, L. K., Mecking, S. and Brookhart M. (1996), Copolymerization of ethylene and propylene with functionalized vinyl monomers by palladium(II) catalysts, *Journal of the American Chemical Society,* 118, 267-268; Killian C. M., Tempel D. J., Johnson, L. K. and Brookhart M. (1996), Living polymerization of α-olefins using Ni^{II}-α-diimine catalysts. Synthesis of new block polymers based on α–olefins, *Journal of the American Chemical Society*, 118, 11664-11665; Mecking, S., Johnson, L. K., Wang L. and Brookhart M. (1998), Mechanistic studies of the palladium-catalyzed copolymerization of ethylene and α-olefins with methyl acrylate, *Journal of the American Chemical Society*, 120, 888-889; WO96/23010 (1996), University of North Carolina at Chapel Hill / Dupont, invs.: Johnson L. K., Killian C. M., Arthur S. D., Feldman J., McCord E., Mclain S. J., Kreutzer K.A., Bennet M. A., Coughlin E. B., Ittel S. D., Parthasarathy A., Tempel D. J. and Brookhart M.
6. Small B. L., Brookhart M., Bennet A. M. A. (1998), Highly active iron and cobalt catalysts for the polymerization of ethylene, *Journal of the American Chemical Society*, 120, 4049-4050; WO98/27124 (1998), Dupont, invs.: Bennet A. M. A.
7. WO97/48735 (1997), EXXON, invs., Canich, J. A. M., Vaughan, G. A., Matsunaga, P. T., Gindelburger, D. E., Shaffer T. D., Squire, K. R.; WO97/38024 (1997), Mitsui, invs., Sugimura, K., Yorozu, K., Suzuki Y., Hayashi T., Matsunaga, S.; WO98/38228 (1998), Targor, inv.: Mecking, S.
8. Mecking, S. (1999), Reactor blending with early/late transition metal catalyst combinations in ethylene polymerization, *Macromolecular Rapid Communications,* 20(3), 139-143.
9. Johnson, L. K., Killian C. M. and Brookhart M. (1995), New Pd(II)- and Ni(II)-based catalysts for polymerization of ethylene and α-olefins, *Journal of the American Chemical Society*, 117, 6414-6415.
10. Huang, J., Feng, Z., Wang, H,, Qian, Y., Sun, J., Xu, Y., Chen, W. and Zheng, G. (2002), Synthesis of hetero-bimetallic metallocene complexes and their catalytic activities for ethylene polymerization, *J. Organomet. Chem.*, 189, 187-194.
11. Kunrath F. A., de Souza R. F., and Casagrande Jr, O. L. (2000), Combination of nickel and titanium complexes containing nitrogen ligands as catalyst for polyethylene reactor blending, *Macromolecular Rapid Communications*, 21, 277-280.
12. Mota F. F., dos Santos R. M., de Souza R. F. and Casagrande Jnr O. L. (2001), Tailoring polyethylene characteristics using a combination of nickel α-diimine and zirconocene catalysts under reactor blending conditions, *Macromoecular Chemistry and Physics*, 202, 1016-1020.
13. Barnhart, R. W. and Bazan, G. C. (1998), Synthesis of branched polyolefins using a combination of homogeneous metallocene mimics, *Journal of the American Chemical Society,* 120, 1082-1083.

14. Quijada, R., Rojas, R., Bazan, G., Komon, Z. J. A., Mauler, R. S. and Galland, G. B. (2001), Synthesis of branched polyethylene from ethylene by tandem action of iron and zirconium single site catalysts, *Macromolecules*, 34, 2411-2417.
15. Komon, Z. J. A., Bu, X. and Bazan, G. C. (2000), Synthesis of butene-ethylene and hexene-butene-ethylene copolymers from ethylene via tandem action of well-defined homogeneous catalysts, *Journal of the American Chemical Society,* 122, 1830-1831.
16. Komon, Z. J. A., Bu, X. and Bazan, G. C. (2000), Synthesis, characterization, and ethylene oligomerization action of $[(C_6H_5)_2PC_6H_4C(O\text{-}B(C_6F_5)_3)O\text{-}^2P,O]Ni(^3\text{-}CH_2C_6H_5)$, *Journal of the American Chemical Society,* 122, 12379-12380.
17. Komon, Z. J. A., Bu, Diamond, G. M., Leclerc, M. K., Murphy, V., Okazaki, M. and Bazan, G. C. (2002), Triple tandem catalyst mixtures for the synthesis of polyethylenes with varying structures, *Journal of the American Chemical Society,* 124, 15280-15285.
18. Jiaxi, W. and Marks, T. J. (2003), Polynuclear catalysis: Enhancement of enchainment cooperativity between different single.site olefin polymerization binuclear catalysts, Abstract ISHHC – 11, July 20 – July 25 Northwestern University, Illinois USA.
19. Liting, L., Metz, M. V., Li H., Chen, M.-C., Marks, T. J., Liable-Sands, L. and Rheingold, A. L. (2002), Catalyst/cocatalyst nuclearity effects in single-site polymerization. Enhanced polyethylene branchingand α-olefin comonomer enchainment in polymerizations mediated by binuclear catalysts and cocatalysts via a new enchainment pathway, *Journal of the American Chemical Society*, 124, 12725-12741.
20. Coates, G.W. (2000), Precise control of polyolefin stereochemistry using single-site metal catalysts, *Chemical Reviews*, 100, 1223-1252.
21. Lieber, S. and Brintzinger, H-H. (2000), Propene polymerization with catalyst mixtures containing different *ansa*-zirconocenes: chain transfer to alkylaluminum cocatalysts and formation of stereoblock polymers, *Macromolecules*, 33(25), 9192-9199.
22. Barazzoni, L., Menconi, F. Ferrero, C., .Moalli, A., Masi, F., Altomare, A., Solaro, R. and Ciardelli, F. (1993), Characterization of supported bimetallic catalysts for ethylene polymerization: the effect of external Lewis bases, *Journal of Molecular Catalysis,* 82, 17-27; Ferrero, C., Masi, F., Menconi, F., Barazzoni, L., Moalli, A., Altomare, A., Solaro, R. and Ciardelli, F. (1993), Effect of protic compounds on ethylene polymerization by Hf-Ti supported catalysts, *Polymer*, 34, 3514-3519.
23. [Chien, J. C. W., Iwamoto, Y., Rausch, M. D., Wedler, W. and Winter, H. H. (1997), Homogeneous binary zirconocenium catalyst systems for propylene polymerization. 1. Isotactic/atactic interfacial compatibilized polymers having thermoplastic elastomeric properties, *Macromolecules*, 30, 3447-3458; Chien, J. C. W., Iwamoto, Y. and Rausch, M. D. (1999), Homogeneous binary zirconocenium catalysts for propylene polymerization. II. Mixtures of isospecific and syndiospecific zirconocene systems, *Journal of Polymer Science, Part A: Polymer Chemistry*, 37, 2439-2445.
24. Beach, D. L. and Kissin, Y. V. (1986), Dual function catalysis for ethylene polymerization to branched polyethylene (II) homogeneous heterogeneous Ziegler-Natta catalyst system, *Journal of Polymer Science, Part A: Polymer Chemistry*, 24, 1069-1075.
25. US5331070 (1994), Phillips Petroleum Company, invs.: Pettijohn, T. M.,Reagan, W. K. and Martin, S. J.

26. Benham, E. A., Smith, P. D. and McDaniel, M. P. (1988), *Polymer Engineering Science*, 28, 1469-1474.
27. US5616529 (1997), Bayer AG, invs.: Ostoja-Starzewski, A. K. -H., Witte, J., Bartl, H., Reichert, K. -H. and Vasiliou, G.
28. Musikabhumma, K., Spaniol, T. P. and Okuda J. (2003), Synthesis of branched polyethylenes by the tandem catalysis of silica supported linked cyclopentadienylamido titanium catalysts and a homogeneous dibromo nickel catalyst having a pyridylimine ligand, *Journal of Polymer Science: Part A: Polymer Chemistry*, 41, 528-544 .
29. Polypropylene Handbook, E. P.Moore Jr Ed, Hanser, Munich, Vienna, New York, 1996, pp. 92-94, 217-222.
30. Ahn, T. O., Hong, S. C., Kim, J. H. and Lee, D. -H. (1998), Control of molecular weight distribution in propylene polymerization with Ziegler-Natta/metallocene catalyst mixtures, *Journal of Applied Polymer Science,* 67(13), 2213-2222; Ahn, T. O., Hong, S. C., Huh, W. S., Lee, Y.-C. and Lee, D-H., Modification of a Ziegler-Natta catalyst with a metallocene catalyst and its olefin polymerization behavior, *Polymer Engineering and Science* (1999), 39(7), 2213-2222.
31. Lisovskii, A., Shuster, M., Gishvoliner, M., Lidor, G. and Eisen, M. S. (1998), Polymerization of propylene by metallocene and Ziegler-Natta mixed catalytic systems, *Journal of Polymer Science, Part A: Polymer Chemistry*, 36(17), 3063-3072.

FREE RADICAL PROCESSES IN WATER DISPERSION : NEW DEVELOPMENTS

BERNADETTE CHARLEUX, CELINE FARCET, CARINE BURGUIERE, and JEAN-PIERRE VAIRON
Unité de Chimie des Polymères – UMR 7610, Université P. et M. Curie, 4 Place Jussieu, 75005 Paris, France

Abstract: The paper describes the connections between the polymerization in dispersed medium and on the so called controlled living polymerization (CRP) taking as a descriptive basis the recent contribution to the domain coming from the authors research group. The synthetic description regards first of all the controlled free radical polymerization in water duispersions giving an overview of the different processes of mediation based either on nitroxide (NMP), atom transfer (ATRP) or reversible transfer (degenerative and addition-fragmentation/RAFT). A more detailed description regards the NMP of styrene with butylacrylate in acquesous emulsion and miniemulsion and the use of well defined amphiphilic block copolymers obtained by CRP as surfactants in emulsion polymerization.

Key words: free radical, emulsion polymerization, living radical polymerization, NMP, ATRP, RAFT, miniemulsuion, amphiphilic copolymers.

1. INTRODUCTION

One of the most recent and striking developments in polymer synthesis is undisputably the so-called « controlled/living radical polymerization » (CRP) which allows via a simple, cheap process, the elaboration of a broad variety of polymers with precisely defined molar masses, distributions, functionalities and architectures.[1,2,3] However, till now, the large majority of studies concerns the polymerizations in homogeneous medium (bulk or solution), whereas the industrial radical polymerization largely proceeds in

F. Ciardelli and S. Penczek (eds.),
Modification and Blending of Synthetic and Natural Macromolecules, 39–46.

aqueous dispersed medium. Thus the transposition of CRP processes towards the elaboration of latexes of well defined polymers is a a key objective. First papers on CRP in dispersed medium appear in the late 90's [4] and the domain is now rapidly developing with many publications in 1999-2003.

This presentation will focus on the linked aspects of the polymerization in dispersed medium and of the CRP. After a brief reminder (tutorial) of the main features of conventional polymerization in dispersed medium (suspension, emulsion, miniemulsion) and on CRP, we shall share the presentation into two separate parts devoted essentially to our own recent contributions to the domain,

- the nitroxide controlled free radical (co)polymerization (NMP) of styrene and butylacrylate in aqueous emulsion and miniemulsion, mediated by a low decomposition temperature nitroxide, the SG1.
- the use of well defined amphiphilic block coplymers of different architectures obtained via CRP, as surfactants in conventional emulsion polymerization.

2. BASICS ON POLYMERIZATION IN DISPERSED AQUEOUS MEDIUM AND ON CONTROLLED/LIVING RADICAL POLYMERIZATION

The characteristics and main differences of the conventional free radical polymerisations in suspension, emulsion and miniemulsion will be briefly presented. As the suspension process behaves like the bulk one, we shall essentially consider the *emulsion* and *miniemulsion* systems[5,6], from the view points of the initial (recipe) and final (latex) stages. The different steps (generation of radicals and initiation, nucleation and particles formation, particles growth) will be successively examined. The advantages and drawbacks of polymerization in water dispersion will be pointed out.

As the Controlled Radical Polymerization will be considered in more detail by another speaker during this session, we shall limit our presentation to an overview of the different processes of mediation based either on nitroxide (NMP), atom transfer (ATRP), or reversible transfer (degenerative, and addition-fragmentation/RAFT).

3. CONTROLLED FREE RADICAL POLYMERIZATION IN WATER DISPERSION

The very first observation of a « living » character for a polymer obtained in emulsion is as old as 1974, long before the emergence of the CRP concept in the 90's. Mikulasova et al noticed this living character for the polymerization and block copolymerization of styrene at 35°C when using a heterogeneous catalyst insoluble in both water and organic phases.[7] After separation of the catalyst, the polymerization proceeded to completion, and molar masses of $\sim 10^6$ $g.mol^{-1}$ were obtained with a very narrow polydispersity. The livingness was attributed to a rapid initiation associated to an efficient segregation of radicals inside the particles drastically limiting the termination. But the experimental technique (process and reactor) is complex, and it is only recently –essentially the past six years- that the NMP, ATRP and RAFT methods, which allow to use conventional devices and process, were applied to CRP in water dispersion.

The NMP has been considered either in suspension[8,9], in emulsion[10-13] (seeded or ab initio) but mostly in miniemulsion[14-27]. Initiating systems can be either bicomponent (conventional radical initiator + nitroxide) or monocomponent (alcoxyamine). Various nitroxides and their alcoxyamines have been used but the efficiency of the control (MM, MWD, low irreversible termination) together with particle size/distribution and latex stability are strongly depending on the nature of the considered nitroxide, its partitioning between organic and water phase and the activation/deactivation pseudo equilibrium constant for a given temperature. Furthermore the occurrence of thermal self-initiation -particularly for styrenics at T>120°C- can lead, as was observed in bulk or solution CRP, to much lower MM than expected and to a broadening of MMD. The miniemulsion procedure, using hydrophobic control agents or hydrophobic initiators (alcoxyamines), appears simpler as the nucleation step is avoided and the number of particles is predetermined. There is no large monomer droplets as in emulsion and the loci of polymerization are exclusively the preformed monomer microdroplets (ϕ = 50-500 nm). Most generally a better control of the CRP is observed in miniemulsion rather than in emulsion. Theoretically, for both water dispersed systems, the observed rate of CRP should be higher than in bulk or solution due to the compartmentalization of the radicals and to the lower rate of irreversible termination, providing small enough particles.[48] In fact the usual size of particles (50 to 500 nm) is larger than the critical value (depending on system) and no significant rate difference is observed between dispersed and bulk CRP systems.

Both direct and reverse ATRP processes have been considered in water dispersion, either in suspension[28,29], emulsion or miniemulsion[30-45]. The common features of NMP and ATRP in bulk or solution still persist in water dispersion, e.g. the kinetics are no more governed by the steady state between initiation and termination observed in conventional radical polymerization but by the activation-deactivation process as irreversible termination becomes negligible with respect to the reversible one and radical concentration is controlled by the persistent radical effect.[47] On the other hand there are obvious differences between NMP and ATRP. It has been shown that both activator (e.g. Cu^{I}) and deactivator (e.g. Cu^{II}) must be present simultaneously at the locus of polymerization (organic phase), and the nature of the ligand plays a key role. The partitioning of the ligated metal between water and organic phases depends upon the hydrophobic character of the ligand, but, in fact, there is always a more or less important concentration of the free metal salt in the water phase. The surfactant is also an essential parameter as it may interfere with the catalyst, and it should not affect the equilibrium between radical and dormant species. It must be cationic or nonionic but only the nonionics with suitable HLB resulted in stable latexes without coagulum. It can also enhance the living character of the system by reducing the diffusion rate of the metal complex to water phase. Again miniemulsion systems are preferred, and correct results were obtained for the homopolymerization, at rather low temperature, of acrylates, methacrylates and styrene.[38] The *in situ* block copolymerization might be considered but the potential secondary reactions of the dormant chain halogenated ends in water medium may lead to a loss of functionality and thus disturb the reinitiation of a second monomer.

The reversible transfer technique -essentially RAFT- has also been considered successfully in emulsion and miniemulsion. The same type of transfer agents as in bulk were used : the poly(methacrylic ester) macromonomers,[49] the dithiocarbonates (xanthates)[50-55] and the very reactive dithioesters[56-65] and dithiocarbamates[66]. The polymerisations controlled by reversible transfer are close to the conventional radical polymerization and should follow the same kinetics whatever the process (bulk, emulsion, miniemulsion). But the reaction in water dispersion appears much more complex than for NMP and ATRP. The rate and the quality of control are depending upon many parameters like the nature of the transfer agent, its transfer constant which must be large enough, the observed retardation effect, the process itself (batch, seeded batch, starved semi-batch, miniemulsion) which should allow to maintain a high [Tr.Agt.]/[M] ratio, the low (but critical) hydrophilicity of the RAFT agent necessary to ensure its transport from the monomer droplets to the locus of polymerization, the nature of the surfactant, etc..[67]

Owing to their complexity it would be too time consuming to detail the above systems in the presentation. We shall limit our talk to a comprehensive illustration using the simpler and « cleaner» miniemulsion (co)polymerizations of styrene and n-butylacrylate in either bi or mono component systems controlled by a « low temperature » nitroxide, the SG1.[23-27]

4. AMPHIPHILIC BI, TRI, AND STARS BLOCK COPOLYMERS PREPARED VIA CONTROLLED RADICAL POLYMERIZATION AND THEIR USE AS SURFACTANTS IN EMULSION POLYMERIZATION

There is presently a growing interest to use well-defined amphiphilic block copolymers as surfactants in emulsion polymerization. Hydrophilic/hydrophobic copolymers with convenient block lengths lead to a better steric (and/or electrosteric) stabilization of particles due to the hydrophilic (and/or polyelectrolyte) surrounding shell and to a lowering of the rate of surfactant exchange (better control of particle number and size).[68] Untill now well-defined amphiphilic block copolymers were usually synthesized using anionic living polymerization[69-72] but the new techniques of controlled radical polymerization, either atom transfer or nitroxides, were recently applied.[73-75] We intend to present here an overview of the CRP synthesis, characterization and surfactant behaviour of amphiphilic bi, tri and star-block copolymers consisting of hydrophilic blocks of poly(acrylic acid) and hydrophobic blocks of polystyrene, which we prepared and used as stabilizers in emulsion polymerization.

The copolymers of various chain structures were prepared either by ATRP (collab. K. Matyjaszewski) or by nitroxide control: AB diblock (A= styrene, B= tert-butylacrylate), ABA and BAB triblock, and three-arm star-block $(AB)_3$ copolymers. Those architectures were obtained in a *two step* procedure, using respectively mono-, di- and tri-functional initiators for the sequential polymerization of the monomers. Blocks lengths and copolymers compositions were varied. Then, the poly(tert-butyl acrylate) blocks were converted into poly(acrylic acid) by hydrolysis, to provide the corresponding amphiphilic copolymers. More recently we were able to synthesize these poly(styrene-*b*-acrylic acid) copolymers via a *direct* NMP method, and short comments will be done on the process.

Series of copolymers (ATRP) with various structural characteristics (composition, molar mass, hydrophilic/hydrophobic ratio) have been synthesized. Their micellization properties and their efficiency as stabilizers in emulsion polymerization of styrene have been studied and correlated with their characteristics and with various parameters like ionic strength, concentration or temperature. All architectures appear to be excellent stabilizers, efficient at much lower concentration ([M]/[S] < 0.5%) than the conventional surfactants. It was shown that the diblock copolymers are able to stabilize more particles than triblocks and star-block copolymers, and the optimal composition was about 10 styrene units and 50 acrylic acid units.[75] Block copolymers with a poly(acrylic acid) content larger than 75 mol%, are involved in both the creation of new particles and in the stabilization of continuously growing polymer/water interfaces in the system, as conventional surfactant does. Block copolymers with a poly(acrylic acid) content lower than 75 mol% act differently, as they participate exclusively in the creation of new latex particles.

5. CONCLUSION

These recent pioneering studies on CRP in water medium are rapidly expanding. The combination of the specificities of polymerization in water dispersion with those of the controlled/living radical polymerization will widely open the field of new latexes with well controlled characteristics on the view point as well of the structure of the polymer (MM, MWD, architectures, block and gradient copolymers, functionalization, etc..) as of the colloidal properties. The new generation of « low temperature » nitroxides (e.g. the SG1) and their alcoxyamines, associated with the miniemulsion process, already allows the elaboration of well defined homo and copolymers of the styrenic and acrylic families without noticeable modification of the conventional experimental process.

REFERENCES

1. Matyjaszewski K (1998). *Controlled Radical polymerization.* Am Chem Soc Symp Series, Washington DC, **685**.
2. Matyjaszewski K (2000). *ControlledLiving Radical Polymerization. Progress in ATRP, NMP and RAFT.* Am Chem Soc Symp Series, Washington DC, **768**.
3. Matyjaszewski K (2003). *Advances in Controlled/Living Radical Polymerization.* Am Chem Soc Symp Series, Washington DC, **854**.
4. Qiu, J.; Charleux, B.; Matyjaszewski, K. (2001) *Prog. Polym. Sci.* ***26***, 2083.
5. Gilbert, R. *Emulsion polymerization: a mechanistic approach.*; Academic press: London, 1995.
6. Sudol, E.; El-Aasser, M., *Emulsion polymerization and emulsion polymers*; Wiley: New York, 1997.
7. Mikulasova D, Chrastova V, Citovicky P (1974). *Eur Polym J*, ***10*** : 551-556.
8. Georges, M. K.; Veregin, R. P. N.; Kazmaier, P. M.; Hamer, G. K. (1993) *Macromolecules*, ***26***, 2987.
9. Schmidt-Naake, G.; Drache, M.; Taube, C. (1999) *Die Angewandte Makromolekulare Chemie*, ***265***, 62.

10. Bon, S. A. F.; Bosveld, M.; Klumperman, B.; German, A. L. (1997) *Macromolecules* ***30***, 324.
11. Marestin C, Noël C, Guyot A, Claverie J, (1998) *Macromolecules*, ***31*** : 4041-4044
12. Cao, J.; He, J.; Li, C.; Yang, Y. (2001) *Polym. J.* ***33***, 75.
13. Lansalot M. ; Farcet C. ; Charleux B. ; Vairon J.P., ; Pirri R. ; Tordo P ; p. 138-151 in Ref. 2
14. Prodpan T, Dimonie VL, Sudol ED, El-Aasser MS (2000). Macromol Symp,***155*** : 1-14.
15. Pan G, Sudol ED, Dimonie VL, El-Aasser MS (2001) Macromolecules, ***34*** : 481-488.
16. Pan G, Sudol ED, Dimonie VL, El-Aasser MS (2002) Macromolecules, ***35*** : 6915-6919.
17. MacLeod PJ, Barber R, Odell PG, Keoshkerian B, Georges MK (2000) Macromol Symp, ***155*** : 31-38.
18. Keoshkerian B, MacLeod PJ, Georges MK (2001). Macromolecules, ***34*** : 3594-3599.
19. Cunningham MF, Xie M, McAuley KB, Keoshkerian B, Georges MK (2002) Macromolecules, ***35***, 59-66.
20. Keoshkerian B, Szkurhan AR, Georges MK (2001). Macromolecules, ***34*** : 6531-6532.
21. Tortosa K, Smith J-A, Cunningham MF (2001). Macromol Rapid Commun, ***22*** : 957-961.
22. Cunningham MF, Tortosa K, Lin M, Keoshkerian B, Georges MK (2002) J Polym Sci Part A: Polym Chem, ***40*** : 2828-2841.
23. Farcet C, Lansalot M, Charleux B, Pirri R, Vairon JP (2000) Macromolecules, ***33*** : 8559-8570.
24. Farcet C (2002) Thèse de l'Université Pierre et Marie Curie, Paris
25. Farcet C, Charleux B, Pirri R (2001). Macromolecules, ***34*** : 3823-3826.
26. Farcet C, Charleux B, Pirri R (2001). Macromol Symp, ***182*** : 249-260.
27. Farcet C, Nicolas J, Charleux B (2002) *J Polym Sci Part A: Polym Chem*, ***40*** : 4410-4420.
28. Lecomte, P.; Drapier, I.; Dubois, P.; Teyssié, P.; Jérôme, R. (1997) *Macromolecules* ***30***, 7631.
29. Nishikawa, T.; Kamigaito, M.; Sawamoto, M.; (1997) *Macromolecules* ***32***, 2204.
30. Makino, T.; Tokunaga, E. (1998) *Polym. Prepr. (Am. Chem. Soc., Div. Polym. Chem.)* ***39***, 288.
31. Chambard, G.; Man, P. d.; Klumperman, B. (2000) *Macromol. Symp.* ***150***, 45.
32. Gaynor, S. G.; Qiu, J.; Matyjaszewski, K. (1998) *Macromolecules*, ***31***, 5951.
33. Matyjaszewski, K.; Qiu, J.; Shipp, D.; Gaynor, S. G. (2000) *Macromol. Symp.* ***155***, 15.
34. Qiu, J.; Shipp, D.; Gaynor, S. G.; Matyjaszewski, K. (1999) *Polym. Prepr. (Am. Chem. Soc., Div. Polym. Chem.)* ***40***, 418.
35. Qiu, J.; Pintauer, T.; Gaynor, S. G.; Matyjaszewski, K.; Charleux, B.; Vairon, J.-P. (2000) *Macromolecules* ***33***, 7310.
36. Jousset, S.; Qiu, J.; Matyjaszewski, K.; Granel, C. (2001) *Macromolecules* ***34***, 6641.
37. Gaynor, S. G.; Qiu, J.; Shipp, D.; Matyjaszewski, K. (1999) *PMSE* ***80***, 536.
38. Matyjaszewski, K.; Qiu, J.; Tsarevsky, N. V.; Charleux, B. (2000) *J. Polym. Sci. Part A: Polym.Chem*, ***38***, 4724.
39. Yoo, S. H.; Lee, J. H.; Lee, J.-C.; Jho, J. Y. (2002) *Macromolecules* ***35***, 1146.
40. Wan, X.; Ying, S. (2000) *J. Appl. Polym. Sci.* ***75***, 802.
41. Wan, X.; Ying, S. (1999) *Polym. Prepr. (Am. Chem. Soc., Div. Polym. Chem.)* ***40***, 1049.
42. Qiu, J.; Ph. D. de Carnegie Mellon University, Pittsburgh, 2000.
43. Matyjaszewski, K.; Shipp, D. A.; Qiu, J.; Gaynor, S. G. (2000) *Macromolecules* ***33***, 2296.
44. Shipp, D.; McMurtry, G. P.; Gaynor, S. G.; Qiu, J.; Matyjaszewski, K. (1999) *Polym. Prepr. (Am.Chem. Soc., Div. Polym. Chem.)* ***40***, 448.
45. Storsberg, J.; Hartenstein, M.; Müller, A. H. E.; Ritter, H. (2000) *Macromol. Rapid Commun.* ***21***, 1342.
46. Qiu, J.; Gaynor, S. G.; Matyjaszewski, K. (1999) *Macromolecules*, ***32***, 2872.

47. Fischer H (1997). *Macromolecules*, ***30*** : 5666-5672
48. Charleux B (2000). *Macromolecules*, ***33*** : 5358-5365.
49. Krstina J, Moad G, Rizzardo E, Winzor CL, Berge CT, Fryd M (1995). *Macromolecules*, ***28*** : 5381-5385.
50. Charmot D, Corpart P, Adam H, Zard SZ, Biadatti T, Bouhadir G (2000) Macromol Symp, ***150*** : 23-32.50.
51. Moad G, Chiefari J, Chong YK, Krstina J, Mayadunne RTA, Postma A, Rizzardo E, Thang SH (2000). *Polym Int*, ***49*** : 993-1001
52. Monteiro MJ, Sjöberg M, Vlist JVD, Göttgens CM (2000) *J Polym Sci Part A: Polym Chem*, ***38*** :4206-4217.
53. Monteiro MJ, de Barbeyrac J (2001) *Macromolecules*, ***34*** : 4416-4423.
54. Smulders W, Gilbert RG, Monteiro MJ (2003) *Macromolecules*, ***36*** : 4309-4318.
55. Monteiro MJ, de Barbeyrac J (2002) *Macromol Rapid Communications*, ***23*** : 370-374.
56. de Brouwer H, Tsavalas JG, Schork J, Monteiro MJ (2000) *Macromolecules*, ***33*** : 9239-9246.
57. Tsavalas JG, Schork FJ, de Brouwer H, Monteiro MJ (2001) *Macromolecules*, **34** : 3938-3946.
58. Monteiro MJ, Hodgson M, de Brouwer H (2000). *J Polym Sci Part A: Polym Chem*, ***38*** : 3864-3874.
59. Lansalot M, Davis TP, Heuts JPA (2002) *Macromolecules*, ***35*** : 7582-7591.
60. Uzulina I, Kanagasabapathy S, Claverie J (2000) *Macromol Symp*, ***150*** : 33-38.
61. Kanagasabapathy S, Sudalai A, Benicewicz BC (2001) *Macromol Rapid Commun*, ***22*** : 1076-1080.
62. Kanagasabapathy S, Claverie J, Uzulina I (1999) *Polym Prepr (Am Chem Soc Div Polym Chem)*, ***40*** : 1080-1081.
63. Butté A, Storti G, Morbidelli M (2001) *Macromolecules*, ***34*** : 5885-5896.
64. Vosloo JJ, De Wet-Roos D, Tonge MP, Sanderson RD (2002) *Macromolecules*, ***35*** : 4894-4902.
65. Prescott SW, Ballard MJ, Rizzardo E, Gilbert RG (2002) *Macromolecules,* ***35*** : 5417-5425.
66. Ferguson CJ, Hughes RJ, Pham BTT, Hawkett BS, Gilbert RG, Serelis AK, Such CH (2002) *Macromolecules*, ***35*** : 9243-9245.
67. Prescott SW, Ballard MJ, Rizzardo E, Gilbert RG (2002) *Aust J Chem*, ***55*** : 415-424.
68. Piirma,I "Polymeric Surfactants", in *Surfactant Science Series*, vol.42, Dekker, M, NY, 1992
69. Jialanella,G.L., Firer,E.M., Piirma,I., (1992) *J. Polym.Sci., A-Polym..Chem.*, ***30***, 1925
70. Valint Jr.P.L., Bock,J., (1988) *Macromolecules*, ***21***, 175
71. Leemans,L., Fayt,R., Ph. Teyssié, (1990) *J. Polym.Sci., A-Polym.. Chem..,* ***28***, 1255
72. Müller,H., Leube,W., Tauer,K., Förster,S., Antonietti,M., (1997) *Macromolecules*, ***30,*** 2288
73. Bouix,M., Gouzi,J., Charleux,B., Vairon,J.P.,. Guinot,P., (1998) *Macromol. Rapid Commun.*, ***19***, 209
74. Burguière,C., Pascual,S., Coutin,B., Polton,A., Tardi,M.,. Charleux,B., Matyjaszewski,K., Vairon,J.P., (2000) *Macromol. Symp.,* ***150***, 39-44 (2000)
75. Burguière,C., Pascual,S., Bui,C., Vairon,J.P., Charleux,B., Davis,K.A., Matyjaszewski,K., Betremieux,I., (2001) *Macromolecules*, ***34,*** 4439-4450

FUNCTIONALIZATION OF POLYOLEFINS IN THE MELT

FRANCESCO CIARDELLI[1,2], MAURO AGLIETTO[1], MARIA BEATRICE COLTELLI[1], ELISA PASSAGLIA[2], GIACOMO RUGGERI[1], and SERENA COIAI[3]

[1]Dipartimento di Chimica e Chimica Industriale, Università di Pisa, Via Risorgimento 35, 56126 Pisa (Italy); [2]ICCOM-CNR Pisa-Station, c/o Università di Pisa, Via Risorgimento 35, 1 50126 Pisa, Italy; [3] Scuola Normale Superiore, Piazza dei Cavalieri, Pisa, Italy

Abstract: Polyolefins with a controlled amount of functional groups can be obtained either by catalytic copolymerization of suitable monomers or by chemical modification of the obtained macromolecules by a free radical reaction in bulk. This paper describes the main aspects of the latter method which is at present the one more extensively used. Thus examples are discussed showing that the modification can be modulated to a great extent depending on feed composition and macromolecular structure. For instance in case of linear polyethylene conditions can be found by which the sole grafting reaction takes place and can be predicted with a simple mathematical model based on a reaction with first order kinetic with respect to the grafting monomer. Under these optimal conditions mixtures of comonomers can be used to introduce in one step two different reactive groups. On the other side it is shown that degradation is favoured by the extensive branching while linearity and double bonds favour crosslinking. Proper coagents are indicated able to limit degradation thus allowing crosslinking and grafting of functional monomers even in case of polypropylene. On the other side the use of transfer agents can hinder crosslinking of highly unsaturated polymers. Selectivity is also observed with preferential conversion of vinyl versus internal double bonds and aliphatic blocks versus styrene blocks.

Key words: functionalization, polyolefins, functionalization kinectics, ethylenepolymers .propylene polymers, styrene polymers, maleic acid derivatives.

F. Ciardelli and S. Penczek (eds.),
Modification and Blending of Synthetic and Natural Macromolecules, 47–71.

1. INTRODUCTION

Polyolefins with a controlled amount of functional groups can be obtained either by catalytic copolymerization of suitable monomers or by chemical modification of the obtained macromolecules. Due to the modest reactivity of the substantially paraffinic structure, polyolefins are in general functionalized by free radical grafting reactions which are more conveniently carried out in bulk at a temperature high enough to grant the proper rheology, accordingly the reactions are carried out in the melt. The process can be further facilitated by introducing through catalysis specific chemical species in the macromolecular chain and these species can help the successive post-modification.

Several examples are reported in the literature describing the copolymerization of 1-alkenes with various monomers in the presence of heterogeneous and homogeneous transition metal catalysts for the preparation of polyolefin macromolecules containing a certain amount of polar groups. The approach has encountered several difficulties due to the significant sensitivity of the active catalytic species to heteroatoms and reactive groups. Also the lewis basicity of the functional comonomer can compete with the olefinic double bond in the complexation to the transition metal.

This problem was partially overcome by sterically protecting the functional groups or by the complexation of the functional monomers with aluminum alkyls[1] which, as also shown in recent contributions[2], can be used with the same catalytic systems used for monoalkenes homopolymeration. However partial deterioration of catalytic performances is observed both for polymerization activity and macromolecular properties. More recently late transition metal catalysts have resulted less sensitive to heteroatoms and more suitable for copolymerization of monoalkenes with polar monomers[3,4]. In these systems however a parallel polymerization mechanism based on free radical propagating species can take place.

The possibility of copolymerizing polar monomers with olefins using transition-metal catalysts was discussed thoroughly in a recent review paper [5]. Several examples are reported where copolymers were obtained but in any case the paper concludes that *"Despite near half-century of work, the smooth incorporation of functional groups into polyolefins still remain a challenging area for further research into catalyst development"*.

In this situation the post modification of polyolefins in order to introduce the desired functional groups remains a very convenient alternative route[6] which is carried out in the melt[7,8] or in solution[9,10] by means of organic peroxides as free radical initiators. The most commonly used functional molecule has been maleic anhydride[11,12] as the high reactivity of the anhydride group in the successive reaction[6,13] is quite convenient for reactive blending application. The reaction in the melt is affected by some limitations due to the very low reactivity of the paraffinic macromolecules[14].

Accordingly it was necessary to work in the melt and with free radicals thus obtaining limited specificity and very low incorporation of polar species. In the present work it is reported that the modification can be modulated to a great extent depending on feed composition and macromolecular structure. On the other side degradation is favoured by extensive branching while linearity and double bonds give crosslinking. Mixtures of comonomers can be used to introduce in one step two different reactive groups and transfer agents can hinder crosslinking. Selectivity is also observed with preferential conversion of vinyl versus internal double bonds and of aliphatic blocks versus styrene blocks[15].

In the literature the most studied functionalizing monomer for polyolefins is maleic anhydride (MAH)[16,17]. Despite the large number of studies on MAH grafting and the commercial success of MAH grafted polyolefins, the chemical mechanism involved in the functionalizing process is not fully understood. Several studies have shown that the reaction pathways depend on the polyolefin molecular structure. When a peroxide is used as initiator, crosslinking or chain scission may occur simultaneously with grafting reaction. Polyolefins with different ethene/propene ratios were grafted with MAH both in the melt and in solution[18]. The results show that the extent of MAH graft, expressed as the ratio E/f, where the extinction coefficient (E) of the anhydride peak at 1785 cm^{-1} is normalized by the film thickness (f), is low ($E/f \cong 0{,}004$) for polyolefins with high propene content and high ($E/f \cong 0{,}020$) when the propene content is below 50% wt. Moreover a decrease of the propene content of the PO results in a transition from degradation to branching/crosslinking as the predominant reaction in the system.

The maleic anhydride reactivity as functionalizing agent has been compared with that of other unsaturated polar molecules such as dibutyl maleate, diethyl fumarate, itaconic anhydride[9]. The functionalization reaction of ethylene-propylene copolymer has been studied in solution in the presence of dicumyl peroxide and dibenzoyl peroxide as radical initiators. In these conditions, maleic anhydride showed the highest reactivity probably because of the higher reactivity of the double bond. The authors observed also the formation of poly(itaconic anhydride) side chains when itaconic anhydride was used as functionalising monomer. Moreover, chain degradation of the rubber could not be avoided. The homopolymerization of maleic anhydride was also observed[11] during the functionalization of low density polyethylene (LDPE) and the authors explained the results considering the electron transfer of PE-MAH radical to maleic anhydride followed by ionic coupling. The presence of dimethyl formamide, as transfer agent, hinders the ionic mechanism leading to the stabilization of PE-MAH radical not directly involved in propagation reactions.

Very recently[19] maleic anhydride grafted polyethylene, [2,3-$^{13}C_2$] MAH-g-PE, which was synthesized with ^{13}C labeled maleic anhydride [2,3-$^{13}C_2$] MAH in solution, was characterized by ^{13}C-NMR spectroscopy in order to make clear the structure of grafted groups. The results reveal that [2,3-$^{13}C_2$] MAH-g-PE has succinic anhydride oligomeric grafts with a terminal unsaturated MAH ring in addition to well-known saturated succinic anhydride oligomeric grafts and that the former grafts are longer but fewer than the latter.

The high reactivity of anhydride groups towards the functional groups of many types of filler[20] and terminal groups of condensation polymers[21] was observed by several authors and the modified polyolefins have been used to obtain composites and blends with excellent mechanical properties. These results drove the research to the optimization of the modification with maleic anhydride in particular conditions for specific applications, for example on the surface of LDPE films[22,23] by photografting polymerization. Moreover the functionalization with MAH of different polyolefins, as polypropylene[24] and block copolymers as acrylonitrile-butadiene-styrene terpolymer[25] and SBS[26] was recently studied.

In previous papers we have reported that polymers of monoalkenes [HDPE[27]; EPR, 17], isotactic and atactic polypropylene[18] can be functionalized in the melt with unsatured dicarboxylic acid esters such as diethyl maleate (DEM): in the presence of a peroxide initiator. The structure of the grafted groups was shown[19] to be that of 2-(diethyl succinate) (DES) (scheme 1).

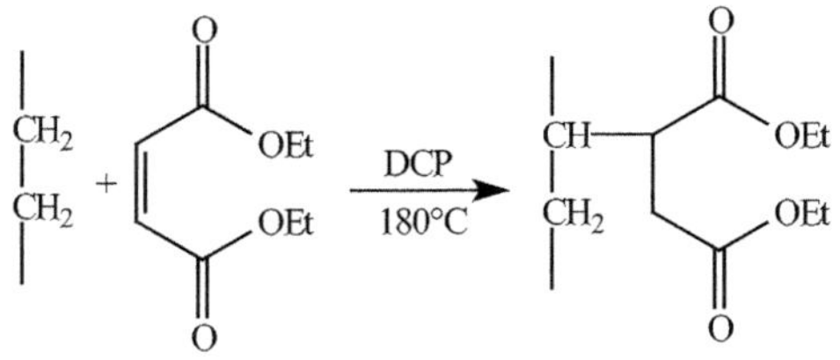

Scheme 1. Grafting of DEM onto HDPE.

Indeed several side reactions can occur due to the reaction conditions (high temperature and free radical mechanism) and the quantitative aspects are not well understood. Therefore a detailed study under well defined conditions was considered very useful in order to grant a better reproducibility of the functionalization degree (FD) and put some light on the whole mechanism.

The experimental data reviewed in this paper derive from experiments performed in our laboratory. Diethyl maleate (DEM) was used as functionalizing agent and dicumyl peroxide (DCP) as free radical generator. The polyolefin structure was varied regularly from LLDPE to EPR and PP. A comparison with the particular behaviour of styrene polymers in these conditions is also reported.

2. KINETICS OF FUNCTIONALIZATION

The functionalization of linear polyethylene (HDPE) in the melt by using dicumyl peroxide (DCP) for the grafting of diethyl maleate (DEM) showed to be a relatively clean reaction bearing to well defined functional groups grafted to the polyolefin backbone[19]. The system therefore was selected as the most suitable for attempting a detailed investigation of the reaction mechanism and kinetics.

The degree of functionalization FD (number of DES groups for 100 monomeric units) was determined with good accuracy on the basis of the comparison of FT-IR spectra of the functionalized polyolefin with standard mixtures of the same polyolefin and polydiethylfumarate of known composition. In many cases an independent control was carried out by NMR.

The general route followed is summarized below.

The IR spectra of the standard mixtures show bands typical of both polymers, the most useful for their high optical density being those at 1738 cm^{-1} (stretching C=O; A1) and at 1460 cm^{-1} (bending $-CH_2$; A2). By plotting the band areas ratio (A1/A2) *vs* the composition of the mixture, in the range of composition going from 14% wt of poly(diethyl fumarate) to 2% wt (corresponding to an FD from 2.8 to 0.3 % by mol), a good correlation was obtained. This suggests that the A1/A2 band area ratio can be safely used for detecting FD of EPR and LLDPE2 (containing 9% of 1-butene) samples functionalized by DEM grafting. Before using directly the A1/A2 values of different functionalized samples a control of the quantitative procedure has been carried out comparing with the results obtained by ^{13}C-NMR spectroscopy. Indeed, a good linear correlation between the ratio A1/A2 of the peak areas and FD, as determined by ^{13}C-NMR (at 12-14 ppm, $-O-CH_2-CH_3$, and at 9-10 ppm, $-CH_3$ of 1 butene), was obtained in the case of functionalized samples derived from LLDPE1 containing 8.7 mol% of propene and 5.2 mol% of 1-butene.

Thus, FD depends linearly on the ratio A1/A2 of the peak areas in the functionalization range 0.2-2.8 mol % of 2-(diethyl succinate) groups grafted on the polyolefin backbone of both LLDPE and EPR. The correlation (eq. 1) can then be used for ethylene polymers with different structure; clearly different values of the coefficient *a* and *b* of eq. 1 are observed for the various samples:

$$FD = a + b(A1 / A2) \qquad (1)$$

According to well-established knowledge about radical reactions, the functionalization reaction with DEM in the presence of DCP can be described on the basis of a series of successive and parallel reactions reported in scheme 1. The simplified mechanism is based on the initial formation of free radicals from the thermal decomposition of DCP, reaction (a), which generates macroradicals by hydrogen abstraction from the macromolecules, reaction (b). The formed macroradicals quickly react with the unsaturated functionalizing monomer (DEM), reaction (e), while chain propagation is promptly interrupted by transfer with the polyolefin chains, reaction (f). Indeed, ^{13}C NMR analysis of the functionalized polyolefins showed that the side-chains contain grafted chains consisting of a single unit of the functional monomer. It is noteworthy that the homopropagation of DEM initiated by RO radicals is to a large extent inhibited as demonstrated by experimental data[14] obtained from the analysis of the functionalization reaction products.

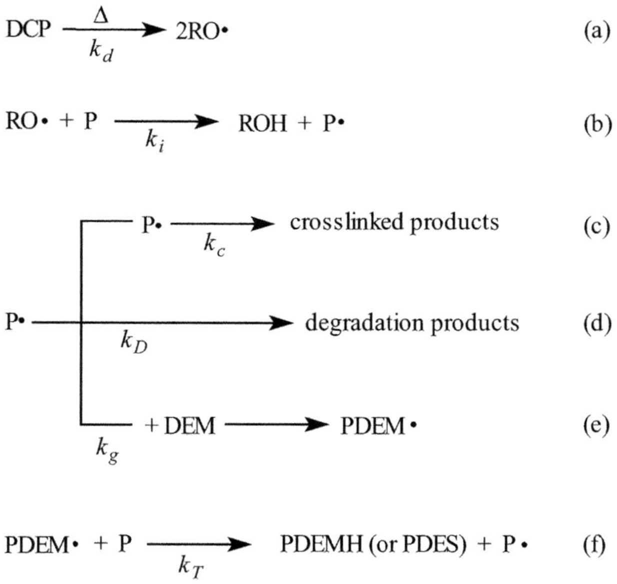

Scheme 2. Simplified reaction mechanism of polyolefin (P) functionalization with diethyl maleate (DEM) and dicumylperoxide (ROOR) as initiator.

According to formal kinetic equations based on the scheme 2, the variation of FD with time (dFD/dt), that is the rate of formation of PDEMH (equivalent to PDES which is a polyolefin with diethyl succinate grafted groups), is given by the rates of v_e+v_f of the successive reactions (e) and (f) of the scheme 2. However, as the reaction (f) is very rapid due to the high reactivity of PDEM· free radicals and the large excess of P over PDEM·, the rate determining step is that of reaction (e). So the time evolution of the functionalization can be determined by equation (2).

$$\frac{dFD}{dt} = k_g[U\bullet][DEM] \tag{2}$$

Where (U)=(P)/(molecular weight of the average monomeric unit) indicates the molar concentration of monomeric units of the unreacted polyolefin.

By introducing the stationary state hypothesis it is possible to write:

$$\frac{dFD}{dt} = k_g[U\bullet][DEM] = k_g\left(\frac{2fk_d[DCP]_0}{k_{TE}}\right)^{1/2}[DEM] \tag{3}$$

Here k_g and k_d are defined in scheme 2, k_{TE} is a rate constant for termination reaction end and f is the initiator efficiency.

Under isothermal conditions and constant chemical composition, it is possible to write:

$$k_g\left(\frac{2fk_d[DCP]_0}{k_{TE}}\right)^{1/2} = k_{FD} \tag{4}$$

Therefore it is possible to write:

$$\frac{\mathrm{dFD}}{\mathrm{d}t} = k_{\mathrm{FD}}[\mathrm{DEM}] \tag{5}$$

Thus one can conclude that FD follows a first order kinetic with respect to the monomer concentration.

A mathematical model based on the proposed kinetic scheme was developed to analyse the experimental values obtained from the DSC runs: it permits the evaluation of the role of secondary reactions such as crosslinking (c) and degradation (d) with respect to the grafting reaction (e).

The modelling of experimental data is based on the solution of eq. 4 in non-isothermal conditions. Starting with the mass balance equation of the monomer (DEM), the reaction rate relevant to the grafting reaction is given by eq. 5 which can be written as:

$$r = -\frac{d[\mathrm{DEM}]}{dt} = k_{\mathrm{FD}}[\mathrm{DEM}] \tag{6}$$

The reaction rate should than be proportional to the heat flow rate determined experimentally by DSC:

$$r = \frac{1}{[\mathrm{DEM}]_0 \, \Delta\mathrm{H}_r} q \tag{7}$$

where ΔH_r is the molar enthalpy of reaction and q the heat flow rate.

The values of the pre-exponential factor and the activation energy were determined by fitting the experimental data, starting with the average values of the data of runs during which degradation and chain extension can be neglected.

The comparison shows a good agreement of the experimental heat flow and the reaction rate curve obtained using the model, relevant to experimental conditions where the grafting process is not affected by secondary reactions (DEM/DCP>7) (Figure 1).

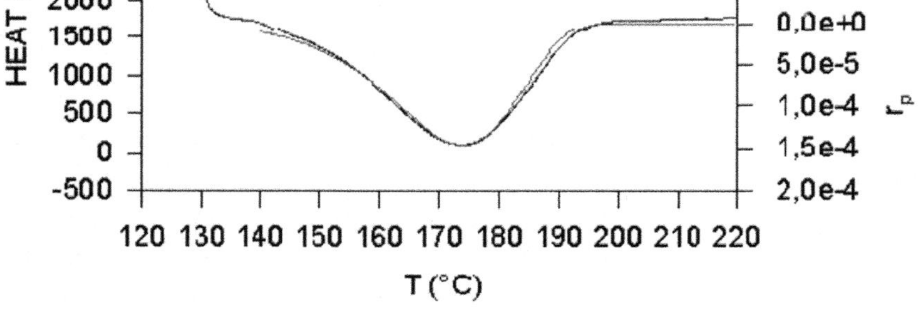

Figure 1. Comparison between DSC curve and reaction rate obtained by model for a sample prepared with a DEM/DCP>7 ratio

The profiles of heat flow and reaction rate versus time are in very good agreement. When the concentration of DPC increases up to DEM/DCP<7, competition between crosslinking and grafting reactions can be observed. Under these conditions the experimental heat flow and the reaction rate curve obtained by the model do not fit well at higher temperatures (figure 2).

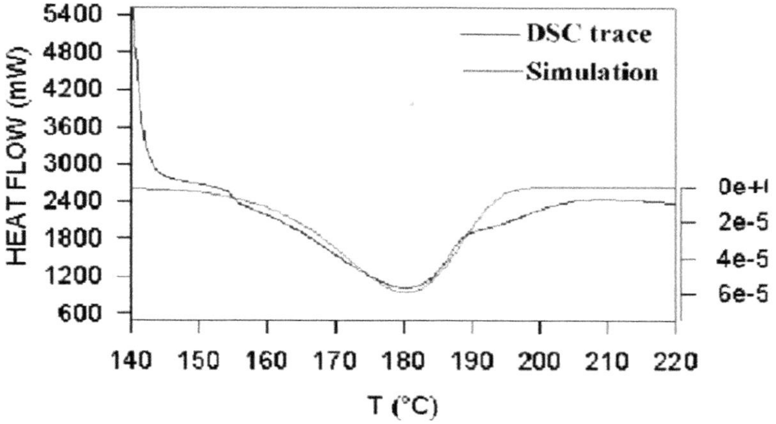

Figure 2. Comparison between DSC curve and reaction rate obtained by model for a sample prepared with a DEM/DCP<7 ratio

3. FUNCTIONALIZATION OF ETHYLENE POLYMERS (LLDPE, VLDPE AND EPM)

Ethylene polymers containing variable amounts of α-olefin (generally propylene) comonomer were investigated in order to evaluate the polymer structure effect on the functionalization process (table 1). The functionalization reactions were carried out in the Brabender mixer with 30 min. residence time and at constant temperature, 190°C for EPM and 200°C for LLDPE samples, while varying the number of branching in the starting polyolefin, the amount of DEM and of dicumyl peroxide (DCP). The starting conditions and the related functionalization degree (FD) are reported in table 1. At the end of the reaction, the polymeric product was treated with boiling acetone to remove low molecular products (DEM, DCP residues and oligomers) formed during the reactions as the starting polyolefins were 100% insoluble in that solvent. In general, the acetone insoluble polymer was soluble in heptane and the FDs reported in table 1 refer to this fraction.

When the toluene extraction left a residue, the FD was calculated respect to the whole acetone insoluble material in order to detect the total amount of grafted groups.

The mass balance cannot be complete with reference to the DEM used as a fraction of the unreacted DEM is lost by volatility during the opening of the Brabender mixer still. Also some DEM could be collected as oligomer or low molecular weight functionalized molecules in the acetone extracted material together with some short polyolefin chains formed upon degradation. The amount of the acetone-soluble fraction was always less than 5% wt of the whole material, except when a very large amount of DEM (>30% wt) was used, thus indicating that in any case degradation to short chains (PM ≅ 1000) and DEM oligomers formation are minor processes. The amount of material extracted with acetone increases with the amount of DEM in the feed. Both GC and MS analyses showed that these fractions contain unreacted DEM, cumyl alcohol and a DEM/peroxide adduct. The cumyl alcohol is formed by decomposition of DCP and transfer reaction to the polyolefin and DEM/peroxide adduct is formed through reaction of DEM with primary radicals. The cross-linking reaction or even chain extension leading to materials insoluble in heptane and toluene seems to be controlled by the DCP/DEM ratio and substantially avoided when working at ratios lower than 0.09. In all other cases, the analysis of the fraction soluble in heptane provides an adequate description of the real extent of the functionalization reaction and can be correlated with the structure of the polyolefin and the starting amount of DEM and DCP.

Table 1. Relative efficiency (E) of free radical derived from DCP at various DCP/DEM ratios.

	C2/C3+C4[a]	*DCP/DEM*	*E*[b]	*FD*	*Acetone ex. % wt*	*Heptane % wt*
LLDPE1	6.2	0.22/1.80	5	1.0	1.2	60.7
LLDPE2	6.2	0.11/1.80	7	0.7	2.8	97.2
LLDPE3	6.2	0.11/3.60	12	1.2	3.1	96.9
VLDPE1	10.1	0.03/0.50	7	0.2	0.4	99.6
VLDPE2	10.1	0.11/1.80	9	1.0	0.6	99.4
VLDPE3	10.1	0.05/1.80	10	0.5	1.0	99.0
EPM1	2.3	0.22/2.70	9	1.9	1.9	98.1
EPM2	2.3	0.11/1.80	12	1.2	1.3	98.7
EPM3	2.3	0.03/0.50	17	0.5	1.0	99.0

a) C2/C3+C4 = ethylene/propylene + 1-butene units ratio; b) E = number of grafted DEM units/moles of DCP.

A comparison between the total amount of DCP and the corresponding FD in all performed experiments (table 1) clearly shows that FD is much larger than the initial moles of DCP confirming the important role of the transfer reaction involving radicals formed after the grafting of the first

DEM unit. Because of this reaction grafted functional groups are monomeric as found by ^{13}C NMR analysis[30] and each primary radical formed through peroxide decomposition can give no more than one grafted group on the polyolefin macromolecule backbone. The relative efficiency (E) of the DCP primary radicals for the grafting reaction can be then evaluated to be dependent on the DCP/DEM ratios and on the secondary/tertiary carbon ratio in the polyolefin macromolecules (table 1). Indeed, E seems to increase by reducing the concentration of DCP, as expected considering the lower probability of termination reactions involving the encounter between two free radicals. A certain increase of the grafting efficiency is observed on going from LLDPE samples to EPM, probably due to larger branching hindering recombination. For obvious reasons, the increase of DEM content in the reaction feed should increase FD and a similar consideration should be valid for DCP, even though its efficiency decreases by increasing the DEM/DCP ratio. Such a dependence of FD on feed composition and polymer structure is clearly evidenced by the results obtained for the various polyolefins under different starting conditions (table 1). In all experiments carried out with variable amounts of DEM, the functionalization degree increases almost linearly. FD also increases with the DCP/DEM ratio but in an asymptotic way and at DCP/DEM ratios higher than 0.09 the formation of cross-linked material is observed. The grafting in any case seems to be positively affected by a larger number of tertiary C-atoms (from EPM to LLDPE) and by a decreased viscosity in the melt (MFI: LLDPE1 = 2.2, LLDPE2 = 1.6). The yield of grafted DEM increases with the DCP/DEM ratio in the same way at any DEM/P value but the efficiency of DCP decreases, probably due to the improved probability of free-radical coupling supported by the formation of cross-linked material at DCP/DEM ratios larger than 0.09. According to the experimental results described here and to the proposed mechanism it is possible to graft in one step two different functional groups by feeding the Brabender reactor with a mixture of two monomers (scheme 3).

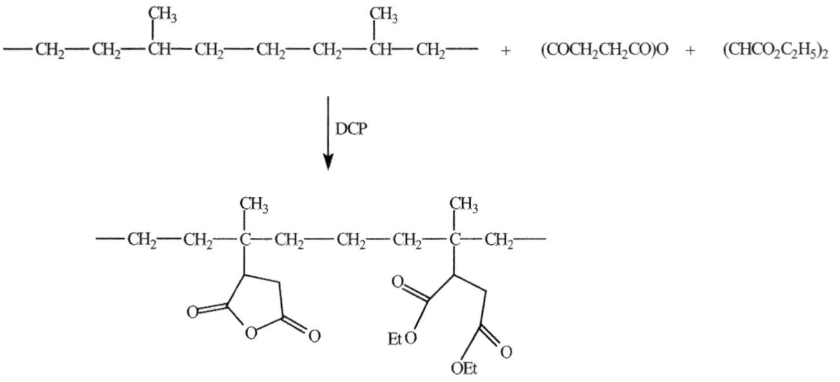

Scheme 3. Contemporary grafting of DEM and MAH onto ethylene polymers.

While the general trend of this reaction followed the previous mechanism giving consistent results, some interference was observed between the two monomers leading to a compositional ratio in the polymer product different from the monomer feed.

In this case parallel experiments were carried out by using both a discontinuous Brabender mixer and a twin-screw extruder. This last resulted more efficient with almost complete grafting yield (table 2).

Table 2. Functionalization of EPM with DEM/MAH mixture and DCP in Brabender mixer or twin-screw extruder

RUN	MIXER	MAH (mol)	DEM (mol)	DCP (mol)	FD_{MAH}	FD_{DEM}	E[a)]
A4	Extruder	0.015	0.0025	0.00045	0.013	0.002	14.9
A5	Extruder	0.029	0.0051	0.00090	0.023	0.002-0.003	14.5
A6	Extruder	0.035	0.0062	0.00125	0.030	0.002-0.003	13.5
B1	Brab.	0.5	1.9	0.12	0.57	0.28	3.7
B2	Brab.	1.9	0.5	0.12	0.28	0.43	3.7
B3	Brab.	0.6	0.6	0.12	0.33	0.35	3.7

a) E = number of grafted DEM and MAH units/moles of DCP.

4. FUNCTIONALIZATION OF ISOTACTIC POLYPROPYLENE

The use of radical iniziators, during the processing of isotactic polypropylene, leads to chain degradation through β-scission reaction[32].

Different reactive monomers and organic peroxides have been employed for PP radical functionalization in the melt. The most studied functionalizing agents are maleic anhydride (MAH)[33,34,35] and glycidyl methacrylate (GMA)[36,37,38]. A careful analysis of the products has been achieved and the results show, in any reaction conditions, the decreasing of the average molecular weight of the polymer as showed by the increasing of the melt flow rate (MFR) with respect to the unprocessed polymer. The grafting reaction occurs by the addition of the terminal macroradical, come from the β-scission reaction, to the monomer double bond. Conditions for a good competition between the desired grafting reaction and the chain scission are difficult to obtain in the melt. The main problem is due to the formation of non stabilized tertiary macroradical. The recent studies on this topic concern the control of the degradation reactions. With this aim molecules or molecules systems (X) able to create a new stabilized macroradical not involved in β-scission, which gives place to transfer and/or coupling reactions[39] (scheme 4) have been proposed.

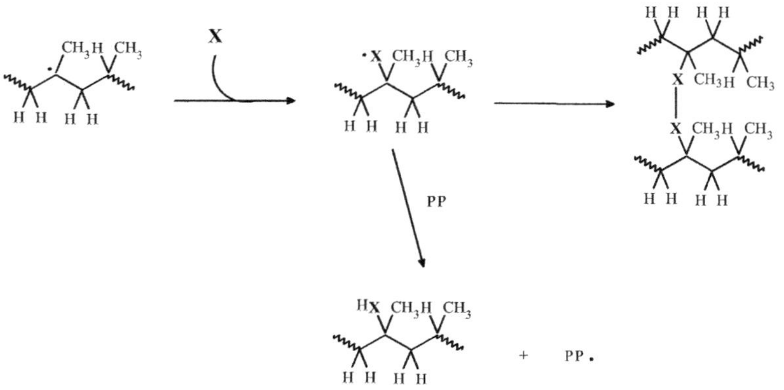

Scheme 4. Free radical reactions of polypropylene with stabilizing the molecules or molecules system (X)

Cartier et al.[40] have studied the radical functionalization reaction of PP with MAH or GMA and a suitable co-monomer such as the styrene (STY). The styrene reacts with macroradicals creating more stabilized radicals which subsequently react with the polar monomer. The reactivity of GMA towards the styryl radical is believed to be greater than its reactivity toward the PP macroradical, hence, the use of this co-monomer produces a synergistic effect on the grafting. In this way, it is possible to limit the degradation and simultaneously to obtain functionalized product.

Recently the chemical crosslinking of propylene polymers has been performed under dynamic conditions using a peroxide and a furan or bis-maleimide based coagent as crosslinking promoter[41] (scheme 5).

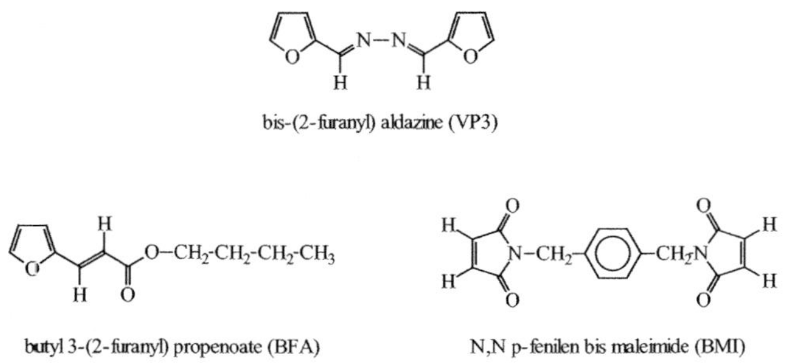

Scheme 5. Coagents employed in PP crosslinking reactions

These coagents are able to react properly with the macroradicals, produced by the primary radicals, thus spacing them from the backbone and promoting chain extension and/or preventing the β-scission (scheme 6):

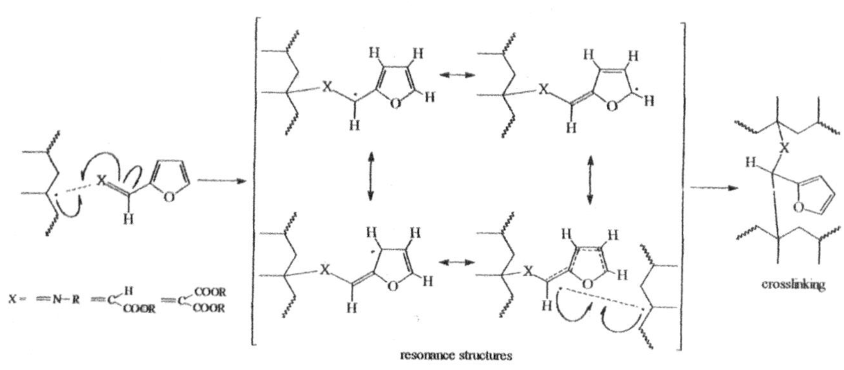

Scheme 6. Reactions of polypropylene monoradicals with furane coagents

The furan derivatives, having the ring conjugated with a double bond substituted with electron attracting groups, promote a high addition rate of the macroradical to the coagent and the formation of stabilized radicals for resonance thus avoiding β-scission. The use of these molecules was then proposed as functionalization coagents able to control transfer reaction, principal reason of PP extensive degradation, and able to introduce onto the polymer backbone suitable polar groups.

In this contest a first attempt was based on the combined use of various furan derivatives in combination with maleic anhydride[42] (scheme 7).

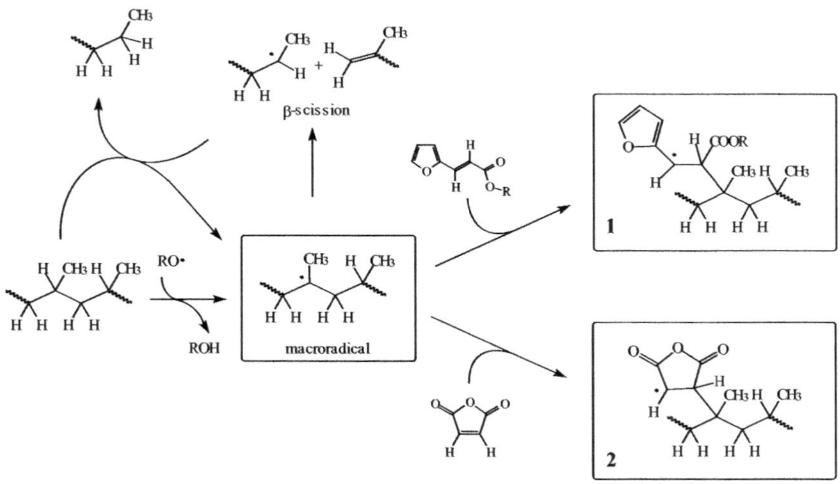

Scheme 7. Reactions of the polypropylene macroradicals with BFA and MAH

Thus with butyl 3-(2-furanyl) propenoate (BFA) and maleic anhydride in the presence of dimethyl peroxide FD 0.7 functional group were obtained with limited degradation as shown by the modest decrease of the torque value during the Brabender mixing. Also coagents bearing two functional groups were prepared and used as functionalizing agents in the presence of peroxides. In this case the functionalization is flanked by crosslinking as the transfer action by MAH is lacking.

5. FUNCTIONALIZATION OF STYRENE POLYMERS

The polystyrene (PS) functionalization can be obtained both during the polymerisation process and by free-radical initiated grafting of unsaturated polar monomers in solution or via melt extrusion of the PS. While for the polyolefins (POs) the functionalization reactions, promoted by free radicals have been largely studied and employed today for the commercial production of modified materials, only few works about PS radical reactions have been reported in literature and principally connected with the preparation of compatibilizing products for PS mixtures with polycarbonates and polyamides.

The PS radical functionalization reaction has been studied in mass and in solution employing maleic anhydride (MAH) as functionalizing monomer and peroxides as radical initiators[43-49]. Although high monomer/polymer molar ratios have been used, the functionalization degrees (FDs) are quite low (FD<1.2% by wt). So, independently from relative reagents concentrations, the PS radical functionalization process is not favoured and this fact is evidenced also by low monomer conversions. Many authors agree in considering that the general extent of PS functionalization reactions involves a selective hydrogen abstraction from the benzylic carbon with formation of a benzylic macroradical more stable than the aliphatic one and therefore less reactive toward the monomer.

In our laboratory a broader research on the PS melt modification has been carried out. Complex reactive systems (from the single initiators, to functionalizing monomers mixtures, co-agents and peroxides) have been employed to increase the functionalization degree: the structural variation as well as the thermal properties of the product have been evaluated with reference to the experimental conditions.

The atactic polystyrene (aPS) reactivity in bulk, at the fluid state, towards free radicals of different chemical structure has been studied varying the molar initiator/PS ratio[50]. In the presence of dicumyl peroxide (DCP), degradation reactions have been observed with a significant variation of the polymer molecular weight (MW) and polymer thermal properties. By treating PS with maleic derivatives (maleic anhydride, diethyl maleate and cetyl maleate monomer) and DCP appreciable FDs have been obtained.

Specially by using diethyl maleate (DEM), the results indicate a certain improvement with respect to the use of maleic anhydride: the FDs ranging from 0.1 to 0.4 % by mol. Moreover, it is possible to obtain an increasing of FD using higher DEM content. Besides, the low DEM conversions confirm the low reactivity of the benzylic macroradical, as reported by Xue et al.[51].

The grafting of diethyl maleate onto Styrene/1-alkenes Block Copolymers (SEBS) can be performed as for ethylene polymers by using monomer and DCP as radical initiator with possible differences in process temperature connected to the different thermal characteristic of the styrene-containing polymers[52]. Experiments with polystyrene, working at 190°C, and also with DEM and DCP, have indicated a lower tendency to functionalization of this polymer with respect to polyolefins. The different behaviour of aromatic units with respect to the aliphatic ones was confirmed by submitting to functionalization under similar conditions a mixture 80/20 by weight of polystyrene with ethylene/1-butene random copolymer. The overall composition of this mixture was analogous to SEBS as far as the aromatic/aliphatic unit ratio and number of branches in the aliphatic blocks are concerned. The analysis of the functionalized product after the separation of polystyrene from LLDPE, by solvent extraction, indicated that only the last contained grafted DEM units. The origin of this high selectivity is not clear yet and different stability of benzyl radical versus purely aliphatic radicals and phase segregation can be claimed. As the same molecular and phase features can be maintained in SEBS like above homopolymer mixture, one could argue that even SEBS was substantially functionalized in the aliphatic blocks. Accordingly solid-state NMR showed the presence of two new peaks in the single pulse excitation (SPE)-MAS spectra of functionalized SEBS, assigned to the ethyl carbons ($-O-CH_2-CH_3$) of diethyl succinate (DES) formed by DEM insertion ^{13}C-NMR-CP/MAS spectra, which allow signals arising from rigid phases to be enhanced, did not display the above absorptions, thus confirming that the functionalization reaction occurs on the polyolefin blocks. Experiments with SEBS and different polymers to DEM and to DCP ratios showed that FD increases linearly with DEM and presents a maximum or at least an asymptotic value at a certain DEM/DCP ratio as observed in the case of polyolefins. Therefore, considering also the different activation of aromatic and aliphatic units, the same kinetic scheme can be proposed as for LLDPE (scheme 8) where DEM propagation is not considered according to previous results.

The primary radicals R· formed by thermal decomposition of DCP are likely to be promptly converted into aliphatic and aromatic blocks. The latter ones disappears as giving rise to transfer reactions. Therefore, substantially only the aliphatic radicals are involved in the grafting. This picture is confirmed by the absence of DEM homopolymer and of DEM grafted on the aromatic units.

A) INITIATION

DCP ⟶ 2R• (1)

R• + DEM ⟶ RDEM • (2)

R• + SEBS ⟶ AlSEBS • (3)

R• + SEBS ⟶ ArSEBS • (4)

AlSEBS • + DEM ⟶ AlSEBSDEM • (5)

B) TRANSFER

RDEM • + SEBS ⟶ RDEMH + AlSEBS (or ArSEBS) • (6)

AlSEBSDEM • + SEBS ⟶ AlSEBSDEMH + AlSEBS (or ArSEBS) • (7)

C) TERMINATION

Al(Ar)SEBS • + Al(Ar)SEBS • ⟶ SEBS-SEBS and SEBS + SEBS(-H) (8)

Al(Ar)SEBS • + RDEM • ⟶ SEBSDEMR (9)

D) DEGRADATION

2Al(Ar)SEBS • ⟶ shorter chain with terminal free radical or double bond (10)

Scheme 8. Reaction mechanism of Styrene/1-alkenes Block Copolymers (SEBS) with diethyl maleate (DEM) and dicumylperoxide (ROOR) as initiator.

It is likely that ArSEBS·radicals, because of the high stability, do not react with DEM to produce a radical of lower stability and rather involved in termination or degradation. The absence of functional groups attached to aromatic units is an indication of the low probability of reaction.

Indeed RDEM·radicals are probably formed in a very small amount as indicated by the absence of the phenyl group in functionalized ethylene polymers. The occurrence of termination by recombination, and degradation, is in keeping with the broadness of molecular weight distribution (MWD) after the functionalization process and with the presence of aromatic low molecular weight polymer in the acetone soluble fraction. Concerning to the functionalization reaction itself, also for SEBS it is based on reaction and it should be linearly dependent on DEM concentration with a lower propagation constant (K_p) due to the lower radical efficiency. On the other side the dependence on DCP concentration cannot be linear but the value must be optimized among the possible reactions. These data suggest that the free radical functionalization of SEBS can be carried under similar conditions as for polyolefins. The use of an adequate DEM/DCP ratio gives a very acceptable level of crosslinking and degradation. Because of the different reactivity of benzyl and aliphatic macromolecular radicals the functional groups concentrate in the aliphatic part.

6. REACTIVITY OF GRAFTED FUNCTIONAL GROUPS WITH MACROMOLECULES

Condensation polymers and polyolefins are thermodynamically immiscible, so the uncompatibilized blends exhibit a coarse morphology that corresponds to very poor mechanical properties. Satisfactory performances in immiscible blends are usually attained by minimizing interfacial tension and improving adhesion between the two phases. One of the most studied method to enhance compatibility consists of adding to the immiscible system a polyolefin with grafted reactive groups that can react in the melt with either hydroxyl [poly(ethylene terefhthalate)] or amino (polyamide) terminal groups, thus yielding comb copolymers able to stabilize the interfaces between the two immiscible phases providing the requested adhesion for properties modulation.

The efficiency of the above reactions at a molecular level depends on different parameters, such as the structure of the condensation polymer and polyolefin, the type and structure of interfacially reactive pre-compatibilizer, the presence and type of catalyst, the original phase segregation, the formation in situ of reactive species and the operative conditions.

The intermolecular reaction and its role in determining the partial compatibility between diethylsuccinate containing linear low-density polyethylene or ethylene propylene copolymer and polyamide 6 (PA6) has been investigated[53] producing blends in the melt using a discontinous mixer. Selective solvent extractions of the obtained blends with formic acid and n-heptane demonstrated the formation of a polyolefin-nylon (PO-PA6) grafted copolymer recovered as final residual fraction. The formation of grafted copolymer has an evident effect on the compatibilization of the two original polymers. Indeed, the differential scanning calorimetry analysis shows a remarkable decrease of temperature and enthalpy of PA6 crystallization. Moreover scanning electron microscopy micrographs show clear evidence of size reduction of PA6 domains associated with improved interface interactions. The reaction among dialkyl succinate groups and amino terminal groups of polyamide produces cyclic imide groups with elimination of alkyl alcohols[54].

The formation of polyolefin-polyamide copolymers during the blending of polyethylene and polyamyde 6 in the melt was observed by many authors[55,56] using polyethylene functionalized with maleic anhydride as compatibilizer. During the blending the partial degradation of polyamide chain was observed by Van Duin et al.[56]. This result is explained by the authors, on the basis of a model compound study, considering the reactivity of water produced during the formation of cyclic imide groups (scheme 9).

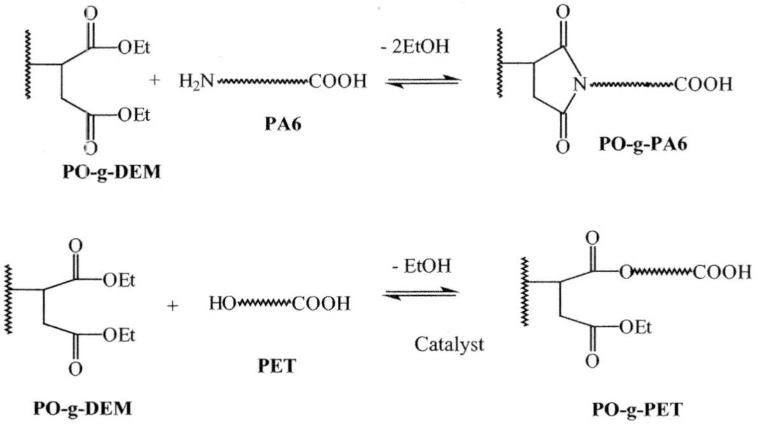

Scheme 9. Reaction of diethyl succinate grafted groups on the polyolefin backbone respectively with Polyamide $-NH_2$ and PET –OH terminal groups

In the case of poly(ethylene terephthalate) (PET) the reaction among hydroxyl terminal groups is reported to happen during the blending with polyolefins functionalized with monomers containing succinic anhydride groups[57], epoxy groups[58], oxazoline groups[59] or isocyanate groups[60]. In particular epoxy groups are able to react with both carboxylic groups and hydroxyl groups.

Considering the possibility of controlling the functionalization of polyolefins with dialkyl maleates and that they are very modestly armful chemicals and being liquids can be easily controlled, it appears of interest to use dialkyl maleate functionalized PE as compatibilizing agent for PO/PET blends[61]. Preliminary data demonstrated that the formation of grafted copolymer during the reactive mixing is rather modest. Therefore the use of a transesterification catalyst appears necessary. Also the crosslinking of the functionalized polyolefin is observed.

Our recent work on this topic showed the necessity to control side reaction in the separate phases. In fact, all the potential transesterification catalysts studied until now lead to PET degradation during blending and in some cases the formation of polyolefin-PET or polyolefin-polyamide copolymer during blending produces an improving of morphologic properties which results in good final properties in terms of mechanical[62,63] and physic properties[64].

7. FINAL REMARKS

The bulk functionalization of polyolefins can be carried out very conveniently by grafting a substituted monomer to the polyolefin macromolecules in the presence of peroxide. The need of temperature large enough to grant for the polymer fluidification and the presence of free radicals give rise to side reactions mainly crosslinking and/ or degradation. In general the functionalization is aimed to improve the compatibility of polyolefins towards more polar polymers, inorganic materials and metals; therefore the functionalization degree (FD) requested is rather low and in the range 0.1-1.5 moles of polar groups per 100 monomeric units. At the same time it is often necessary that the process does not affect the original chains to a large extent. Indeed the change of the original macromolecule structure may hinder future application. The data presented in this paper show that the chemical control of the process can be governed acting on the reaction and feed conditions which have to be selected taking into account the structure of the polyolefin and its rheological behaviour .

The reactivity of the grafted functionalities with the typical polycondensation polymers indicates that these groups react with the polymer end groups bearing to the formation of the grafted polymer which acts as the compatibilizer but the extent of this reaction is generally limited due to the low mobility of both types of reactive groups bound to long molecules that are not able to reach the interfacial area between the two polymeric phases.

REFERENCES

1. Giannini, U., Brueckner, G., Pellino, E., Cassata, A. (1967) Stereospecific polymerisation of monomers containing oxygen and nitrogen with Ziegler-Natta catalysts, *Journal of Polymer Science, Polymer Letters Edition* **5**(6), 527-533
2. Goretzki, R., Fink, G. (1999) Homogeneous and heterogeneous metallocene/MAO-catalyzed polymerization of functionalized olefins, *Makromolecular Chemistry and Physics* **200,** 881-886
3. Endo, K. (1999) Polymerization of tert-butyl methacrylate with $Ni(acac)_2$-methylaluminoxane catalyst, *Makromolecular Chemistry and Physics* **200**, 1722-1725
4. Novak, B. M., Kim, J.; Stewart, J.; Schlitzer, D. (2000) Introducing topological chirality and functionality into stiff-chain polymers, *ACS Polymer Preprints* **41**, 893
5. Boffa, L. S., Novak, B. M., (2000) Copolymerization of polar monomers with olefins using transition-metal complexes, *Chemical Reviews* **100**, 1479-1494

6. Hu, G. H.; Sun, Y. J.; Lambla, M. (1996) Devolatilization: a critical sequential operation for in situ compatibilization of immiscible polymer blends by one-step reactive extrusion, *Polymer Engineering and Science* **36**, 676-684
7. Greco, R., Maglio, G., Musto, P. (1989) Bulk functionalization of ethylene-propylene copolymers I. Influence of temperature and processing on the reaction kinetics *Journal of Applied Polymer Science* **33**, 2513-2527
8. Sun, Y., Hu, G., Lambla M. (1995) Free radical grafting of glycidyl methacrylate onto polypropylene in a co-rotating twin screw extruder, *Journal of Applied Polymer Science* **57**, 1043-1054
9. De Vito, G., Lanzetta, N., Maglio, G., Malinconico, M., Musto, P., Palumbo, R.,(1984), Functionalization of an amorphous ethylene-propylene copolymer by free radical initiated grafting of unsaturated molecules, *Journal of Polymer Science: Polymer Chemistry Edition* **22**, 1335-1347
10. Raval, H., Singh, Y. P.; Mehta, M. H.; Devi, S. (1991) Grafting of low-density polyethylene with butyl acrylate: synthesis and characterization, *Polymer International* **24**, 99-104
11. Gaylord, N. G., Mishra, M. (1983) Nondegradative reaction of maleic anhydride and molten polypropylene in the presence of peroxides , *Journal of Polymer Science, Polymer Letters Edition* **21**, 23-30
12. Wu, C. H., Su, A. C. (1991) Functionalization of Ethylene-Propylene rubber via melt mixing, *Polymer Engineering and Science* **31**, 1629
13. Greco, R., Malinconico, M., Martuscelli, E., Ragosta, G., Scarinzi, G. (1988) Rubber modification of polyamide-6 effected concurrently with caprolactam polymerization: influence of blending conditions and degree of grafting of rubber, *Polymer* **29**, 1418-1425
14. Ciardelli, F., Aglietto, M., Passaglia, E., Ruggeri, G. (1998) Molecular and mechanistic aspects of the functionalization of polyolefins with ester groups, *Macromolecular Symposia* **129**, 79-88
15. Ciardelli, F., Aglietto, M., Passaglia, E., Picchioni, F. (2000) Controlled functionalization of olefin/styrene copolymers through free radical processes, *Polymer for Advanced Technologies.* **11**, 371-376
16. Moad, G. (1999) The synthesis of polyolefin graft copolymers by reactive extrusion, *Progress in Polymer Science* **24**, 81-142
17. Al-Malaika, S. (1997) *Reactive modifiers for polymers,* Academic & Professional, London, UK
18. Machado, A. V., Covas, J. A., Van Duin, M. (2001) Effect of polyolefin structure on maleic anhydride grafting, *Polymer* **42**, 3649-3655
19. Yang, L., Zang, F., Endo, T., Hirotsu, T. (2002) Structural characterization of maleic anhydride grafted polyethylene by 13C NMR spectroscopy, *Polymer* **43**, 2591-2594
20. Wang, D., Wilkie, C. A. (2003) In-situ reactive blending to prepare polystyrene–clay and polypropylene–clay nanocomposites, *Polymer Degradation and Stability* **80**, 171-182
21. Tedesco, A., Barbosa, R. V., Nachtigall, S. M. B., Mauler, R. S. (2002) Comparative study of PP-MA and PP-GMA as compatibilizing agents on polypropylene/nylon 6 blends, *Polymer Testing* **21**, 11-15
22. Deng, J. P.,Wan, T. Y., Ranby B. (2002) Melt-photografting polymerization of maleic anhydride onto LDPE film, *European Polymer Journal* **38**, 1449-1455
23. Jianping, D., Wantai, Y. (2003) Surface photografting polymerization of vinyl acetate, maleic anhydride, and their charge-transfer complex. IV. Maleic anhydride, *Journal of Applied Polymer Science* **87**, 2318-2325

24. Dong, O., Liu, Y. (2003) Styrene-assisted free-radical graft copolymerization of maleic anhydride onto polypropylene in supercritical carbon dioxide, *Journal of Applied Polymer Science* **90**, 853-860
25. Qi, R., Oian, J., Zhou, C. (2003) Modification of acrylonitrile-butadiene-styrene terpolymer by grafting with maleic anhydride in the melt. I. Preparation and characterization, *Journal of Applied Polymer Science* **90**, 1249-1254
26. Aimin, Z., Chao, L., (2003) Chemical initiation mechanism of maleic anhydride grafted onto styrene–butadiene–styrene block copolymer, *European Polymer Journal* **39**, 1291–1295
27. Aglietto, M., Bertani, R., Ruggeri, G., Ciardelli, F. (1992) Radical bulk functionalization of polyethylenes with ester groups, *Makromolecular Chemie* **193**, 179-186
28. Aglietto, M., Ruggeri, G., Luppichini, E., D'alessio, A., Benedetti, E. (1993) Functionalized polyolefins Materials Engineering, *Materials Engeneering* **4**, 253-260
29. Ruggeri, G., Aglietto, M., Petragnani, A., Ciardelli, F. (1983) Polypropylene functionalization by free radical reactions, *European Polymer Journal* **19**, 863-866
30. Aglietto, M., Bertani, R., Ruggeri, G., Segre, A. L. (1989) Functionalization of polyolefins: structure of functional groups in polyethylene reacted with ethyl diazoacetate, *Macromolecules* **22**, 1492-1493
31. Passaglia, E., Corsi, L., Aglietto, M., Ciardelli, F., Michelotti, M., Suffredini, G. (2003) One step functionalization of ethylene-propylene random copolymer with two different reactive groups, *Journal of Applied Polymer Science* **87**, 14-23
32. Gaylord, N. G., Mehta, M., Mehta, R. (1987) Degradation and cross-linking of ethylene-propylene copolymer rubber on reaction with maleic anhydride and/or peroxides, *Journal of Applied Polymer Science* **33**, 2549-2558
33. Garcia-Martinez, J. M., Laguna, O., Collar, E. P. (1997) Role of reaction time in batch process modification of atactic polypropylene by maleic anhydride in melt, *Journal of Applied Polymer Science* **65**, 1333-1347
34. Garcia-Martinez, J. M., Laguna, O., Collar, E. P. (1998) Chemical modification of polypropylenes by maleic anhydride: influence of stereospecificity and process conditions, *Journal of Applied Polymer Science* **68**, 483-495
35. Garcia-Martinez, J. M., Laguna, O., Areso, S., Collar, E. P. (1998) Functionalization of atactic polypropylene by succinyl-fluoroscein: A two step process of chemical modification in the melt, *Journal of Applied Polymer Science* **70**, 689-696
36. Sun, Y. J., Hu, G. H., Lambla, M. (1995) Melt free-radical grafting of glycidyl methacrylate onto polypropylene, *Die Angewandte Makromoleculare Chemie* **229**, 1-13
37. Sun, Y. J., Hu, G. H., Lambla, M. (1995) Free radical grafting of glycidyl methacrylate onto polypropylene in a co-rotating twin screw extruder, *Journal of Applied Polymer Science* **57**, 1043-1054
38. Hu, G. H., Sun, J., Lambla, M. (1996) Effects of processing parameters on the in situ compatibilization of polypropylene and poly(butyleneterephthalate) blends by one-step reactive extrusion, *Journal of Applied Polymer Science* **61**, 1039-1047
39. Fritz, H. G., Cai, Q., Boelz, U. (1993) Polypropylen-modifikation durch reaktive kunststoffaufbereitung, *Kunststoffe* **86**, 439-444
40. Cartier, H., Hu, G. H. (1998) Styrene-assisted melt free radical grafting of glycidyl methacrylate onto polypropylene, *Journal of Polymer Science: Part A: Polymer Chemistry* **36**, 1053-1063
41. Romani, F., Corrieri, R., Braga, V., Ciardelli F. (2001) Monitoring the chemical crosslinking of propylene polymers through rheology, *Polymer* **43**, 1115-1131
42. Ciardelli, F., Passaglia, E., Coiai, S. (2003) Procedimento di funzionalizzazione radicalica controllata di una poliolefina, Italian Patent Application n° TO2003A000478

43. Li, H-M., Chen, H-B., Shen, Z-G., Lin, S. (2002) Preparation and characterization of maleic anhydride-functionalized syndiotactic polystyrene, *Polymer* **43**, 5455-5461
44. Jo, W.H., Park, C.D., Lee, M.S. (1996) Preparation of functionalized polystyrene by reactive extrusion and its blend with polyamide 6, *Polymer* **37**, 1709-1714
45. Kesselmeier, R., Moritz, H-U. (1998) Grafting maleic anhydride onto polystyrene during reactive extrusion, 6th International Workshop on Polymer Reaction Engineering **134** 207-216
46. Liliac, W.D., Lee, S. (1999) Analysis of the solid phase copolymerization grafting process, *Korean Journal of Chemical Engineering* **16**, 275-284
47. Zhang, X., Lim, J.G., Baik, J.H., Kim, H.J. (2001) Method of preparing monomer-grafted syndiotactic polystyrene with polarity, WO Patent 01/92352
48. Ishihara, T. (1988) Process of mixing melts amorphous polyester and a graft modified polystyrene and composition thereof, US Patent 4, 775,717
49. Park, I., Barlow, J.W., Paul, D.R. (1991) Terminal anhydride functionalization of polystyrene *Journal of Polymer Science: Part A* **29,** 1329-1338
50. Passaglia, E., Coiai, S., Ricci, L., Ciardelli, F. (2003) Polystyrene modification by reactive processing in the bulk, *Atti XVI Convegno di Scienza e Tecnologia delle Macromolecole, Pisa, 22-25 Settembre*
51. Xue, T.J., Jiang, D., Wilkie, C. A. (1996) Grafting of vinyl monomers onto polymers containing styrene, *212th ACS National Meeting, Orlando, Florida 25-29 August 1996 (Conference Paper)*
52. Passaglia, E., Ghetti, S., Picchioni, F., Ruggeri, G. (2000) Grafting of diethyl maleate and maleic anhydride onto styrene-b-(ethylene-co-1-butene)-b-styrene triblock copolymer (SEBS), *Polymer* **41**, 4389-4400
53. Passaglia, E., Aglietto, M., Ruggeri, G., Picchioni, F. (1998) Formation and compatibilizing effect of the grafted copolymer in the reactive blending of 2-diethylsuccinate containing polyolefins with poly-ε-caprolactam (nylon-6), *Polymers for Advanced Technologies* **9**, 273-281
54. Greco, R., Lanzetta, N., Maglio, G., Malinconico, M., Martuscelli, E., Palumbo, R., Ragosta, G., Scarinzi, G. (1986) Rubber modification of polyamide 6 during caprolactam polymerization: influence of composition and functionalization degree of rubber, *Polymer* **27**, 299-308
55. Jiang, C., Filippi, S., Magagnini, P. (2003) Reactive compatibilizer precursors for LDPE/PA6 blends. II: maleic anhydride grafted polyethylenes, *Polymer* **44**, 2411–2422
56. Van Duin, M., Borggreve, R. J. M. (1998) Graft formation and chain scission in blends of polyamide-6 and -6.6 with maleic anhydride containing polymers, *J. of Polymer Science, Part A: Polymer Chemistry* **36,** 179-188
57. Sanchez-Solis, A., Calderas, F., Manero., O. (2001) Influence of maleic anhydride grafting on the rheological properties of polyethylene terephthalate–styrene butadiene blends, *Polymer* **42,** 7335-7342
58. Hu, G. H., Cartier, H. (1999) Styrene-assisted melt free radical grafting of glycidyl methacrylate onto an ethylene and propylene rubber, *Journal of Applied Polymer Science* **71**, 125-133
59. Park, S. H., Park, K. Y., Suh, K. D. (1998) Compatibilizing effect of isocyanate functional group on polyethylene terephthalate/low density polyethylene blends, *Journal of Polymer Science, Part B: Polymer and Physics* **36,**447-453
60. La Mantia, F. P., Scaffaro, R., Colletti, C., Dimitrova, T., Magagnini, P., Paci, M., Filippi, S. (2001) Oxazoline functionalization of polyethylenes and their blends with polyamides and polyesters, *Macromolecular Symposia* **176**, 265-278

61. Aglietto, M., Coltelli, M.B., Savi, S., Lochiatto, F., Ciardelli, F., Giani, M. (2004) Post-consumer polyethylene terephthalate (PET)/polyolefin blends through reactive processing, *Journal of Material Cycles and Waste Management*, in press
62. R.A. Kudva, R.A., Keskkula, H., Paul, D.R. (1999) Morphology and mechanical properties of compatibilized nylon 6/polyethylene blends, *Polymer* **40**, 6003–6021
63. Pawlak, A., Perkins, W. G., Massey, F. L., Hiltner, A., Baer, E. (1999) Mechanical Properties of Poly(ethylene terephthalate) Modified with Functionalized Polymers, *Journal of Applied Polymer Science* **73**, 203–219
64. Yeh, J.T., Chao, C. C., Che, C. H. Effects of processing conditions on the barrier properties of polyethylene (PE)/modified polyamide (MPA) and modified polyethylene (MPE)/polyamide (PA) blends, *Journal of Applied Polymer Science* **76**, 1997-2008

HYPERBRANCHED ARCHITECTURES
Synthesis-Characterization-Properties

HERMIS IATROU, MARINOS PITSIKALIS, and NIKOS HADJICHRISTIDIS
Chemistry Department, University of Athens, Panepistimiopolis Zografou, 15771, Athens

Abstract: Recent developments in the science and technology of polymeric materials have allowed the synthesis and characterization of well-defined complex macromolecular architectures. The use of these polymers provided the insight necessary to establish the structure-property relationship. Branching is known to affect the dilute solution and melt properties of the polymers, however a precise prediction of this effect is not easily assessable, mainly due to the random fashion by which branching takes place. Consequently elaborated synthetic methods allowing the synthesis of well-defined structures with high molecular and structural homogeneity and predetermined sites of branching are necessary. The synthesis, characterization and properties of a variety of well-defined hyperbranched polymers, such as multifunctional stars, comb polymers, single and double graft, block-graft, α-ω branched copolymers, dendritic-like structures and polymacromonomers, synthesized in our Laboratory, will be discussed.

Key words: Multifunctional stars, combs, single and double grafts, block-grafts, α-ω branched copolymers, dendritic, polymacromonomers, synthesis, characterization, solution properties, bulk properties.

1. INTRODUCTION

Recent developments in the science and technology of polymeric materials have allowed the synthesis and characterization of well-defined complex macromolecular architectures. The use of these polymers provided the insight necessary to establish the structure-property relationship. Branching in polymers is probably the most important structural variable allowing the fine tuning of polymers characteristics and properties. Branched

F. Ciardelli and S. Penczek (eds.),
Modification and Blending of Synthetic and Natural Macromolecules, 73–89.

polymers contain branch points, ie. atoms or small groups from which two or more polymer chains emanate. Branching is known to affect the dilute solution and melt properties of the polymers, however a precise prediction of this effect is not easily assessable, mainly due to the random fashion by which branching takes place. Consequently elaborated synthetic methods allowing the synthesis of well-defined structures with high molecular and structural homogeneity and predetermined sites of branching are necessary. Among the different synthetic methods anionic polymerization is the most powerful technique providing the best control over the molecular and structural characteristics of the polymers.

Hyperbranched polymers are highly branched macromolecules but their structure is not regular or highly symmetrical. The importance of these materials in affecting the dilute solution and bulk properties of the polymers has been recognized since 1952 by Flory. The synthesis, characterization and properties of a variety of well-defined hyperbranched polymers, such as multifunctional stars, comb polymers, single and double graft, block-graft, α-ω branched copolymers, dendritic-like structures and polymacromonomers, synthesized in our Laboratory, will be discussed.

2. SYNTHESIS-CHARACTERIZATION-PROPERTIES

2.1 Multifunctional Stars

Highly functionalized polybutadiene (PBd) stars having 64 and 128 arms[1] have been synthesized by using carbosilane dendrimeric molecules as linking agents, having equal number of Si-Cl groups with the corresponding stars. The stars were prepared by reacting the functional dendrimers with monodisperse polybutadienyllithium living chains. As an example the reaction scheme for the synthesis of the 64-arms stars is given in Scheme 1.

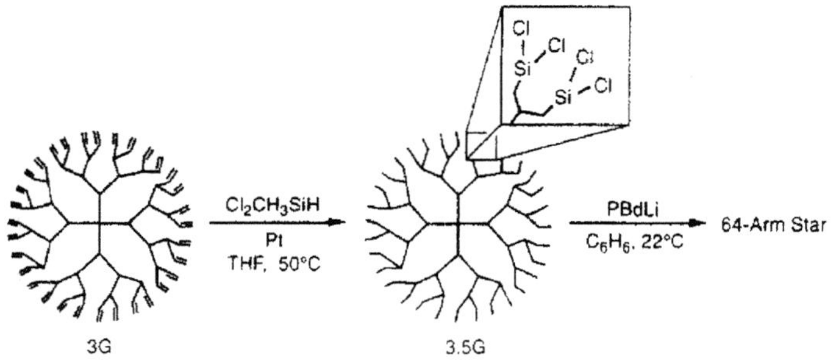

Scheme 1.

The arm molecular weight of the stars prepared was ranging between 6400 and 72000. They were characterized by low angle laser light scattering (LALLS) and Size Exclusion Chromatography (SEC). The dilute solution properties were determined in cyclohexane at 25 °C and dioxane at 26.5 °C, a good and a Θ solvent, respectively, for polybutadiene. The obtained R_G, A_2, D_o and [η] indicate that the isolated stars behave like hard spheres. The ratio of the hydrodymamic radius over the radius of gyration obtained was slightly higher than $(5/3)^{1/2}$. It was found that the ratio $g=\langle R_G^2\rangle / \langle R_G^2\rangle_{lin} \propto f^{-0.5}$, which is in good agreement with the scaling model developed by Gast et al (*Macromolecules 1991, **24**, 1670*) for polymeric micelles, and as a consequence it was concluded that stars with many arms represent good models for polymeric micelles.

Carbosilane dendrimers with functionality 16 were used in order to synthesize star copolymers of the A_8B_8 type, called "Vergina" copolymers[2], where A is polystyrene (PS) and B polyisoprene (PI). The synthetic approach involved the incorporation of 8 PS arms first, by reacting living PS arms, prepared by anionic polymerization, with the linking agent in a molar ratio 8:1, followed by reaction of the remaining 8Si-Cl groups of the macromolecular linking agent $(PS)_8(SiCl)_8$ with an excess of living PI chains. The Vergina stars were characterized by using membrane osmometry (MO), LALLS, SEC-UV and NMR spectroscopy. The characterization results revealed that the copolymers exhibited high degree of molecular and compositional homogeneity.

Highly functionalized PS stars, having 4 to 644 arms, have been prepared by using living radical (TEMPO) polymerization[3]. The stars were synthesized by reacting 2,2,6,6-tetramethylpiperidine-1-yloxy (TEMPO) - capped polystyrene (PS-T) with divinylbenzene. The characterization of the PS-T and the final stars was carried out by SEC, LALLS and viscometry. In the same work an asymmetric $PS_nPS'_n$ star was prepared, consisting of n arms of different PS and PS' arms. The stars exhibited polydispersity indices from 1.15-1.56. The reactions used are shown in Scheme 2.

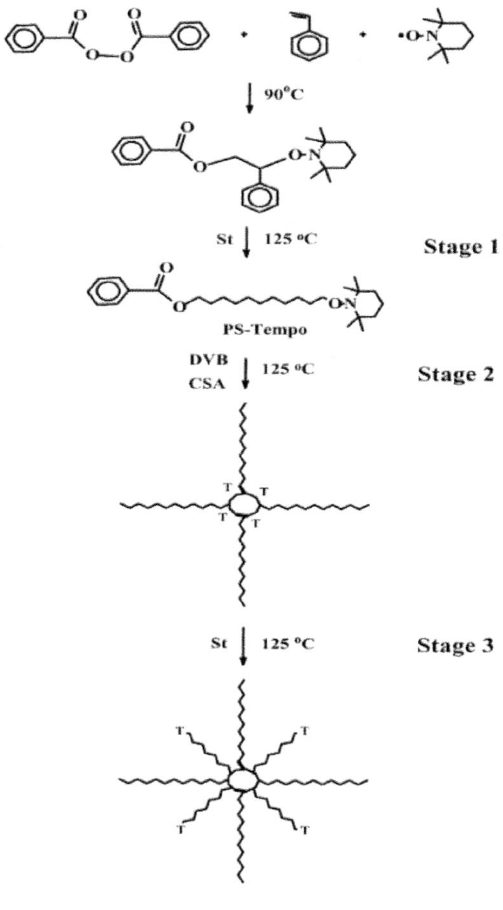

Scheme 2.

2.2 Comb and Graft Architectures

Model multigraft copolymers of Bd and styrene (St) with randomly placed trifunctional, for the poly(Bd-*g*-St), and tetrafunctional branch points for the poly(Bd-*g*-St_2) (Scheme 3) have been synthesized[4]. The synthetic approach involved the hydrosilylation of PBd (~92 wt % 1,4) with $HSiMe_2Cl$ (trifunctional) and $HSiMeCl_2$ (tetrafunctional), followed by reaction of the resultant silicon-chloride groups along the backbone with polystyryllithium (Scheme 4) .

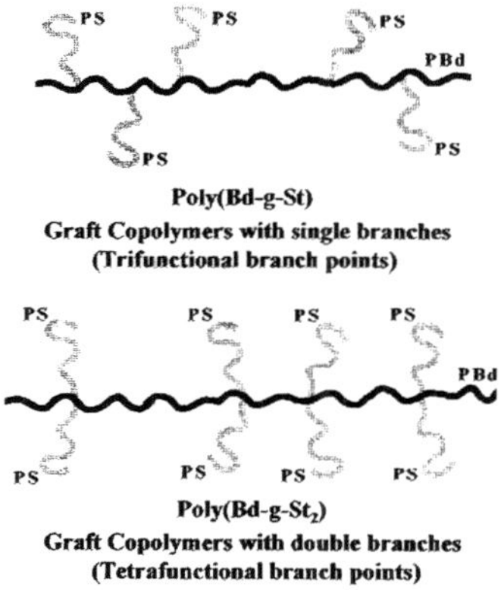

Scheme 3.

The extent of hydrosilylation was controlled by appropriate adjustment of the [silane]/[C=C] ratio. The reaction sequence used for the synthesis of the multigraft copolymers is given in Scheme 4.

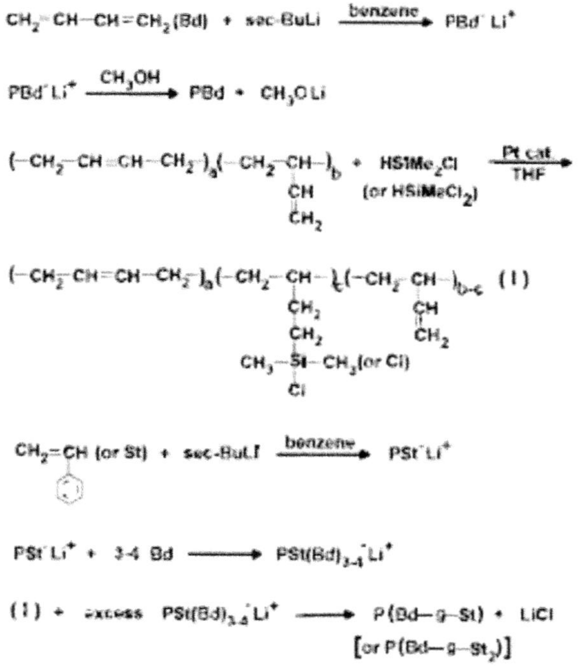

Scheme 4.

Characterization carried out by size exclusion chromatography (UV and RI detection), LALLS and NMR spectroscopy indicated that the synthesized molecules are highly homogeneous concerning the molecular weight and the composition.

Regularly spaced comb polystyrenes and graft polyisoprene/polystyrene copolymers with two branches at each junction point, so called "Centipedes"[5] have been presented recently (Scheme 5). The synthetic approach involved the selective replacement of two chlorines of $SiCl_4$ by PS, followed by step-growth polymerization of the produced $(PS)_2SiCl_2$ with α,ω-dilithium PS or 1,4 PI. The polydispersity indexes of the final centipedes was reduced to 1.2-1.3 by fractionation.

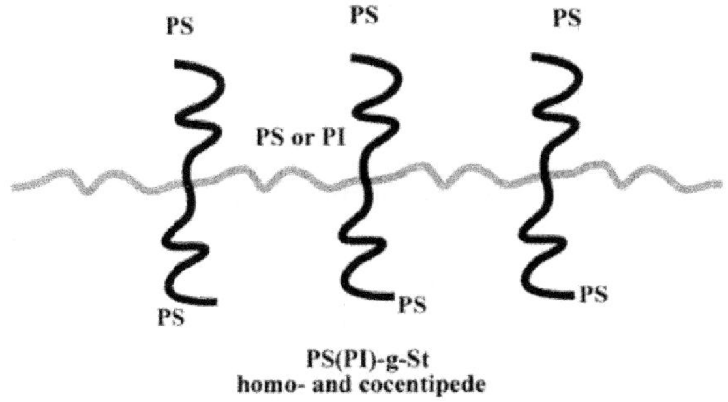

Scheme 5.

Tensile properties and morphology of tetrafunctional multigraft copolymers were investigated by transmission electron microscopy (TEM), small-angle X-ray scattering (SAXS), and tensile testing[6]. It was found that the copolymers with 22 vol % PS and seven branch points show a surprising high strain at break at about 2100 %, about double that of commercial thermoplastic elastomers such as Kraton (Figure 1). With increasing number of branch points strain at break and tensile strength increases. The morphology of the 22 % PS has a wormlike microphase separated structure with much lower long range order than other thermoplastic elastomers like Kraton.

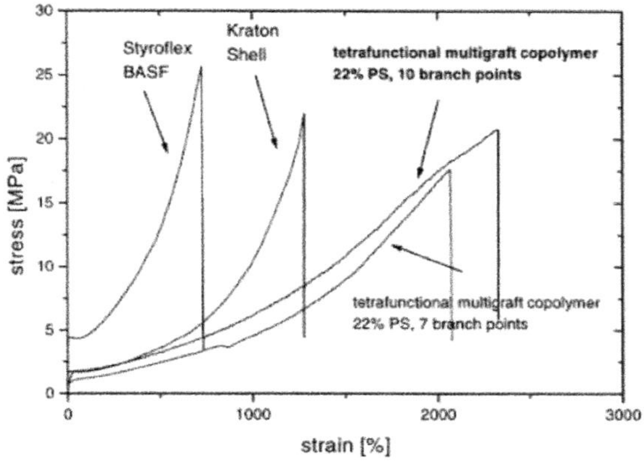

Figure 1. Mechanical properties of multigraft copolymers (MG-4-7-21 and MG-4-10-21) compared to commercial TPE's Kraton (20% PS) and Styroflex (58% PS), revealing the exceptional large strain at break of tetrafunctional multigraft copolymers.

The comb centipedes were also analyzed using a SEC system with a multiangle light scattering and refractive index detector to determine the relationship between the *z*-average mean square radius of gyration $<S^2>$ and the weight-average molecular weight M_w.[7] Tetrahydrofuran (THF) was chosen as the solvent. The relation fell considerably below that of linear polystyrene. It was found that as the ratio *r* of the molecular weight of the side chain to that of the connector increased, the $<S^2>$ vs M_w relation was lowered (Figure 2). Comparison of the experimental relationship with theoretical predictions for flexible discrete chains suggested that the main chains of the two types of polymer are stiffened by the crowding of the side chains.

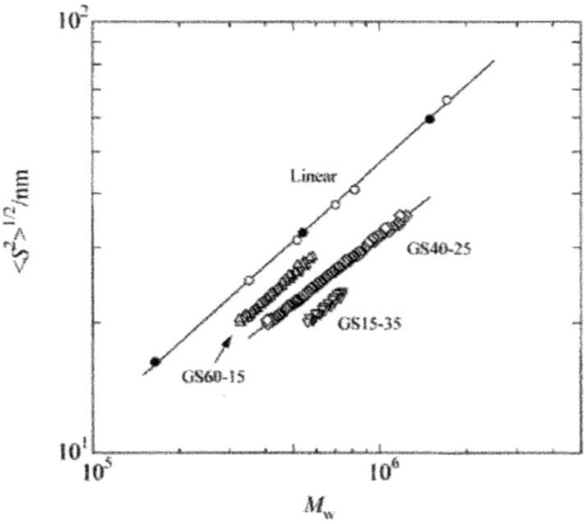

Figure 2. Molecular weight dependence of $<S^2>$ for linear polystyrene and the indicated polystyrene centipedes in THF.

2.3 α,ω Branched Architectures

Miktoarm stars of the $(PS)(PI)_5$ and $(PI)_5PS(PI)_5$[8] type were prepared by using hexafunctional chlorosilane linking agent. The reactions used for the synthesis of the $(PS)(PI)_5$ stars are given in Scheme 6. A stoichiometric amount of PSLi was added slowly to the linking agent, in order to avoid the replacement of more than one chlorine by the PS. Nevertheless, the formation of a small amount of $(PS)_2S\text{-}Cl_4$ could not be avoided. The remained Si-Cl groups were reacted with an excess of PILi. The side products were removed by fractionation.

$$\text{Is} + sec\text{-BuLi} \xrightarrow{\text{benzene, 25 °C}} \text{PILi}$$

$$\text{S-Cl}_6 + \text{PILi } (\sim 10\% \text{ excess}) \xrightarrow{\text{benzene, 25 °C}} (\text{PI})_5\text{-S-Cl} + 5\text{LiCl}$$

$$\text{St} + \text{DLi} \xrightarrow{\text{benzene, 25 °C}} \text{LiPSLi}$$

$$\text{LiPSLi} + \text{excess } (\text{PI})_5\text{-SCl} \xrightarrow{\text{benzene, THF (3\%), 25 °C}} (\text{PI})_5\text{PS}(\text{PI})_5 + 2\text{LiCl}$$

Scheme 6.

The reaction sequence used for the synthesis of the α,ω-branched copolymer $(PI)_5PS(PI)_5$ is given in Scheme 7.

$$\text{St} + sec\text{-BuLi} \xrightarrow{\text{benzene, 25 °C}} \text{PSLi}$$

$$\text{PSLi} + \text{S-Cl}_6 \xrightarrow{\text{benzene, 25 °C}} \text{PS-S-Cl}_5 + \text{LiCl}$$

$$\text{Is} + sec\text{-BuLi} \xrightarrow{\text{benzene, 25 °C}} \text{PILi}$$

$$\text{PS-S-Cl}_5 + \text{excess PILi} \xrightarrow{\text{benzene, 25 °C}} \text{PS}(\text{PI})_5 + 5\text{LiCl}$$

Scheme 7.

The synthetic route involved the reaction of a small excess over a five-fold amount of living PILi in respect to the chlorosilane linking agent. The formed $(PI)_5\text{-}S\text{-}Cl$ was then reacted with an α,ω-difunctional PS, leading to the formation of the $(PI)_5PS(PI)_5$ miktoarm star copolymer.

A series of well-defined dumbbell copolymers (Scheme 8) of the $(PBd)_{32}PS(PBd)_{32}$[9] type has been synthesized. The copolymers exhibited the same PS connector and 32 PBd branches of varying molecular weight on each end.

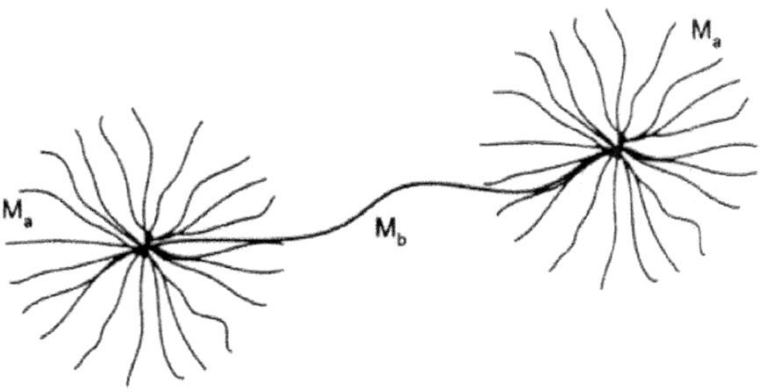

Scheme 8. Schematic representation of a Dumbbell Macromolecule. It consists of a main (connector) chain with molecular weight M_b and with each of its two ends attached to star polymers with the same high functionality, each having arm molecular weight M_a.

The synthetic route of the stars involved the preparation of a 1,2PBd-*b*-PS-*b*-1,2PBd triblock copolymer, by coupling a living diblock copolymer of the 1,2PBd-b-PS type with dichlorodimethylsilane. The small 1,2PBd end blocks that exhibited exactly the same molecular weight, was further hydrosilylated with $MeSiCl_2H$, incorporating the $-SiCl_2$ groups along the PBd chains. These groups were reacted with living PBd chains leading to the formation of the model dumbbell copolymers. The reactions used are given in Scheme 9.

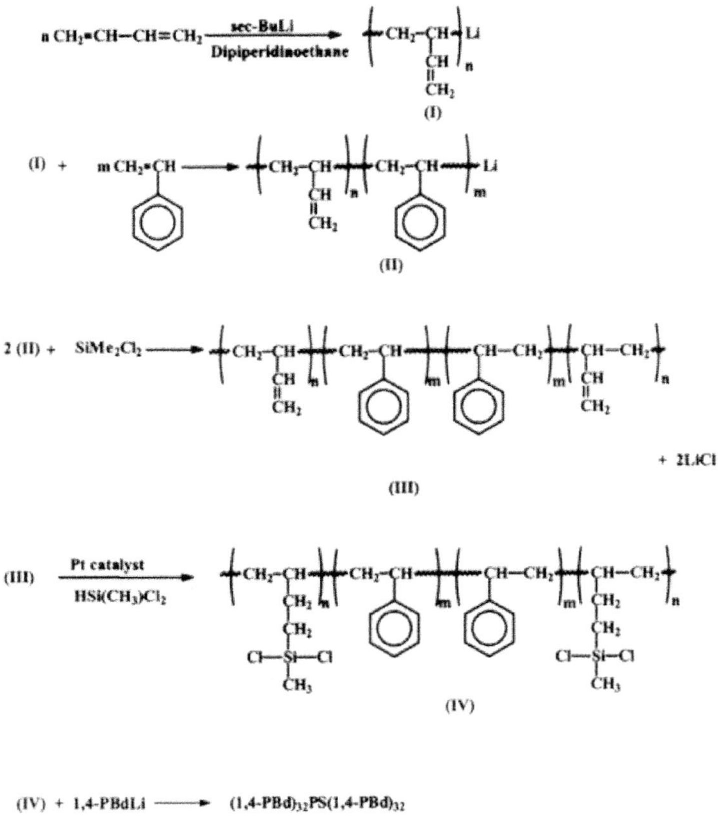

Scheme 9. Key reactions for the synthesis of the dumbbell copolymers $(PBd)_{32}PS(PBd)_{32}$

The combined characterization results (MO, LALLS) indicated that the copolymers were well-defined. Their viscoelastic measurements indicated that for the copolymers exhibiting the three larger branches, i.e. 5.0 x 10^3, 1.5 x 10^4 and 4.3 x 10^4, the dynamics present a starlike behavior, similar to those of a symmetric 64- or 32-arms stars with the same arm molecular weight. For the copolymer with the short branch, the influence of the backbone fraction is not negligible, and should be considered in the assessment of the dynamics (Figure 3).

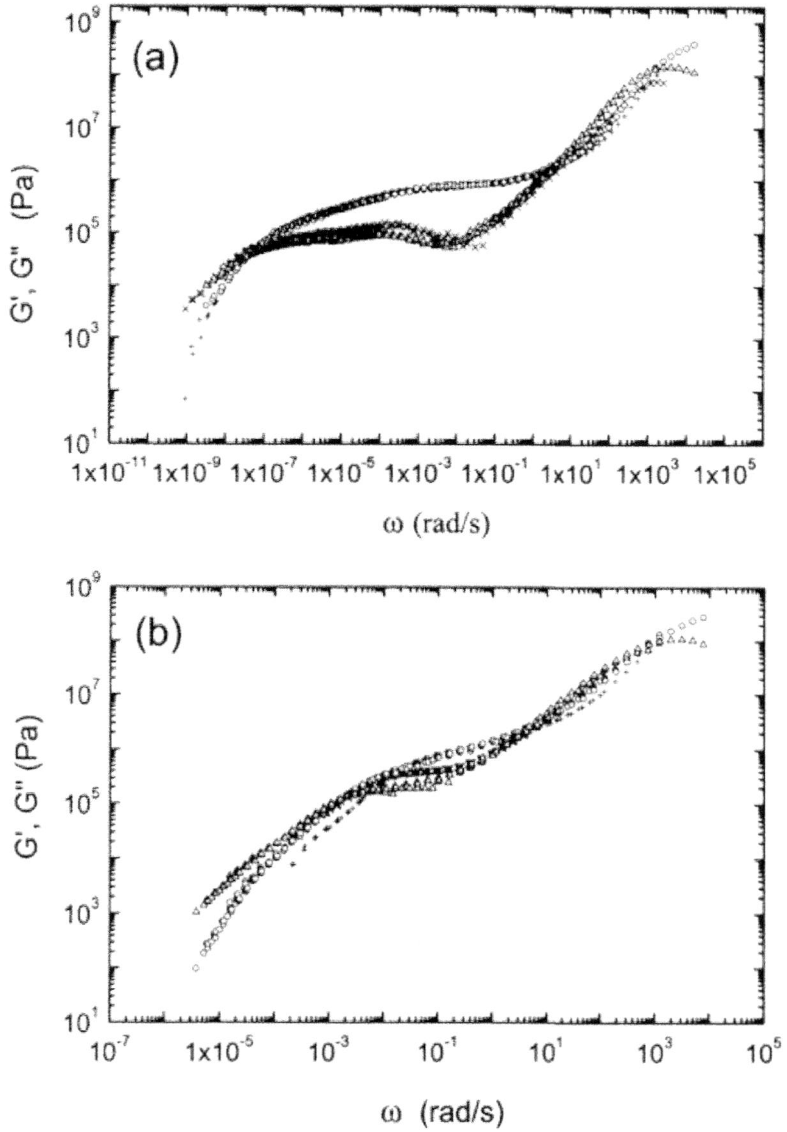

Figure 3. Comparison of linear viscoelastic spectra of dumbbells (G', ○; G' ', Δ) and respective multiarm star polymer (G', +; G' ', ×): DB43000 and 32-PBdStar37 (a); DB5000 and 64-PBdStar05 (b).

2.4 Block-Graft Copolymers

Model block-double graft co- (BDG) and terpolymers of the poly(S-*b*-(1,2Bd-*g*-X_2)] type, where X is S, I, Bd (in case of BDG) or S-*b*-I (in the case of terpolymers) have been recently synthesized[10]. The synthetic approach involved the preparation of a PS-b-1,2PBd, followed by the hydrosilylation reaction of the vinyl bonds with $HSiMeCl_2$. The resulted macromolecular linking agent was then reacted with living polymer X, leading to the formation of the desired poly(S-*b*-(1,2Bd-*g*-X_2)] terpolymer. The reactions used are given in Scheme 10.

$$\text{styrene} + \text{sec-BuLi} \xrightarrow[\sim 25^0\text{C}]{\text{cyclohexane/benzene}} \text{PS}^{(-)}\text{Li}^{(+)}$$

$$\text{PS}^{(-)}\text{Li}^{(+)} + \text{butadiene} \xrightarrow{\text{DIPIP, } 5^0\text{C}} \text{P(S-}b\text{-1,2Bd)}^{(-)}\text{Li}^{(+)}$$

$$\text{P(S-}b\text{-1,2Bd)}^{(-)}\text{Li}^{(+)} + CH_3OH \longrightarrow \text{P(S-}b\text{-1,2Bd)}$$

$$\text{P(S-}b\text{-1,2Bd)} + HSiMeCl_2 \xrightarrow[\text{THF, 65-67}^0\text{C}]{\text{Pt catalyst}} \text{P[S-}b\text{-(1,2Bd-}g\text{-SiMeCl}_2)] \quad (I)$$

$$\text{X} + \text{sec-BuLi} \xrightarrow[25^0\text{C}]{\text{benzene}} \text{PX}^{(-)}\text{Li}^{(+)}$$

$$I + \text{PX}^{(-)}\text{Li}^{(+)} \xrightarrow[25^0\text{C}]{} \text{P[S-}b\text{-(1,2Bd-}g\text{-X}_2)]$$

Where X : S, Bd, I, S-*b*-I.

Scheme 10.

Extensive characterization results revealed that the terpolymers exhibit high degree of molecular and compositional homogeneity.

The morphological characteristics (Figure 4) and the mechanical properties (Figure 5) of the block-double-graft co- and terpolymers were investigated[12] by transmission electron microscopy (TEM), small-angle X-ray scattering (SAXS), and tensile testing. It was found that when the branches are polydienes, the BDG molecules form the same morphologies as their linear diblock analogues. In this case phase separation occurs between the PS domain and a combined diene microdomain formed by the backbone and the polydiene branches. In BDG materials in which the branches are polystyrene-polyisoprene diblock copolymers lamellae are obtained at a total PS volume fraction close to 0.50. It was found that the domain spacing of these materials is mainly determined by the molecular weights of the diblock branches instead of the backbones.

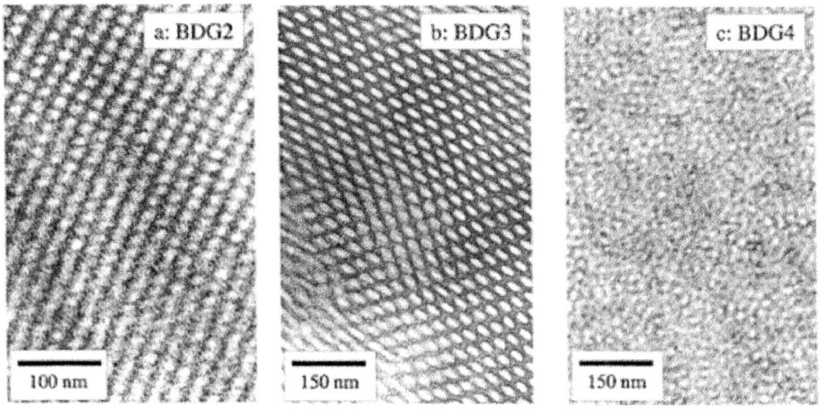

Figure 4. Morphology of BGD copolymers

A lamellae-forming BDG terpolymer with an average of three tetrafunctional junction points per molecule exhibited characteristic thermoplastic elastomer properties (Figure 5) with a stress at break of 32 MPa and strain at break 1000 %. This was attributed to the chain conformation in the microphase separated state, in which the PBd blocks of the backbone bridge the PS domains through multiple junction points, resulted in enhanced elastomeric properties. It was found that the parameters that influence the mechanical properties of these BDG materials are the existence of the PS backbone, the molecular weight of the branches and the number and functionatity of branch points on the 1,2 PBd part of the backbone.

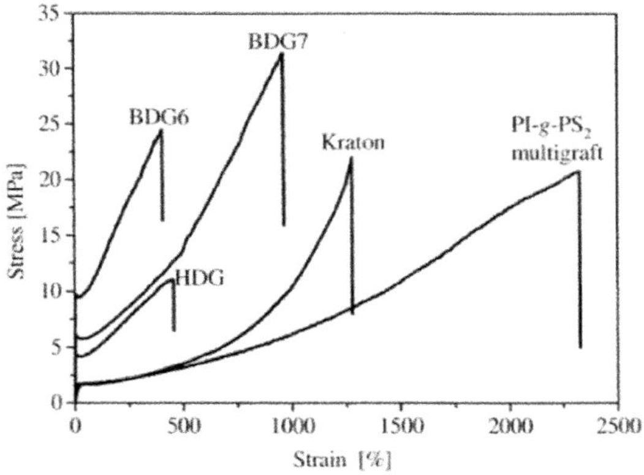

Figure 5. Stress-strain curves for (1) BDG6, 9 junction points, and the branch molecular weight is 14 000 g/mol; (2) BDG7, 3 junction points, and the branch molecular weight is 32 800 g/mol; (3) HDG, 9 junction points, and the branch molecular weight is 12 500 g/mol; (4) Kraton D1101; and (5) PI-*g*-PS$_2$ multigraft copolymer with 9 junction points at branch molecular weight of 13 000 g/mol.

2.5 Dendridic Architectures

Well-defined polymeric second-generation dendritic polymers of I and S, of the $(S_2I)_3$, $(SI'I)_3$, $(I''I'I)_3$ and $(I'_2I)_4$ (Scheme 11) type have been synthesized[12] by using anionic polymerization and controlled chlorosilane chemistry.

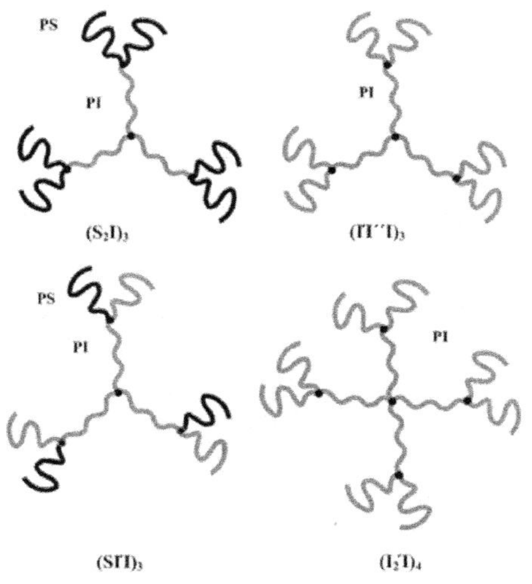

Scheme 11

The synthesis was performed in 4 steps. In the first a selective reaction of a living chain with the chlorosilane group of 4-(chlorodimethylsilyl)styrene was performed to produce a macromonomer. The second step involved the addition of a second living chain (same or different) to the double bond of the macromonomer. In the third step the polymerization of I was conducted with the anionic sites created in the second step and the last step involved the linking reaction with trichloromethyl or tetrachloro silane. The reaction sheme is given in Scheme 12.

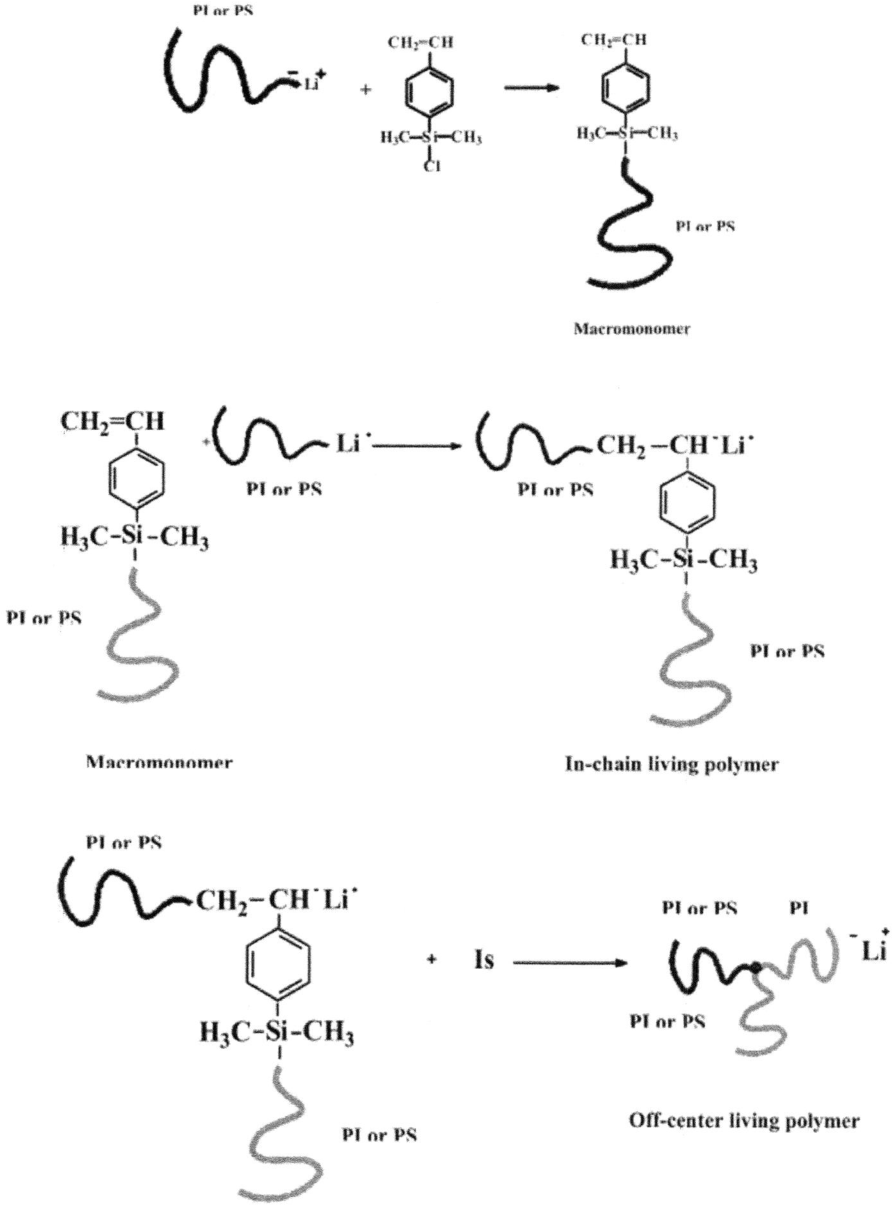

Scheme 12.

Combined characterization results has shown that the dendritic macromolecules exhibit high degree of molecular and compositional homogeneity.

2.6 Polymacromonomers

Anionic polymerization combined with the high vacuum techniques was used for the synthesis of polymacromonomers[13], i.e. polymers in which in every monomeric unit one other polymeric chain is attached. The macromonomers were prepared in the same way shown in scheme 11. The initiator used for the polymerization of the macromonomers was s-BuLi. The macromonomers were either PI, PBd or poly(S-b-I). Characterization results of the polymacromonomers indicate that the polymacromonomers synthesized were well-defined.

REFERENCES

1. Roovers, J., Zhou, L.-L., Toporowski, P., van der Zwan, M., Iatrou, H. Hadjichristidis, N (1993) Regular star polymers with 64 and 128 arms. Models for polymeric micelles, *Macromolecules*, **26**, 4324-4331.
2. Avgeropoulos, A., Poulos, Y., Hadjichristidis, N., Roovers, J. (1996) Synthesis of Model 16-Miktoarm (Vergina) Star Copolymers of the A_8B_8 Type, *Macromolecules,* **29**, 6076-6078.
3. Tsoukatos, T., Pispas, S., Hadjichristidis, N. (2001) Star-branched polystyrene by nitroxide living free-radical polymerization, *J. Polym. Sci, Part A: Polym. Chem.* **39**, 320-325.
4. Xenidou, M. and Hadjichristidis, N. (1998) Synthesis of model multigraft copolymers of butadiene with randomly placed single and double polystyrene branches, *Macromolecules*, **31**, 5690-5694.
5. Iatrou, H., Mays, J., Hadjichristidis, N., (1998) Regular comb polystyrenes and graft polyisoprene/polystyrene copolymers with double branches (centipedes). Quality of (1,3-phenylene)bis(3-methyl-1-phenylpentylidene)dilithium initiator in the presence of polar additives, *Macromolecules*, **31**, 6697-6701
6. Weidisch, R., Gido, S., Uhrig, D., Iatrou, H., Mays, J., Hadjichristidis, N. (2001) Tetrafunctional multigraft copolymers as novel thermoplastic elastomers *Macromolecules*, **34**, 6333-6337.
7. Nakamura, Y., Wan, Y., Mays, J., Iatrou, H., Hadjichristidis, N. (2000) Radius of gyration of polystyrene combs and centipedes in solution, *Macromolecules*, **33**, 8323-8328.
8. Velis, G., Hadjichristidis, N. (1999) Synthesis of model $PS(PI)_5$ and $(PI)_5PS(PI)_5$ nonlinear block copolymers of styrene (S) and isoprene (I), *Macromolecules*, 32, 534-536.
9. Houli, S., Iatrou, H., Hadjichristidis, N., Vlassopoulos, D. (2002), *Macromolecules*, **35**, 6592-6597.

10. Velis, G., Hadjichristidis, N. (2000) Synthesis of model block-double-graft copolymers and terpolymers of styrene (S), butadiene (Bd), and isoprene (I): poly[S-*b*-91,2Bd-*g*-X_2)] (X: S, Bd, I, S-*b*-I), *J. Polym. Sci, Part A: Polym. Chem.*, **38**, 1136-1138.
11. Zhu, Y., Weidish, R., Gido, S., Velis, G., Hadjichristidis, N. (2002) Morphologies and Mechanical Properties of a series of block-double-graft copolymers and terpolymers, *Macromolecules*, **35**, 5903-5909
12. Chalari, I., Hadjichristidis, N. (2002) Synthesis of well-defined second-generation dendritic polymers of isoprene (I) and styrene (S): $(S_2I)_3$, $(S_2I'I)_3$, $(I''I'I)_3$, and $(I'_2I)_4$, *J. Polym. Sci, Part A: Polym. Chem.*, **40,** 1519-1526.
13. Pantazis, D., Chalari, I., Hadjichristidis, N. (2003) Anionic polymerization of styrenic macromonomers, *Macromolecules*, **36**, 3783-3785.

EFFECTS OF INTRAMOLECULAR COMPOSITION AND TOPOLOGY ON INTERMOLECULAR STRUCTURE AND BULK PROPERTIES OF COMPLEX POLYMER SYSTEMS:

Experimental Characterization and Computer Simulation

TADEUSZ PAKULA

Max Planck Institute for Polymer Research, Postfach 3148, 55021 Mainz, Germany ; E-mail: pakula@mpip-mainz.mpg.de

Abstract: The experimental characterization and computer simulation of the intermolecular structure and bulk properties of complex polymer systems is described. The reported experiments concern the structure analysis by means of various scattering methods and the analysis of the dynamics performed through dynamic mechanical measurements. The simulation methods used are applicable to various size scales of the macromolecules and in particular at the nanometer scale. The simulations methods being based on cooperative macromolecular rearrangements in condensed state are very convenient for representing the stricture of dense systems consisting of complex polymers. The results of the simulation provide a significant supplement to the experimental investigation.

Key words: complex polymers, bulk properties, simulation methods, macromolecular architecture, topology, X ray diffraction, X ray scattering, light scattering, neutron scattering.

1. INTRODUCTION

Progress in technologies related to chemistry and physics with increasing success allows manipulate atoms and molecules to create novel structures with desired properties. A lot of systems have been developed in which a

F. Ciardelli and S. Penczek (eds.),
Modification and Blending of Synthetic and Natural Macromolecules, 91–121.

controlled molecular architecture leads to self-organization of molecules to various supra-molecular structures with interesting properties[1,2]. Nevertheless, the knowledge about the phenomena controlling such self assembling processes and about the correlation between structures and properties is still far away from a state which could be considered as satisfactory, i.e. allowing predictions and synthesis of molecular structures necessary to achieve desired molecular assembles with expected functions or properties.

Some examples of self assembling macromolecular structures are presented in Figure 1, in which the topology of molecular skeletons varies horizontally and the atomic constitution along chains is varied vertically. Structures, to which some kinds of these molecules can organize in condensed states, are schematically shown in Figure 2 where the hierarchical nature of order, extending often over many decades of the size scale, is illustrated.

The self-assembling systems are generally complex fluids, in which various kinds of interactions and a predetermined atomic order within molecules involve spontaneously, under certain external conditions, a specific supramolecular order. Generally, all kinds of interactions present in a given system, like excluded volume, intra-molecular bonds, dipoles, charges or incompatibility of molecular segments, can contribute to the self organization processes. By a specific molecular design of atomic constitution and topology of bond skeletons of molecules, contributions of various interactions can be controlled to a large extend.

The broad size range of structures in such systems involves a broad variety of related relaxations[3], which contribute to the dynamics extending over an extremely broad time range sometimes exceeding the observation possibilities. The dynamic spectrum of such materials becomes very important for understanding the correlation between parameters of the molecular and supramolecular structures on one hand, and macroscopic properties on the other hand. Analysis of such correlations appears to be extremely difficult and requires usually application of many experimental techniques for characterization of both the structure and the dynamics.

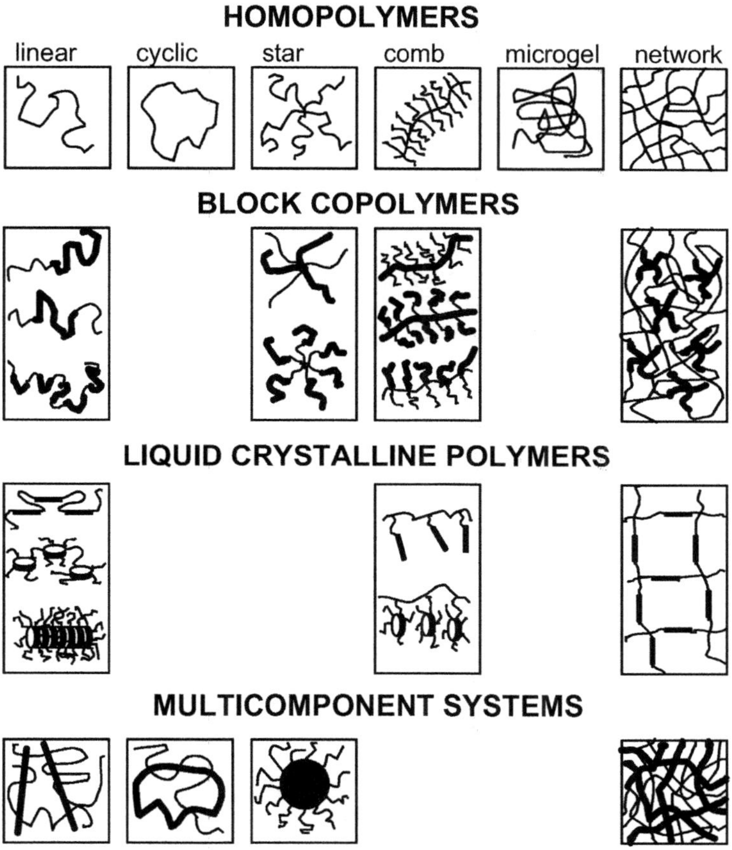

Figure 1. Examples of macromolecular architectures with various topology and constitution

In this chapter, structure analysis by means of various scattering methods and analysis of the dynamics by means of the dynamic mechanical measurements will be illustrated and discussed.

A theoretical description of correlations between molecular parameters and resulting supramolecular structures, on one hand, and between the structure and macroscopic properties, on the other hand appears to be extremely difficult. The problem demands consideration of a large number of independent parameters while most of theoretical problems can successfully be solved only when the number of parameters is considerably reduced. Therefore, there is a hope, that computer simulation can be more successful than an analytical theory in describing properties of such complex systems. In the later case there are, however, drastic limitations concerning size and time scales. In order to overcome this problem, various simulation methods have been developed, which can operate successfully, to some extend, in correspondence to various size scales of real systems, as illustrated in Figure 3.

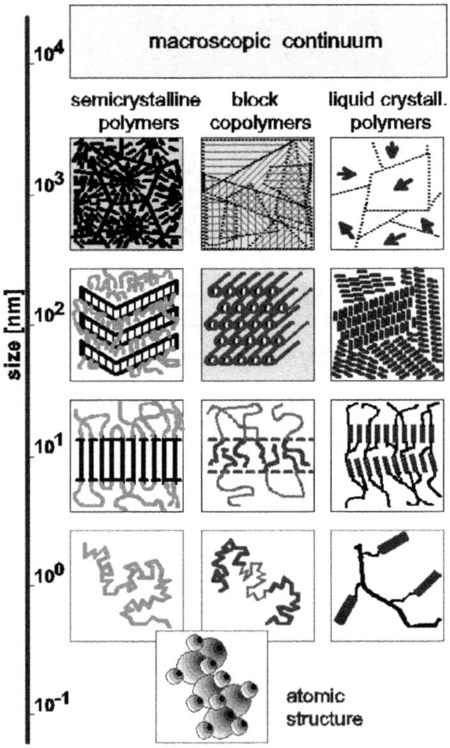

Figure 2. Examples of supramolecular hierarchies extending over many orders of magnitude of the size scale in various classes of polymer materials

The simulation methods can, from this point of view, be classified in the following groups: (1) methods which consider molecular structures with atomic details and with interatomic interactions, (2) methods which consider coarse grained molecular objects and drastically simplified interactions between them, (3) microstructural methods considering the material as consisting of microphases which are regarded as continuous and finally (4) methods (theories) considering the material as a continuum.

Whereas at the two extremes i.e. at the atomic and the continuum scales, the situation can be considered as to some extend satisfactory, serious problems appear still in the intermediate scales. The problems are of versatile nature but the main consist in a limited spatial resolution of the models, in independent application of methods corresponding to various size scales and in a lack of methods bridging them uniquely. A lot of effort is still required to construct a hierarchy of interrelated simulation methods which in analogy to the hierarchical nature of the material structures will correlate the properties of the extreme scales.

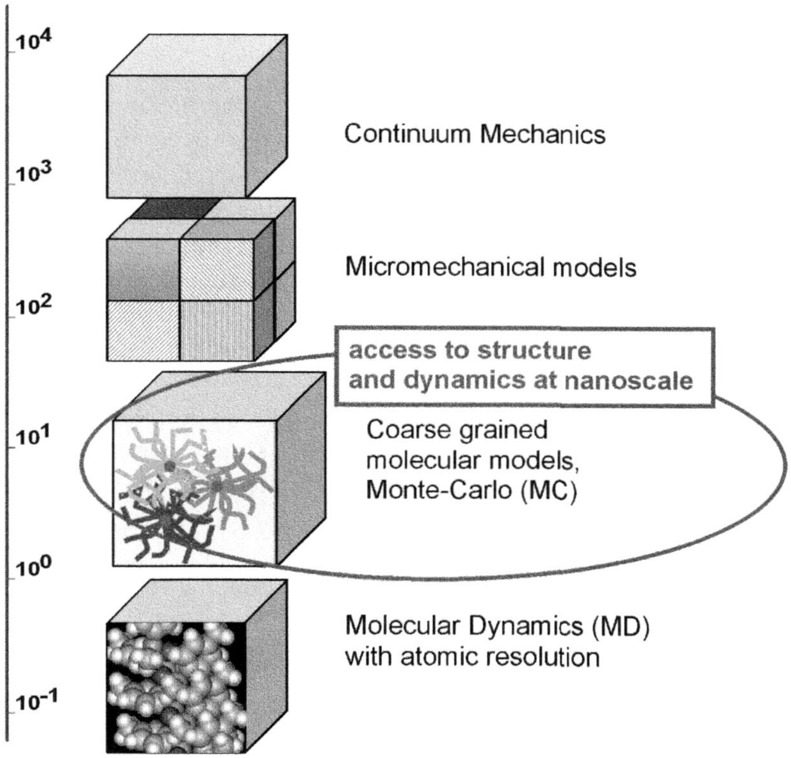

Figure 3. Various modeling methods applicable at various size scales of the macromolecular systems

In this chapter, examples of simulation results will be shown which are obtained by means of methods particularly suitable to represent the macromolecular systems at the nanometer scale. These results are considered as information about structure and dynamics of the studied systems which is supplementary to information coming from experimental investigation. The simulation methods used are based on cooperative macromolecular rearrangements taking place in condensed states and are unique in representing structure and dynamics of dense systems consisting of complex polymers.

2. CHARACTERIZATION METHODS

2.1 Structure and Scattering

Very characteristic for polymeric materials is the hierarchical nature of the structural organization, which often extends over many orders of magnitude on the size scale (see Figure 2). Comprehensive analysis and understanding of these structures and their functions requires application of experimental methods which are sensitive to the ordering beginning at the scale of interatomic distances and extending up to the scale of macroscopic dimensions. Various radiation scattering techniques are applied in order to detect and characterize structures at various size scales (see Figure 4 for examples):

- X –ray diffraction sensitive for interatomic spatial correlations within the sub-nanometer range (crystallinity, unit cells, structure in mesophases)
- Small angle X-ray scattering indicating structures in the size range 1-60 nm characteristic for semicrystalline polymers, copolymers, nanocomposites etc.
- Wide and small angle light scattering extending the possibility of structural analysis up to the micrometer range (all optically turbid materials have structures within this size range).
- Neutron scattering sensitive within the nanometer size scale and with a unique possibilities of controlling contrast of macromolecular elements e.g. by deuteration.

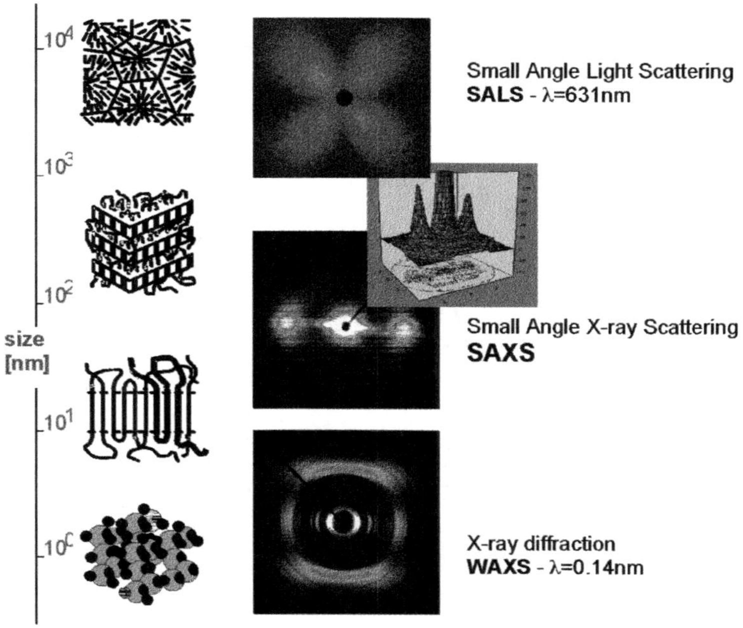

Figure 4.. Illustration of scattering patterns recorded by means of various techniques using radiation of different wave lengths (e.g. light, x-rays) in order to detect structures of various sizes

2.2 Dynamics and Mechanical Spectroscopy

Formally, the properties of linear viscoelastic materials are described by the storage G' and loss G" moduli, representative for the elastic and viscous components of material behavior, respectively. Both real and imaginary parts of the complex modulus are functions of frequency with a characteristic behavior at frequencies around the relaxation frequency of structural units in a simple viscoelastic system. An example of such behavior is illustrated schematically in Figure 5 by using the G'(ω) and G"(ω) dependencies determined for the simple Maxwell model[3].

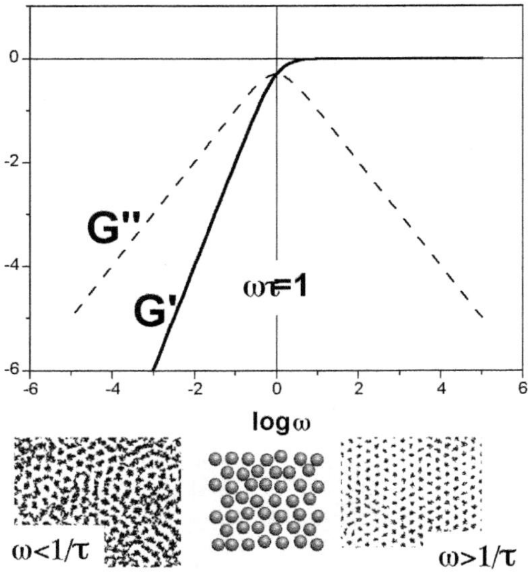

Figure 5 Frequency dependencies of the real (G') and imaginary (G'') parts of the modulus for the simplest relaxation model (Maxwell) with a single relaxation time (τ). Different behavior of elements in the systems of spherical particles below and above the frequency of 1/τ is illustrated by trajectories of their centers.

In the high frequency range the plateau of G' and the low G'/G" ratio (G'>>G") both are characteristic for the elastic behavior (Hookean). The elastic shear modulus describing the material properties in this range becomes nearly constant.

In the low frequency limit, the behavior typical for the Newtonian flow is seen, for which G">>G' and where G' and G" obey the characteristic proportionalities to ω^2 and ω, respectively. Here, the system can be characterized by the complex viscosity, η*(ω), which assumes values equal to the kinematic viscosity, η, only if G" >> G'. Only in this range, the viscosity is a meaningful material property and the complex viscosity can approach here the value of the zero shear viscosity. The two limiting types of behavior are separated by the intermediate cross-over region in which the system can be regarded as typically viscoelastic. The point at which G' and G" cross each other determines the place (ω on the frequency scale) which is related to the relaxation time of the structural units constituting the system ($\tau=1/\omega_r$). In the simple example of the relaxation represented by the Maxwell model, the G" assumes a maximum and the lines logG' and logG" vs. logω extrapolated from low to higher frequencies cross each other at the same frequency directly related to the relaxation time.

Figure 5 gives an example of a possible structural interpretation of the mechanically observed relaxation, which can be considered for molecular or colloidal systems. For the dense system of spheres, which can be regarded as representing molecules or particles, there is the solid-like state at higher and the liquid-like state at low frequencies. The nature of these states is illustrated in the figure by particle trajectories. In the high frequency regime, they indicate only vibrational motions and quasi localized positions of particles between the neighbors, whereas, at low frequencies, displacements exceeding particle sizes make the system viscous - changing neighbors becomes possible. According to such a model, the relaxation is related to the position correlation of particles and the mechanical response can be considered as a Fourier transform of the position autocorrelation function with the characteristic relaxation time necessary for particles to escape from the surroundings of the initial neighbors.

3. EXPERIMENTAL RESULTS

Viscoelastic spectra for real polymers are usually more complex than the spectrum for the idealized case shown in Figure 5.

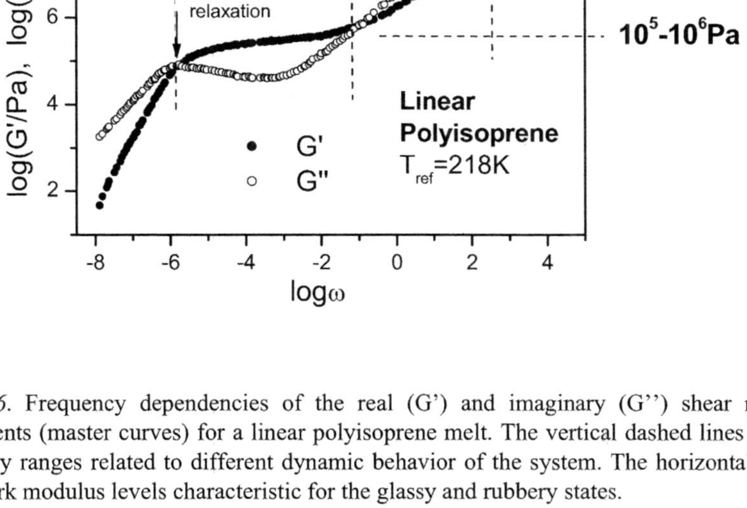

Figure 6. Frequency dependencies of the real (G') and imaginary (G'') shear modulus components (master curves) for a linear polyisoprene melt. The vertical dashed lines indicate frequency ranges related to different dynamic behavior of the system. The horizontal dashed lines mark modulus levels characteristic for the glassy and rubbery states.

The complexity concerns the behavior of real systems at the single specific relaxation, as well as, the versatility of relaxation processes in complex systems. One of the simplest examples is the behavior of a linear polymer melt, in which the viscoelastic spectrum indicates two characteristic relaxations related to motion of monomer units at high frequencies and the relaxation of polymer chains usually at much lower frequencies. An example of such a spectrum is shown in Figure 6. The two relaxation processes separate the diverse dynamic states of the system which are related to drastically different mechanical macroscopic behavior. There are the well known glassy, rubbery and melt states as illustrated in Figure 6 for the polyisoprene linear polymer[4].

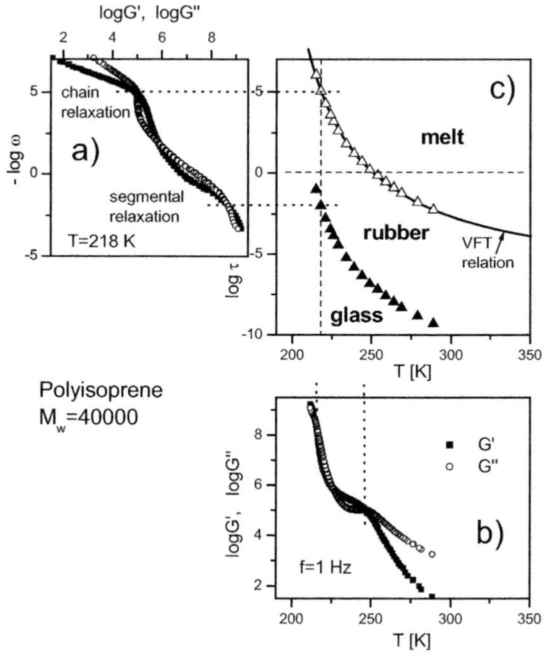

Figure 7. Correlation between frequency and temperature dependencies of the viscoelastic spectra for a linear polyisoprene melt. In (a) and (b), the frequency and temperature dependencies are shown, respectively. In (c), the temperature dependencies of the segmental and chain relaxation times are shown.

The results like these presented in Figure 6 can also be recorded as a function of temperature in isochronal experiments. The relation between these two types of results is illustrated in Figure 7, where the two dependencies (a) and (b) are interconnected by the temperature dependence of the relaxation times (c) of both the segmental and the polymer chain relaxations.

The two dashed lines (vertical and horizontal) indicate sections through the diagram, which can be directly related to results presented in Figures 7a and 7b. As indicated in Figure 7c the temperature dependencies of the segmental end chain relaxation rates separate ranges of qualitatively different mechanical behavior of the polymer. In some cases, these dependencies can be described by phenomenological relations[3].

3.1 Complex Topology

3.2 Star polymers

An example of the behavior of multiarm stars in the melt is illustrated in Figure 8, where the mechanical results indicate three relaxation processes from which the two at higher frequencies are typical for polymeric bulk systems and are attributed to segmental and arm relaxations, whereas, the slowest is a new one and is believed to be related to rearrangements within the supramolecular order documented by means of the small angle X-ray scattering result presented in the insert[5].

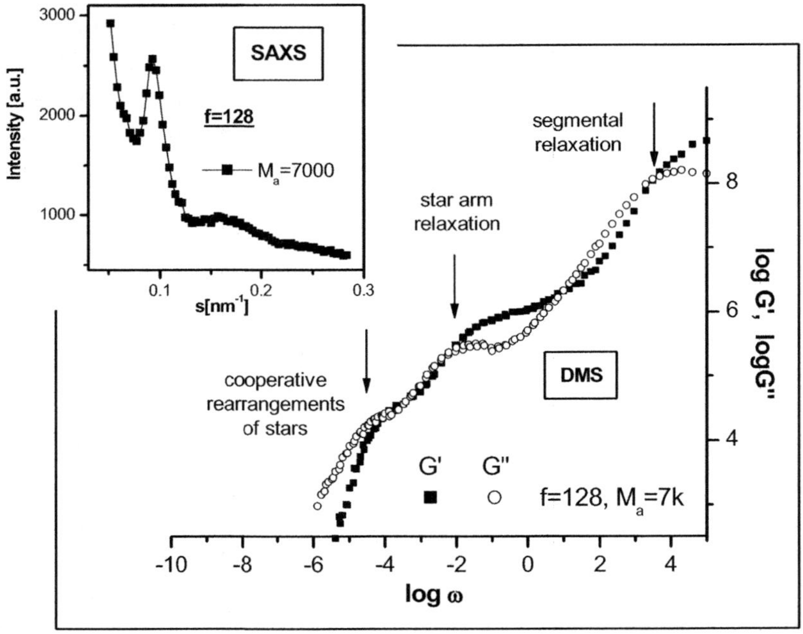

Figure 8. An example of the small angle X-ray scattering (SAXS) and the mechanical spectroscopy results for a polybutadiene melt of multiarm stars. A scattering peak indicating star ordering and various modes in the mechanical relaxation can be seen.

It is interesting to notice that there is a new dynamic state in this system between the relaxation of the arms and the structural relaxation. In this state, the melt of multiarm stars shows a new elastic plateau, with the G' level which is considerably lower than that characteristic for the entangled linear polymers.

3.3 Brush polymers

Comb or brush like polymers are similar to polymer stars because of having a large number of arms but the arms are here anchored to linear backbone, usually one side chain per backbone monomer unit. A certain kind of order exists in the melts of this kind of polymers, as well. Figure 9 shows results of the X-ray scattering experiments performed both in the wide and small angle range for a melt of PnBA brush macromolecules[6]. The small angle peak (a) indicates correlation distances of about 10 nm and results from interchain backbone-backbone correlation or from correlated positions of Br atoms decorating the ends of PnBA arms in the brush polymer. Both these correlations should be related to the thickness of the brush which depends first of all on the side chain length. The two wide angle peaks (b) and (c) are attributed to the correlation distances between cores of the nBA side chains and to the amorphous hallo of this polymer, respectively.

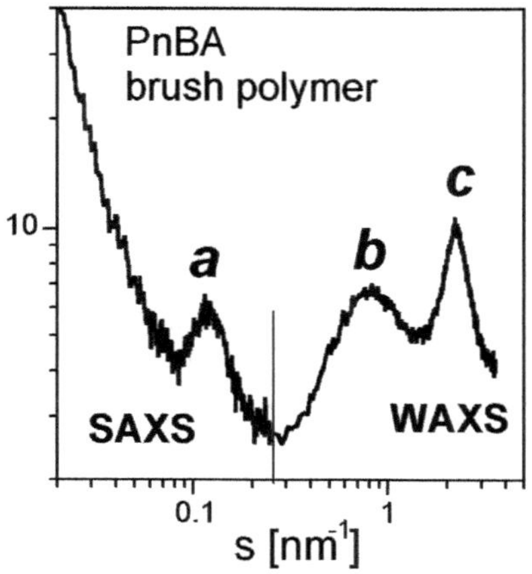

Figure 9. Small and wide angle X-ray scattering intensity distributions recorded for the melt of PnBA brush polymer in the molten state. The scattering peaks a, b and c indicate backbone-backbone distance correlation between side chain cores an the amorphous halo, respectively. [Please provide a caption for this figure]

Viscoelastic properties determined for the melt of brush-like macromolecules are presented in Figure 9 as master curves of G' and G'' determined for the reference temperature of 254K[6]. The results indicate again a presence of three relaxation ranges: the high frequency relaxation corresponds to segmental motion, the intermediate relaxation is attributed to the reorientation rates of the side chains and the slowest process is the global macromolecular relaxation in this system. The last controls the melt shear flow and the zero shear viscosity. The rate and nature of this relaxation must be dependent on the length of the backbone. For short backbone chains these macromolecules can behave similarly to stars, for which translational motion dominates the slow dynamics but for longer backbone chains the reorientation possibilities should become slower than translation[7].

It has been recently demonstrated, that cross-linking in such systems can create a new class of bulk materials having properties considerably different form these which the conventional cross-linked polymers have[8,9]. The lightly cross-linked brush polymers constitute elastomers with the shear modulus which can by some orders of magnitude be lower than for the conventional polymeric elastomers.

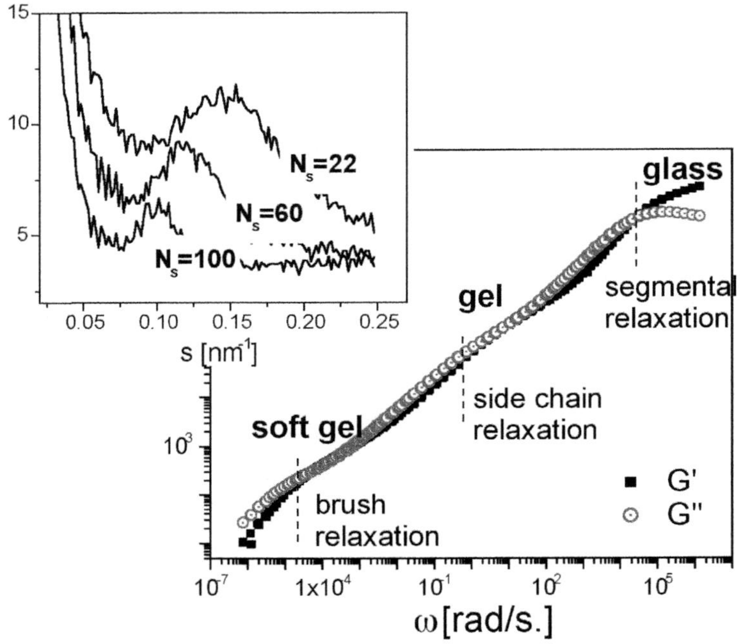

Figure 10. Frequency dependencies of the storage (G') and loss (G'') shear moduli (master curves at T_{ref}=254K) for the melt of brush polymers. The insert: Small angle X-ray scattering from melts of p(nBA) brush polymer with various side chain length.

3.4 Copolymers

3.4.1 Copolymers with various composition profiles

Various distributions of monomers of different type along polymer chains create various types of copolymers, which can considerably differ by their supramolecular structure and resulting properties. Random copolymers and block copolymers with homopolymer blocks constitute two extreme cases of such materials. Copolymers with more sophisticated distributions of monomers such as tapered block copolymers[10,11] and gradient copolymers[12,13] are also known but the structure-properties relationships have not so much been investigated as in the characteristic extremes.

Besides creation of new relaxations and new properties resulting from the microphase separated structure, there is a possibility in copolymers to influence considerably the rates of segmental relaxation by combining monomers of different flexibility at various compositions within the polymer chains. Figure 11 shows an example of effects of copolymer composition on the mechanical spectra for p(nBA/MMA) nearly random copolymers[14]. It is seen, that the whole viscoelastic spectra can be considerably shifted along the temperature (frequency) scale by variation of composition in the copolymers.

The copolymerization allows combining monomers which in blends would appear strongly incompatible and would not be able to form structurally stable bulk materials.

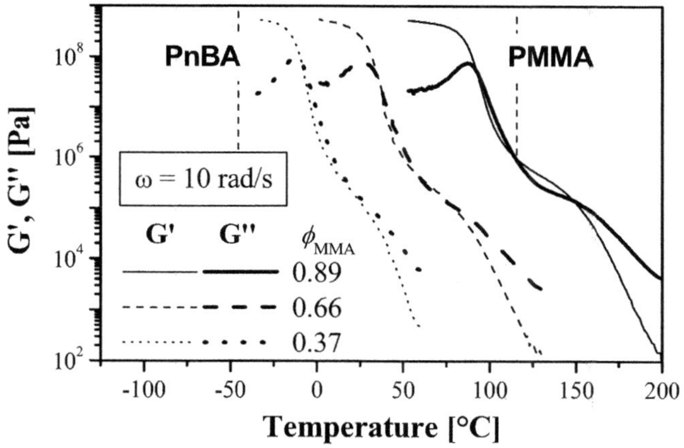

Figure 11. Temperature dependencies of the storage (G') and loss (G'') shear moduli for the copolymer samples of various compositions. The vertical dashed lines indicate Tg's for the homopolymers.

In the case of a control of nearest neighbor sequences, as achieved for the alternating copolymers, the material remains homogeneous and main effect obtained is related to the segmental mobility of the system and consequently to the localization of the glass transition on the temperature scale. In the other cases, when the control of composition extends to the size scale of whole macromolecules, the microphase separation process and the resulting microstructures are influenced.

3.4.2 Copolymer micelles

Spherical block copolymer micelles with glassy core and flexible corona provide interesting model hairy particles with usually very good defined dimensions. When mixed with linear polymer compatible with the corona, they show a phase behavior very similar to that of hard sphere colloidal particles constituting, in this way, a very interesting model system. We consider here spherical micelles of polyisoprene-b-polystyrene block copolymers dispersed in a polyisoprene matrix[15]. Such micelles show ordering which depends on the micelle concentration, as illustrated in Figure 12c. Mechanical relaxation spectroscopy of such copolymer/homopolymer blends revealed three relaxation processes which are marked in Figs. 12a and 12b by vertical dashed lines. The fast processes have been attributed to the polymer typical segmental and chain relaxations. The mechanical results indicate that with the increasing order a new slow relaxation process is created, which also here can be attributed to structural relaxation[15]. For the system containing 50% micelles this structural relaxation becomes so slow that the G' plateau at the level of 10^3 Pa extends to the lowest detectable frequencies.

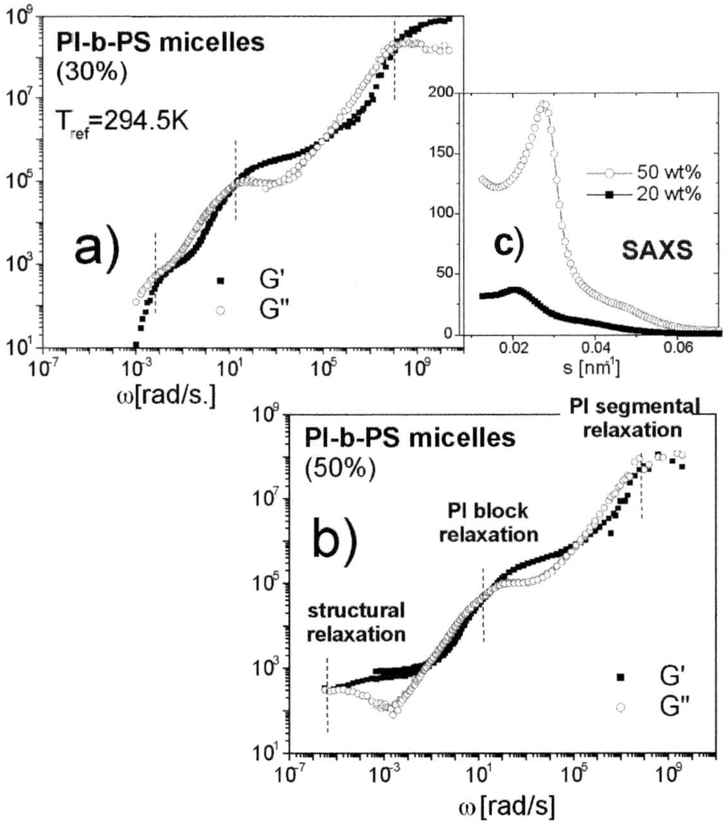

Figure 12. Temperature dependencies of the storage (G') and loss (G'') shear moduli for the copolymer samples of various compositions. The vertical dashed lines indicate Tg's for the homopolymers.

4. COMPUTER SIMULATION

There is a large number of different methods (algorithms) used for simulation of polymers on the coarse grained molecular scale[16-19]. Models of polymers considered in this range disregard usually details of chemical constitution of macromolecules and represent them as assemblies of beads connected by non breakable bonds. In order to speed up recognition of neighboring beads, the simplified polymers are often considered on lattices with beads occupying lattice sites and bonds coinciding with lines connecting neighboring sites. The methods used for simulation of the lattice polymers can be considered within two groups: first including algorithms, which can operate only in systems with a relatively large fraction of lattice sites left free and the other with algorithms suitable for lattice systems, in which all lattice sites are occupied by molecular elements. Whereas, the systems considered within the first group should be regarded as lattice gases, the systems treated within the second group of methods can be considered as lattice liquids. This reflects in differences of mechanisms of molecular rearrangements used within these two groups to move the systems through the phase space in order to reach equilibrium. The later problem concerns the physical nature of molecular rearrangements in dense polymer systems and is related to the microscopic mechanism of motion in molecular liquids. Unfortunately, the most popular picture is that a molecule or a molecular segment in the case of polymer, needs a free space in its neighborhood in order to make a translational step beyond the position usually occupied for a quite long time. Most of simulation methods assume this picture and consequently a relatively large portion of the space in form of free lattice sites has to be left free in order to allow a reasonable mobility[16-18]. There is only one simulation method, which assumes a cooperative nature of molecular rearrangements on the local scale and which does not require such a reserve space to allow the molecular mobility. The method, which uses the mechanism of cooperative rearrangements for polymer systems is the Cooperative Motion Algorithm (CMA) suggested originally in[19] and presented in improved form in subsequent publications[20-22]. The mechanism of this kind has been formulated recently also for low molecular liquids, as based on assumptions taking into account both a dense packing of molecules interacting strongly due to excluded volume and a condition of preservation of local continuity of the system. The later version of the microscopic mechanism, called the dynamic lattice liquid (DLL) model, has been described in detail[23].

The aim of this chapter is to present possibilities of simulation of many static and dynamic properties of dense complex polymer systems, as well as, to show some examples of application of the CMA method for polymers of various complex architectures.

4.1 The Model and Generation of Model Systems

The model is aimed to mimic the behavior of a macromolecular liquid-like state i.e., primarily, a polymer melt but also a solution of macromolecules in a low molecular solvent. A number of simplifications are assumed in the model, in order to achieve fast simulation, which however, as we believe, do not influence the basic physical features of the system. The assumptions concern both the structure and the dynamics. The type of structural simplification is illustrated in Figure 13. The models usually neglect the chemical specificity of the macromolecules and represent polymer fragments (monomers or so) by an equivalent sphere with the size approximately corresponding to the polymer segment volume. Such spheres are considered as beads which are connected by bonds to represent various polymer topologies from which the simplest is the linear one but some others as these presented in Figure 2 are of great interest. In order to describe the dynamic simplification we have to consider a qualitative description of the behavior of a simple molecule in a liquid. It can be characterized by the trajectory of motion, which consists of oscillatory components about some quasi-equilibrium positions and of the occasional translational movements between subsequent quasi-localized states, as illustrated in Figure 14. The oscillatory motion is caused by the fact, that in the liquid, each molecule is hemmed to a substantial extend by neighboring molecules forming so called "cage" and can escape from the cage only when at least one of the surrounding molecules moves away at the same time. This, however, is a rare event taking place occasionally after many unsuccessful attempts and involves a larger scale rearrangement in the local system. Under such conditions the translational steps can occur only in a strictly cooperative way so that each moving element replaces its nearest neighbor, as illustrated in Figure 14. Such rearrangements have to take place along self avoiding closed paths in order to preserve the local continuity of the system. The arrangement of molecules in the system is in this way unstable. Each molecule can wander through the system changing neighbors during the translational motion steps. Algorithms implementing these considerations have been described and applied in various former publications[19-28].

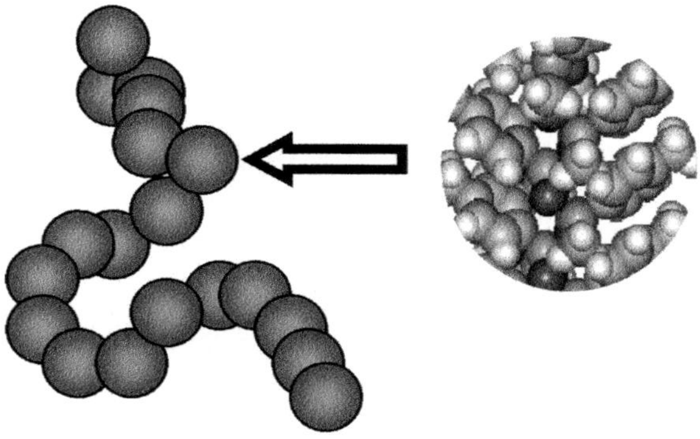

Figure 13. Illustration of a structural simplification of a poly(methyl-phenyl siloxane) chain to a series of bounded beads having sizes corresponding to the sizes of polymer segments .

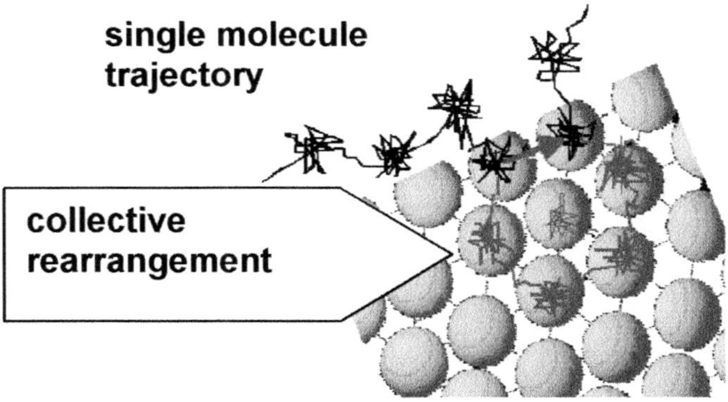

Figure 14. Illustration of the dynamic simplification in a dense system of beads occupying all sites in a lattice. The rearrangement consists in cooperative displacement of system elements along closed trajectories within which each element replaces one of the nearest neighbors.

Because of the specific property of the CMA simulation consisting in ability of treatment of dense systems on a lattice, the methods of generation of initial states of model systems are versatile and very effective. Two of them are worth while to be mentioned: (1) melting of initially ordered dense systems and (2) polymerization performed in the systems which mimic real conditions of polymer synthesis[24]. The first method provides model systems, which can be perfectly regular, for example, systems with exactly uniform lengths of polymers. The model systems obtained by polymerization have usually more realistic parameters with the characteristic distributions of polymer sizes. Examples of model systems generated in these two ways are illustrated in Figures 15 and 16.

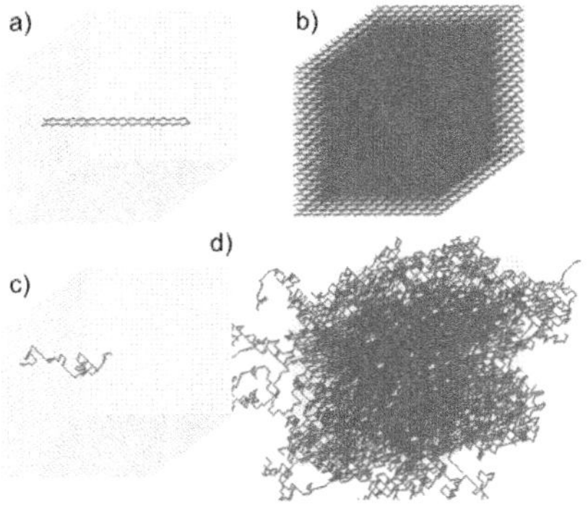

Figure 15. Illustration of generation of an amorphous model system with linear chains: (a) the single chain in an ordered state on the fcc lattice fitting to the system dimensions, (b) the system filled completely with chains having ordered conformations, (c) and (d) single chain and the system after equilibration, respectively.

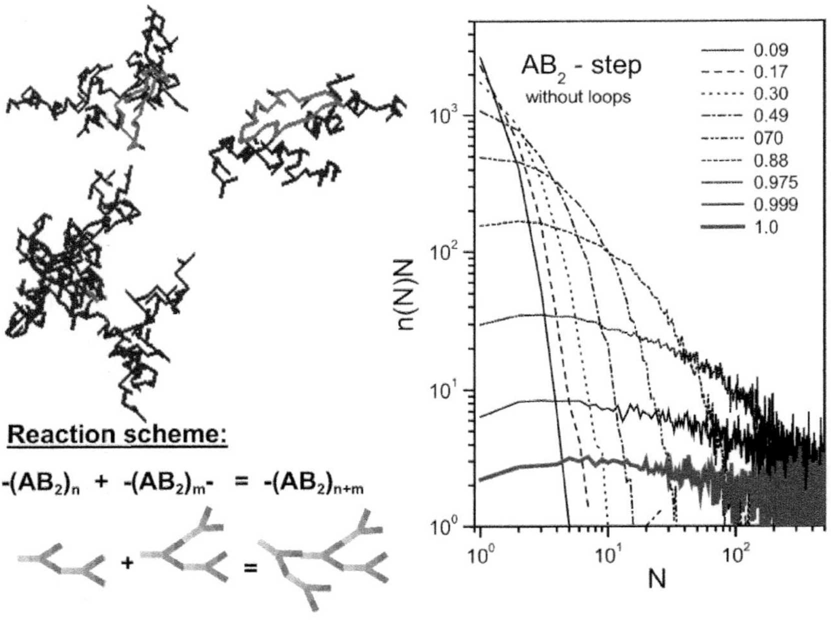

Figure 16. Illustration of evolution of the molecular mass distribution of hyperbranched polymers when simulated by a step polymerization process in a bulk system of branched monomers. Some structures of polymers obtained are shown.

5. EXAMPLES OF SIMULATION RESULTS

5.1 Melts of Branced Macromolecules

Model systems representing melts of polymer stars and polymer brushes with various numbers of arms and various arm lengths were simulated using the CMA[7,22,25,26]. Results of simulated dense systems of this kind of macromolecules have shown that the structure develops due to strong excluded volume effect between structural elements on the macromolecular scale. This leads to additional dynamic relaxation modes, which although taking place on the macromolecular size scale, seem to have many similarities with the cooperative rearrangements considered in the dynamics of small molecules in the simple liquid. An illustration of such behavior is presented in Fig. 17. The pair correlation functions of star centers of mass indicate the ordering increasing with increasing number of arms. This structure involves considerable changes in the dynamics of multiarm star melts with respect to melts composed of stars with a low number of arms. The flow of such systems becomes controlled not by the star arm relaxation but by an additional slow relaxation process which is attributed to cooperative rearrangements of stars within the ordered state. This involves a slowing down of the decay of the position correlation of such stars (Fig. 18) which for a large number of arms becomes considerably slower than the relaxation of star arms constituting in this way an extra slow mode in the translational relaxation.

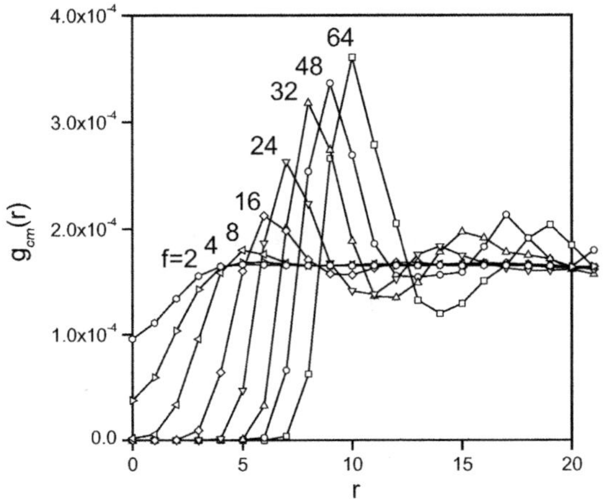

Figure 17. Illustration of increasing correlation between positions of centers of mass of stars in the melt with increasing number of arms (between f=2 and f=64)..

Both the experimental and simulation results have shown that the degree of order in star polymer melts depends on both molecular parameters: the number and the length of arms[26]. It can be supposed, that the degree of order in the studied systems is mainly controlled by the ratio of the core radius to corona thickness which evidently depends on these parameters. Therefore, for stars with a large number of short arms the highest degree of ordering should be expected, whereas, stars with long arms could show a limited order even when the number of arms is high.

The type of order observed in the multiarm star melts can be described as liquid like on the macromolecular scale. Neither in real nor in simulated systems, has any clear signature of lattice formation been detected. This results probably from the deformability and related form fluctuations of the star coronas. They consist of flexible arms and remain soft spheres even when the number of arms is large and the core radius is large. The latter is well confirmed by the dynamic properties of the studied systems. Similar behavior has been observed in various other systems, such as microgels or polymer melts filled with spherical copolymer micelles. In all these cases, the ordering was related to the excluded volume interaction between compact but deformable macromolecular elements in a dense system.

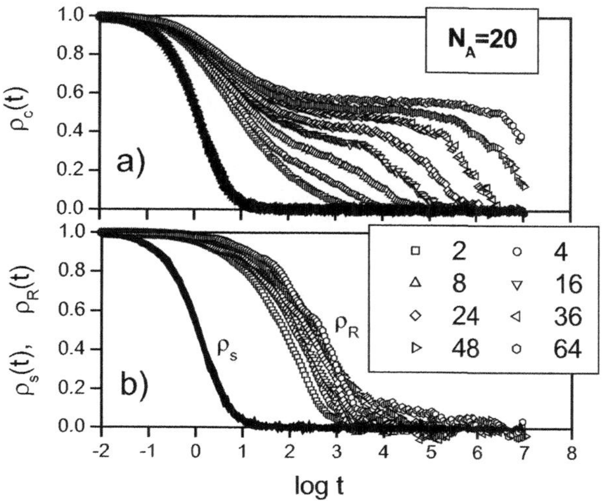

Figure. 18. Illustration of the main relaxation processes in simulated star polymer melts for stars with various arm numbers: ρ_s - segmental relaxation, ρ_R - star arm relaxation and ρ_c - star position correlation. The relaxation rate of segmental relaxation is not influenced by the number of arms, the arm relaxation depends only weakly on the arm number.

Qualitatively, the simulation results concerning dynamics seem also to be in a good agreement with the experimental observations[26]. The suggested assignments of the relaxations detected by means of the viscoelastic measurements appear to be reflected in the simulation. Two relaxations, one related to segmental motion and the other to star arm relaxation, are observed in all systems. The relaxation rate of the first one is independent of star structure parameters both in experiments and in the simulation. On the other hand, the star arm relaxation has been observed to be considerably dependent on the arm length but essentially independent of the arm number. The most interesting effect observed both in the simulated and in the real systems was the additional slow relaxation process appearing in systems with clear ordering of stars. The analysis of the simulation results concerning dynamics, as well as, a direct observation of star motions in these systems (Fig. 19) leads to the conjecture that the slow process can be related to translational cooperative rearrangements of stars within the ordered state which are of the same character as these suggested for the rearrangements in low molecular liquids. This might be a cooperative process on the macromolecular scale with the mechanism analogous to that postulated for liquids[23].

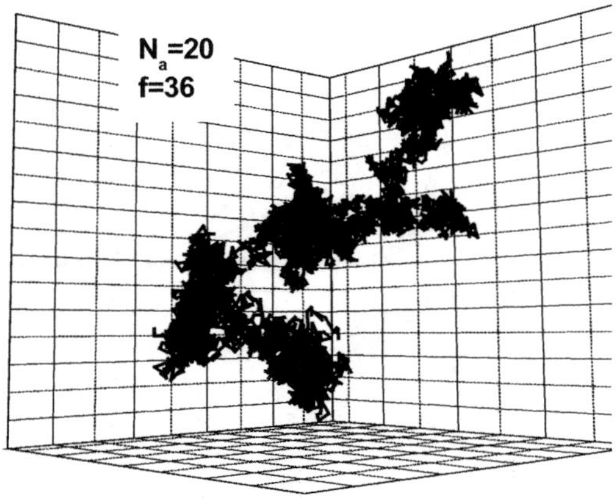

Figure. 19. A typical star center of mass trajectory recorded for a multiarm star in the simulated melt. The trajectory consists of blobs related to a longer residence of the star at some well distinguishable places which are regularly distributed along the trajectory and are connected by thinner trajectory fragments related to faster displacements between the localized states.

5.2 Copolymers

Various kinds of copolymers have been simulated and characterized by a large number of quantities providing information about the thermodynamic properties, the structure and the dynamics. Examples of temperature dependencies of the thermodynamic quantities recorded during heating of the initially microphase separated systems are shown in Figure 20. The following quantities have been determined: (1) the energy of interaction of a monomer E_m determined as the average of interactions of all monomer pairs at a given temperature[27] and (2) the specific heat calculated via the fluctuation of energy of states sampled during simulation of the system at constant temperature. The temperature, at which a step-wise change in the energy and corresponding peak in the specific heat is observed, is regarded as the temperature of the order-disorder transition, T_{ODT}. In Figure 20, the behavior of linear copolymers with various distributions of comonomers along chains is compared. The results indicate a high sensitivity of the systems to changes of the comonomer distribution by other parameters (chain length and composition) kept constant. The observed range of variation of the temperature of the ODT is very broad. Similar results have been obtained also for other types of comonomer distributions[12,27].

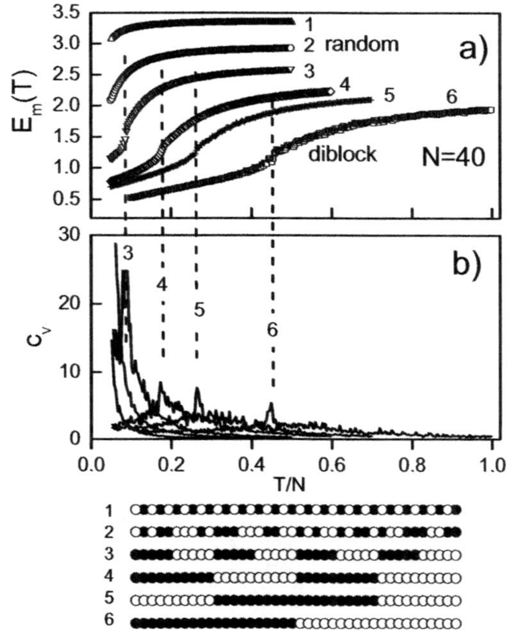

Figure 20. Temperature dependencies of (a) the average interaction energy per monomer and (b) the specific heat, for copolymer systems with various distributions of monomers along the chain (illustrated in the bottom of the Figure): (1) alternating copolymer, (2) random copolymer, (3) and (4) multiblock copolymers, (5) triblock and (6) diblock copolymer. Vertical dashed lines indicate temperatures of the microphase separation in systems (3) to (6).

The nature of the transitions corresponding to structural changes in copolymers can be well established from an analysis of distributions of local concentrations which are directly related to the free energy. An example of such distributions for a symmetric diblock copolymer, in a broad temperature range, is shown in Figure 21, by means of contour lines of equal composition probability projected on the composition-temperature plane. Such contour plots reflect many details of the thermodynamics and structure of the system. It is easily seen that, at high temperatures, the system can be considered as homogeneous because locally the most probable concentration corresponds to the nominal composition in the diblock. This is changed at temperatures close to T_{ODT} where at first a plateau and later two maxima corresponding to two coexisting phases are detected. At T_{ODT}, a sudden change transforms the system to a state with well defined microphases indicated by the most probable local concentrations corresponding to pure components. These results indicate three characteristic ranges of thermodynamic behavior of the system assigned as (1) disordered, (2) weakly segregated and (3) strongly segregated regimes appearing with decreasing temperature. Structures of simulated systems corresponding to these regimes are illustrated in Figure 21 by assuming different colors for copolymer constituents (black and gray for A and B, respectively).

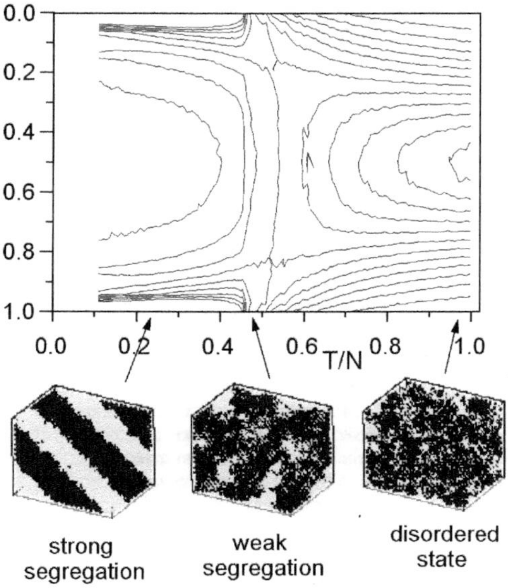

Figure 21. Temperature dependence of concentration distributions in small volume elements consisting of the nearest neighbors of each chain segment. Characteristic structures corresponding to various temperature ranges are illustrated.

In order to get information about dynamic properties of simulated systems the relaxations of various structural elements are monitored with time at equilibrium states corresponding to various temperatures: (1) the mean squared displacement of monomers, (2) the mean squared displacement of the center of mass of chains, (3) the single bond autocorrelation function and (4) the end-to-end vector autocorrelation function.

In Figure 22, various correlation functions, specified above, are shown for the diblock copolymer system at high and low temperatures. At high temperature, the system behaves like a homogeneous melt. All correlation functions show a single step relaxation. The fastest is the bond relaxation and the slowest is the chain relaxation described by the end-to-end vector autocorrelation function. The relaxation of the block is faster than the whole chain relaxation by approximately a factor of two. Such relations between various relaxation times in the disordered state of the copolymer can be regarded as confirmed experimentally for some real systems, in which the dielectric spectroscopy allows distinction of the different relaxation modes [28]. At low temperatures, drastic changes can be noticed for the dynamics of the block copolymer. At temperatures T/N<0.45 (Fig. 22c) the diblock system is in the microphase separated regime and most of the correlation functions determined show bifurcation of the relaxation processes into fast and slow components. The fast components of chain, block and concentration relaxations are attributed to the almost unchanged in rate, but limited, relaxation of chains when fixed at the A-B interface and the slow component indicates the part of relaxation coupled to the relaxation of the interface within uniformly ordered grains with the lamellar morphology. The concentration relaxation becomes the slowest one in such state of the system. The dynamic behavior of diblock copolymers, as simulated by CMA is presented in detail and discussed in[28], where the spectra of various relaxation modes have been determined in order to compare simulation results with dielectric spectra determined for real copolymer systems in the vicinity of the microphase separation transition.

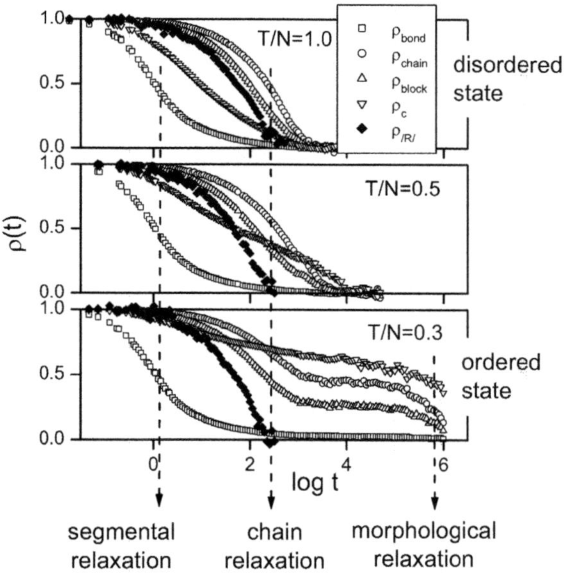

Figure 22. Various correlation functions determined at various temperatures for a symmetric diblock copolymer. The two extreme temperatures correspond to edges of the temperature range within which the system has been simulated. The high temperature (T/N=1.0) corresponds to the homogeneous regime and the low temperature (T/N=0.3) to the strongly segregated limit. The intermediate temperature (T/N=0.45) is only slightly higher then the temperature of the microphase separation for this system.

The above effect can also be well seen when temperature dependencies of relaxation times of various structural elements are compared. The characteristic bifurcation of the chain relaxation in the phase separated regime results in two relaxation times, which can be detected below the microphase separation temperature. Similar effects as these for chain relaxation can be observed in temperature dependencies of other relaxations considered here, i.e. the block or concentration relaxation.

Block copolymer systems have been simulated also in a stationary flow under applied stress[27]. The relaxed states of simulated model systems after cessation of flow with various stresses are shown in Figure 23. It is interesting to notice that different structure orientations are obtained depending on the shear stress under which the system was previously deformed. The results obtained have shown that, in simulated systems, similar morphological rearrangements as observed experimentally can be obtained when systems are subjected to shear flow with various shear rates. It is demonstrated that interaction of copolymer constituents with walls can considerably influence the observed structure orientations. This interaction can control the structure at low shear rates, as well as, the structure to which a system, disordered under high shear rate, relaxes after cessation of flow.

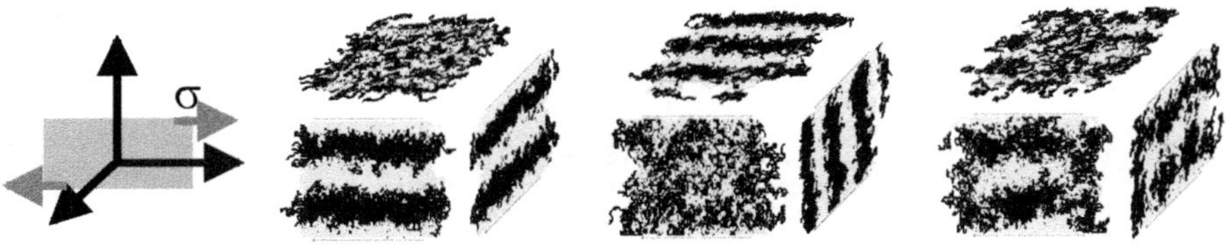

σ=0.2 σ=0.2 σ=0.5

Figure 23. The system subjected to flow between walls having properties of component A, i.e. interacting repulsively with component B, at T/N=0.3, is presented. Morphologies in the simulated block copolymer system after cessation of flow are shown by means of projections of polymer chains on three planes XY, XZ and YZ. Copolymer constituents A and B are plotted in different colors, black and gray, respectively. Results for T/N=0.3 (below the microphase separation temperature) and various shear stresses are considered.

6. CONCLUSIONS

New complex macromolecular architectures are of particular interest when they lead to new properties of materials. A large variety of macromolecules differing by topology of bond skeletons and by distributions of monomers of different type have been synthesized. In many cases, such macromolecules constitute self-assembling systems in which a supramolecular order can result in a modified dynamics and consequently in new properties.

In the simplest case of linear polymers, joining of monomeric units into linear chains results in a dramatic change of properties. A system consisting of non bounded monomers can usually be only liquid like or solid (e.g. glassy) but the polymer can additionally exhibit a rubbery state with properties, which make these materials extraordinary in a large number of applications. This new state is manifested by the very slow relaxation of polymer chains in comparison with the fast motion of monomers, especially when the chains become so long that they can entangle. The relaxation processes of smallest polymer units and of the whole polymers separate the behavior of materials in the temperature-time space into glassy, rubbery and flowing melt states as illustrated in Figures 6 and 7.

For complex macromolecular architectures, the mechanical behavior is a little bit more complex and we believe that the knowledge collected about such systems, including examples presented in this chapter, can be concluded by the diagram presented in Figure 24.

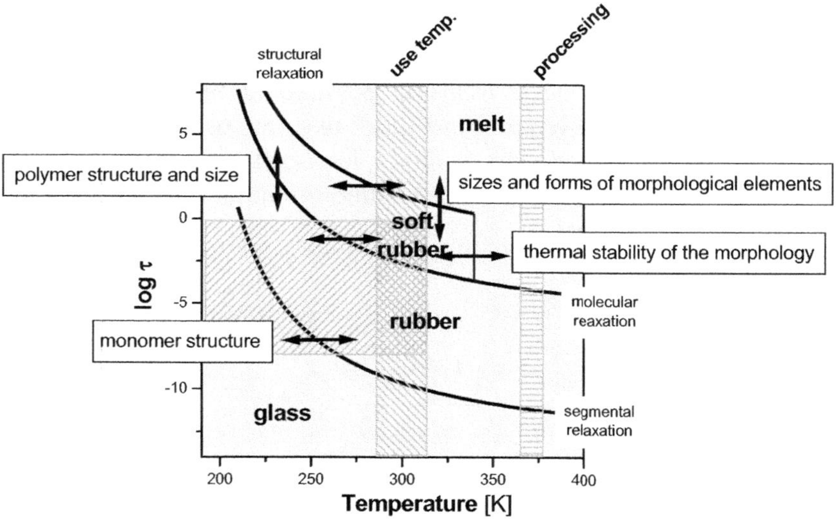

Figure 24. Schematic temperature dependences of relaxation times corresponding to various structural units in polymer bulk materials with a supramolecular ordering resulting from complex polymer architectures. The vertical lined areas indicate interesting temperature ranges where the materials can be used or processed and the horizontal lined area indicates the for applications relevant time range. The arrows show directions of possible shifts of relaxation processes by variation of parameters given in the rectangular frames.

The diagram is analogous to that shown in Figure 7c and shows temperature dependences of relaxation times corresponding to various characteristic structural units in the systems. The relaxation time dependencies separate the states of different mechanical behavior. New and characteristic for the here discussed materials with supramolecular structures, is the existence of a state with long relaxation times related to these structures i.e. the microstructure in copolymers, the ordering in colloidal or star-like systems and the networking by entanglements or cross-linking in the systems of brush molecules. The new state is characterized by a modulus plateau usually considerably lower than the conventional rubbery plateau of linear polymers and can be considered as soft rubbery state. The ranges of different behavior depend in such systems on many structural parameters. Possible variations are schematically illustrated in the diagram by vertically or horizontally oriented arrows depending on whether the given relaxation can be shifted along the time or temperature scale when caused by a variation of the parameters specified in the rectangles. This illustrates qualitatively a complexity of interrelation between molecular parameters and the macroscopic behavior.

It has also been shown that the simulation of such complex systems, when performed using an appropriate simulation method, not only can mimic the structure and dynamic behavior of real systems but in addition can bring insight on a molecular level. This is possible through the selection of parameters and the monitoring of quantities that are not always accessible in the experiments. The modeling methods described are able to monitor the systems from substrates to products including synthesis and processing as the following diagram illustrates.

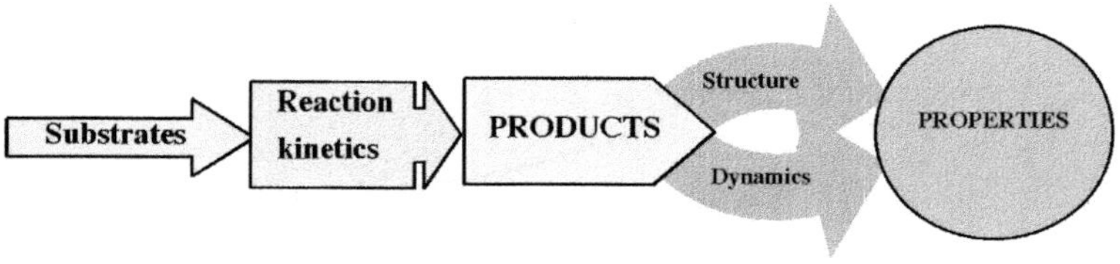

Main advantages of these methods are: physically reasonable simplifications of the dynamics and structures, applicability to dense systems, flexibility in representation of various molecular topologies and high computational efficiency. The CMA method is super efficient on traditional sequentially working hardware (e.g. even a single PC).

REFERENCES

1. Hadjichristidis, N,, Pispas, S. and Floudas, G. A. (2003) *Block Copolymers, Synthetic Strategies, Physical Properties and Applocations*; Wiley-Interscience, New Jersay
2. Matyjaszewski, K.; Davis, T. P., Eds. (2002) *Handbook of Radical Polymerization*; Wiley Interscience: Hoboken

3. Pakula, T. (2003) Dielectric and Mechanical Spectroscopy – a Comparison, in F. Kremer and A. Schönhals (eds.) *Broadband Dielectric Spectroscopy*, Springer, Berlin, pp. 598-623
4. Pakula, T., Geyler, S., Edling, T., Boese, D.,(1996) *Rheol Acta* 33,631
5. Pakula, T., Vlassopoulos, D., Fytas, G., Roovers, J., (1998) *Macromolecules* 31,8931
6. Pakula, T., Minkin, P., Beers, K.L., Matyjaszewski, K., (2001) *Mater. Sci. Eng.* 84,1006
7. Pakula, T., Harre, K., (2000) *Comput. Theor. Polym. Sci.* **10**,197
8. Pakula, T., Minkin, P., Matyjaszewski, K., (2003) Ch. 26, in K. Matyjaszewski (ed.) *Advances in Controlled/Living Radical Polymerization,* ACS Symposium Series 854, Washington, pp. 366-382
9. Neugebauer, D., Zhang, Y., Pakula, T., Matyjaszewski, K. (2003) *Macromolecules* – in press
10. Tsukahara, Y., Nakamura, N., Hashimoto, T., Kawai, H. (1980) *Polymer J.* 12,455.
11. Hashimoto, T.; Tsukahara, Y.; Tachi, Y.; Kawai, H. (1983) *Macromolecules* 16,648
12. Pakula, T., Matyjaszewski, K., (1996) *Macromol. Theory Simul.* 5, 987
13. Matyjaszewski, K., Ziegler, M.J., Arehart, S.V.,Greszta, D., Pakula, T., (2000) *J. Phys. Org. Chem.* 13,775
14. Okrasa, L., Pakula, T., Matyjaszewski, K., - paper in preparation
15. Gohr, K., Pakula, T., Tsutsumi, K., Schärtl, W., (1999) *Macromolecules* **32**,7156(1999)
16. Wall, F.T., Mandel, F., (1975) *J. Chem. Phys.* 63,4592
17. Carnesin, I., Kremer, K., (1988) *Macromolecules* 21,2819
18. Binder, K., Kremer, K., (1988) *Comput. Phys. Rep.* 7,261
19. Pakula, T., (1987) *Macromolecules* 20,679(1987)
20. Geyler, S., Pakula, T., Reiter, J., (1990) *J. Chem. Phys.* 92,2676
21. Reiter, J., Edling, T., Pakula, T., (1990) *J. Chem. Phys.* 93,837
22. Gauger, A., Pakula, T., (1995) *Macromolecules* 28,190
23. Pakula, T., Teichmann, J., (1997) *Mat. Res. Soc. Symp. Proc.* 495,211; Teichmann, J. (1996) Ph.D. Thesis, University of Mainz; Pakula, T.,(2000) *J. Mol. Liquids* 86,109; Polanowski, P., Pakula, T. (2002) *J. Chem. Phys.* 117,4022
24. Pakula, T., Müller, A., Matyjaszewski, K., *Paper in preparation.*
25. Pakula, T., (1998) *Comput. Theor. Polym. Sci.* 8,21
26. Pakula, T., Vlassopoulos, D., Fytas, G., Roovers, J., (1998) *Macromolecules* 31,8931
27. Pakula, T., Floudas, G., (2000) Ch. 6 in *Block Copolymers* F.J. Balta Calleja and Z. Roslaniec (eds.) Marcel Dekker Inc., N.Y. pp. 123-177; Pakula, T. (1998) *J. Macromol. Sci. – Phys.* B37,181;
28. Pakula, T., Karatasos, K., Anastasiadis, S.H., Fytas, G. (1997) *Macromolecules* 30,8463;

PREPARATION OF NANOCOMPOSITES WITH WELL-DEFINED ORGANIC (CO)POLYMERS

KRZYSZTOF MATYJASZEWSKI

Center for Macromolecular Engineering, Department of Chemistry, Carnegie Mellon University, 4400 Fifth Avenue, Pittsburgh, PA 15213

Abstract: Nanocomposites with well-defined organic polymers are hybrid materials with large interphase between two incompatible organic and inorganic segments. The well-defined organic segments are generated by controlled/living polymerization. Due to its simplicity, robustness and tolerance to impurities, radical polymerization is preferred. Atom transfer radical polymerization has been most often used, since it is perhaps most versatile. Inorganic component can comprise colloidal spherical particles, flat layered structures or macroscopically flat wafers but also irregular inorganic surfaces. The nancomposites are formed by covalent linkage between the incompatible organic and inorganic parts. Both grafting from surfaces and onto surfaces was employed for that purpose. Spherical polymer brushes were synthesized by the ATRP of styrene, *n*-butyl acrylate and methyl methacrylate from initiators covalently attached to the surface of colloidal silica and silsesquioxanes. In a similar way core-shell colloids were prepared by block copolymerization using ATRP. Polymer brushes were formed by grafting from and onto flat silicon surfaces and functionalized carbon black.

Key words: nanocomposites, organic/inorganic hybrids, controlled radical polymerization, atom transfer radical polymerization, ATRP

1. INTRODUCTION

Controlled/living polymerizations enable the synthesis of well-defined (co)polymers of controllable dimension and composition.[1,2] In the past decade the use of controlled/living radical polymerization (CRP) to prepare

F. Ciardelli and S. Penczek (eds.),
Modification and Blending of Synthetic and Natural Macromolecules, 123–134.

nanostructured materials has gained significant attention toward this endeavor, as a wider range of functional monomers can be incorporated into polymeric systems.[3,4] The combination of various controlled/living radical systems with inorganic components such as growth from/onto flat and colloidal as well as irregular surfaces, formation of novel nanocomposites is among the most active areas of CRP. The computer based search on May 31, 2003 using SciFinder Scholar on topics related to general area of CRP (using entries of controlled radical polymerization, living radical polymerization and atom transfer radical polymerization) revealed nearly 5,000 publications.

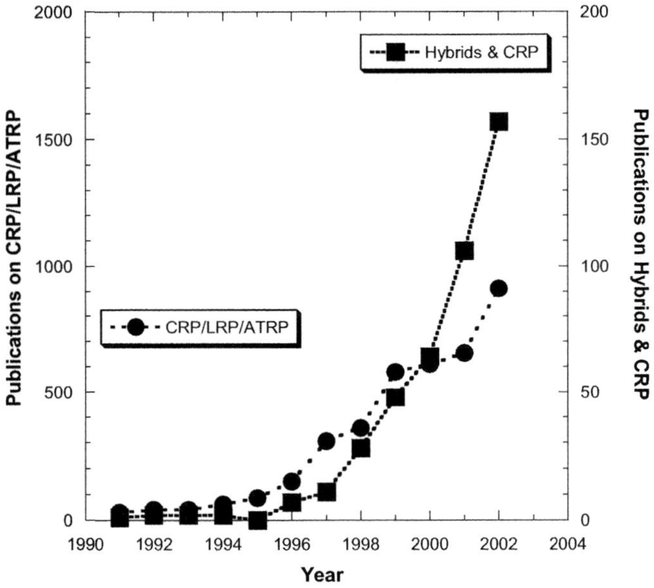

Figure 1. Number of publications devoted to controlled radical polymerization, living radical polymerization and atom transfer radical polymerization (left axes and circles) and to sub-areas of CRP related to hybrids, surfaces and nanocomposites (right axes and squares), according to SciFinder Scholar, May 31, 2003.

532 of them where identified in the sub-areas of CRP related to hybrids, surfaces and nanocomposites. However, although they constitute only 12 % of all papers devoted to CRP, the dynamics of their increase exceeds that of general CRP. Figure 1 illustrates this dynamics and may suggest that the interest in controlled radical polymerization as a tool to prepare various nancomposites will be increasing within the next several years.

The covalent attachment of well-defined organic segments to inorganic elements allows preparation of various novel materials which include segmented copolymers, networks, layered or dispersed silicates and brushes from various surfaces. These systems comprise different inorganic components such as linear or crosslinked polymers, clays and glass, silicon, gold and even compounds semiconductors. The covalent linkage can be achieved by block-copolymerization, grafting through-, from- and onto-, co-crosslinking, etc. as illustrated in Scheme 1.

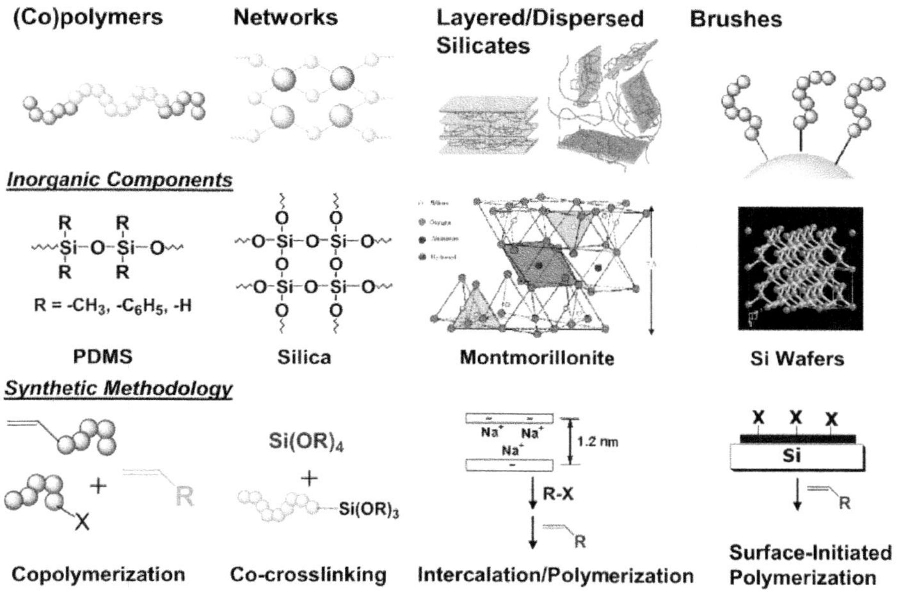

Scheme 1.

In particular, atom transfer radical polymerization (ATRP)[5-7] was demonstrated to be a versatile technique to synthesize well-defined (co)polymers and organic/inorganic hybrid materials.[8] By the ATRP of various monomers from preformed polymeric, surface and attached or colloidal initiators, discrete macromolecules possessing nanometer scale morphologies were prepared. Some of them are presented in Scheme 2 below. We will discuss in this paper some examples of organic/inorganic hybrids prepared in our laboratory by using ATRP.

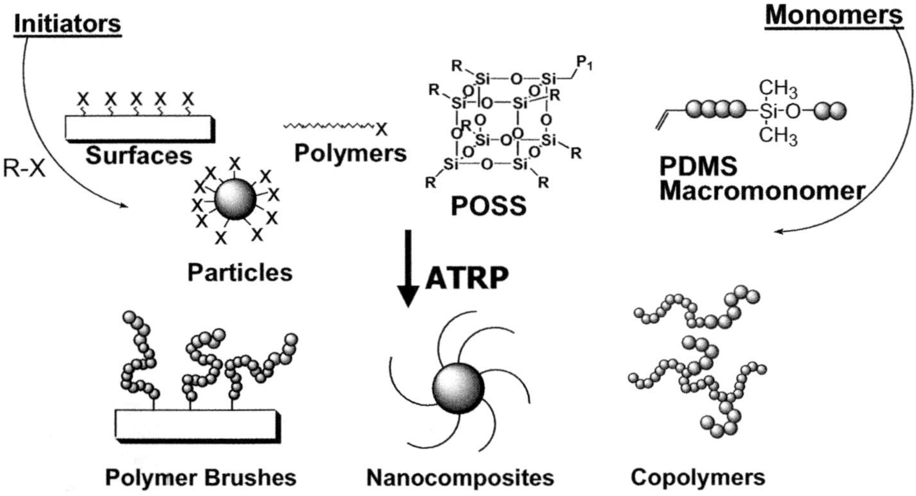

Scheme 2.

2. BRUSHES FROM COLLOIDAL PARTICLES

Hybrid nanoparticles containing an inorganic core and tethered glassy homopolymers polymers have been conducted by the ATRP of styrene and (meth)acrylates) from colloidal initiators.[9-12] These materials serve as an interesting example of spherical brushes, since a high grafting density of polymer chains can be achieved on a curved nanoparticle surface. An advantage of using preformed, colloidal initiators is that brush length can be systematically controlled by changing the ratio of monomer to initiating sites ($[M]_o/[I]_o$) while maintaining a core of static dimensions. This is in direct contrast to block copolymer micelles, where variation of shell length requires altering the relative composition and DP_n of the respective segments. In the synthesis of silica-*graft*-polystyrene (SiO_2-*g*-pS) hybrid nanoparticles, the ATRP of S was performed using 2-bromoisobutyrate functional silica colloids using varying target DP_n's. SiO_2-*g*-pS hybrid nanoparticles possessing molar masses of tethered pS in the range of M_n = 5,000 to 32,000 g/mol were prepared and characterized both in the solid state and in solution using transmission electron microscopy (TEM) and dynamic light scattering (DLS), respectively. TEM images of SiO_2-*g*-pS colloids revealed the formation of submonolayer patches where interparticle spacing increased as a function of tethered pS molar mass (Figure 2). DLS of SiO_2-*g*-pS in toluene was conducted to determine hydrodynamic radii (R_h) of discrete colloids. Comparison of R_h for hybrid nanoparticles of varying size with SEC values of cleaved pS after hydrofluoric acid treatment of SiO_2-*g*-pS colloids revealed a linear relationship. The linearity observed in plotting R_h vs. $M_{n\ cleaved\ pS}$ indicated that tethered chains in toluene were in an extended conformation, presumably due to the high density of grafted chains. Calculations based on differential refractometry measurements of dn/dc values of colloidal initiator and SiO_2-*g*-pS were performed to determine number of tethered chains (N_t) per colloidal initiator ($N_t = 1850$).

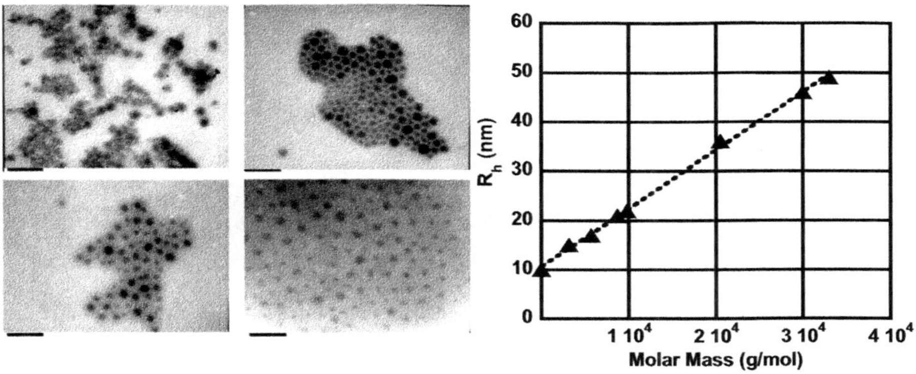

Figure 2. (left) TEM images of SiO_2 colloidal initiator and SiO_2-*g*-pS hybrid nanoparticles ($M_{n\ tethered\ pS}$ = 5,230, 14,960, 32,670 g/mol), black bar = 100 nm; (right) Plot of hydrodynamic radius (R_h in toluene, 25 °C) of SiO_2-*g*-pS versus molar mass of cleaved pS determined from SEC. Ref.[11].

The synthesis of spherical brushes possessing a rubbery shell of tethered polymers was achieved by the ATRP of *n*-butyl acrylate (BA) from 2-bromoisobutyrate functional silica colloids. SiO_2-*g*-pBA colloids ($M_{n\ tethered\ pBA}$ = 17,990; M_w/M_n = 1.28) were then spin cast onto a mica surface and imaged using tapping-mode AFM. Similar to images of pBA molecular brushes, [13] individual colloids with a corona of tethered pBA extending on the mica substrate were observed in both height and amplitude AFM images (Figure 3). In AFM height images, silica cores were discernable as tall features (white spots) surrounded by tethered pBA. Strikingly different features were observed in phase images, where silica cores (yellow spot) were embedded in a dense shell of collapsed pBA (red corona), followed by extension of single chains from the outer edge of the pBA corona (blue outer shell). Effective diameters of the SiO_2-*g*-pBA colloid were assessed from AFM and indicated the following: $D_{eff\ spherical\ brush}$ = 235 nm, $D_{eff\ silica\ core}$ = 22 nm.

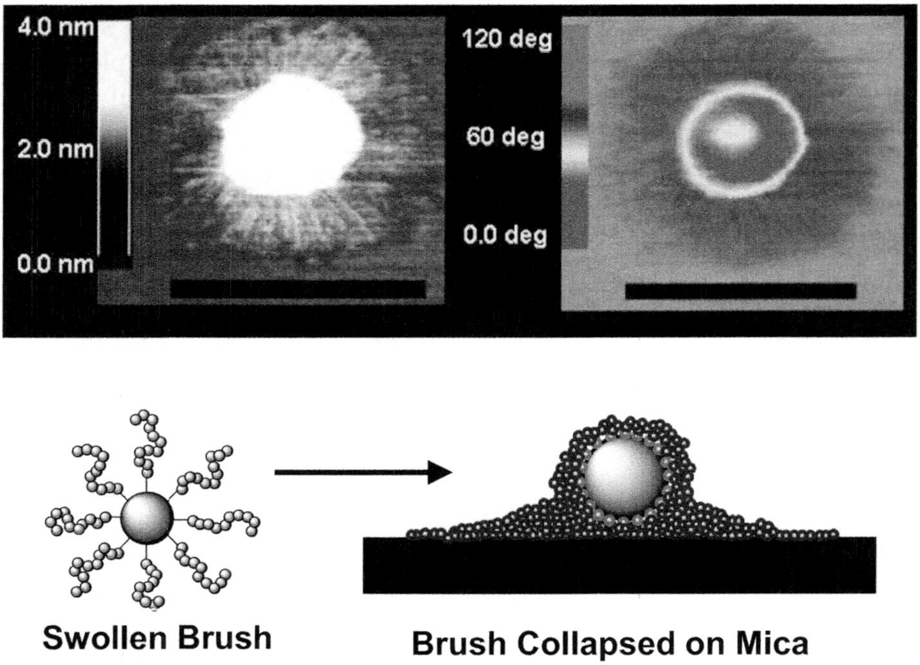

Figure 3. a) AFM height (left) (silica cores-white spots) and b) phase (right) images of SiO_2-*g*-pBA hybrid nanoparticles (silica core-yellow spot, pBA-dark corona), black bar = 250 nm. SEC of cleaved pBA indicated M_n = 17,990; M_w/M_n = 1.28; c) schematic of SiO_2-*g*-pBA colloid in solution conformation versus after spin coating onto a mica surface. Ref. [14].

Carbon black is a commercially important filler composed of primary particles with diameters of 10-75 nm that are fused into aggregates that range from 50-500 nm in size. Carbon black is extensively used in various industries as reinforcing agent for rubbers and as a pigment for coatings, inks and toners. A key requisite in these systems is the efficient dispersion of carbon black in polymer blends, or dispersed media. It has been found that dispersion of carbon black in resins and solvents can be improved by grafting polymers onto the surface of carbon black, providing steric stabilization against flocculation.

The synthesis of carbon black grafted with pBA chains was conducted via surface initiated ATRP from functional carbon black surfaces.[15] Analysis of pBA chains cleaved from the cores indicated that polymers with tunable molar mass and low polydispersity (M_n = 17,990; M_w/M_n = 1.28) were obtained. AFM of pBA coated carbon black particles cast onto mica (Figure 4) confirmed the efficient grafting of pBA to the central core with morphologies consistent with those observed for other pBA spherical brushes (cf. Figure 3).

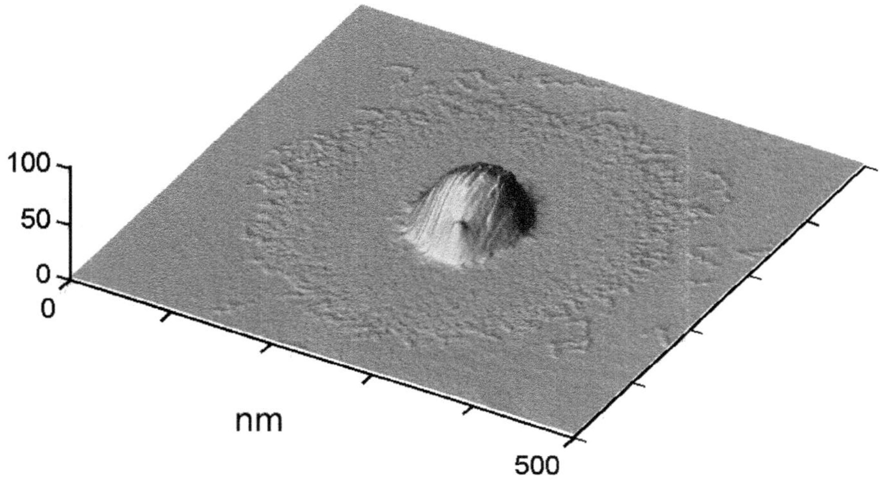

Figure 4. Tapping mode AFM image of poly(*n*-BA)bound carbon black . SEC of cleaved pBA indicated $M_n = 17{,}990$; $M_w/M_n = 1.28$ Ref. 15]

3. CORE-SHELL COLLOIDS

The synthesis of covalently bonded layered core-shell colloids was successful by the consecutive polymerization of styrene and (meth)acrylates from nanoparticle surfaces using ATRP (Scheme 3). From this approach, block copolymers containing both glassy and rubbery segments have been grafted to inorganic cores. Previously, the synthesis of polysilsesquioxane-*graft*-(polystyrene-*block*-poly(benzyl acrylate)) ($SiO_{1.5}$-*g*-(pS-*b*-pBzA)) hybrid nanoparticles was reported, followed by morphological characterization of ultrathin films cast onto mica using AFM.[11] SEC of cleaved pS ($M_n = 5{,}230$; $M_w/M_n = 1.20$) and pS-*b*-pBzA ($M_n = 26{,}760$; $M_w/M_n = 1.40$) confirmed that block copolymers were indeed prepared. Phase images rendered in tapping-mode AFM revealed the presence of distinct domains, tentatively assigned to each component of the colloid. Additionally, multilayered colloids were designed by the ATRP of BA from a $SiO_{1.5}$ colloid followed by chain extension from the $SiO_{1.5}$-*g*-pBA colloid with styrene. In this system, hard and soft segments were incorporated, where a rubbery hard colloidal core (D_{eff} = 19 nm) was surrounded be an inner shell of pBA (DP_n = 108) and an outer shell of glassy pS (DP_n = 410). AFM phase images also observed discrete domains of each component confirming that discrete segments were incorporated to the colloid after each ATRP reaction.

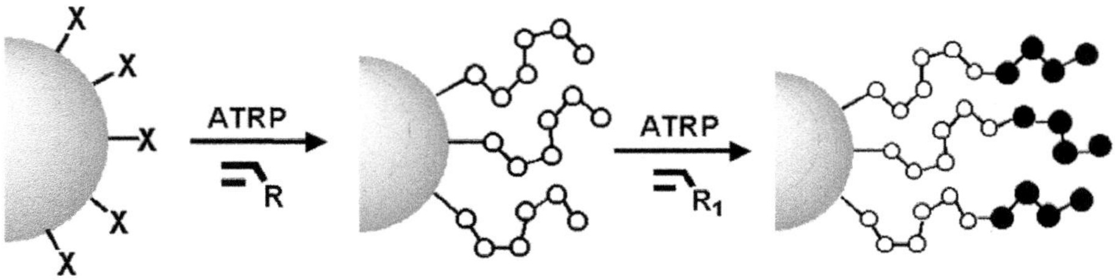

Scheme 3.

4. ABC TRIBLOCK COPOLYMER BRUSHES

The synthesis of copolymer brushes was conducted by the preparation of a surface reactive ABC triblock followed by covalent attachment to a silicon wafer. In the synthesis of the ABC triblock copolymer, a pDMS macroinitiator ($M_{n\ SEC}$ = 6,200; M_w/M_n = 1.19) was prepared by the living anionic ring-opening polymerization and hydrosilation.[19] The ATRP of styrene was then performed from the pDMS macroinitiator to prepare a pDMS-*b*-pS diblock copolymer (M_n = 66,730; M_w/M_n = 1.38). Chain extension of the diblock copolymer with 1-(dimethoxymethylsilyl)propyl acrylate (DMSA) yielded a triblock copolymer ($M_{n\ NMR\ pDMS\text{-}b\text{-}pS\text{-}b\text{-}pDMSA}$ = 156,700) capable of covalently bonding to a silicon wafer while also containing rubbery and glassy segments (Scheme 4).[16]

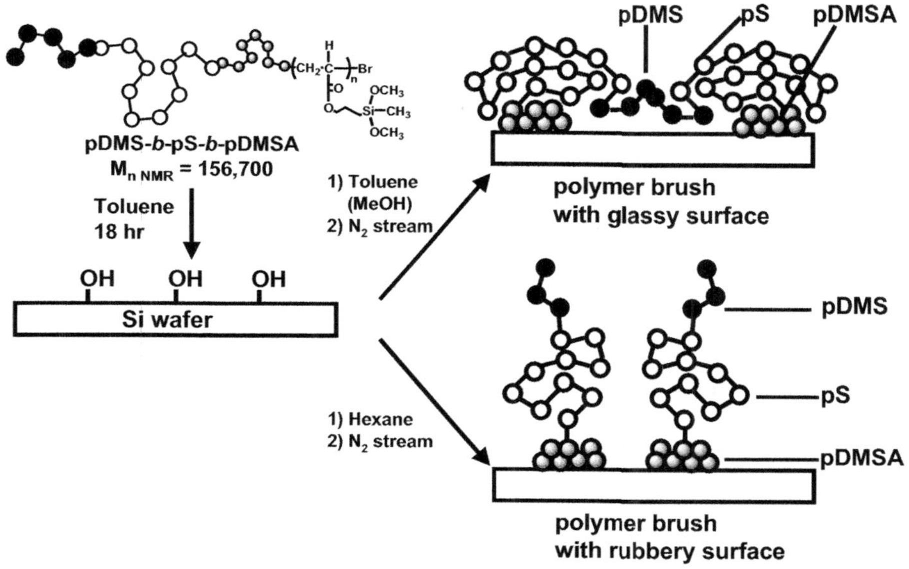

Scheme 4.

Surface properties of the copolymer brush could be reversibly controlled to present either pS, or pDMS segments by treatment of the ultrathin film with various solvents. When the brush was immersed in a good solvent for both pDMS and pS segments (e.g., toluene) the brush surface topography as assessed from tapping-mode AFM indicated speckled features due to some phase separation of pDMS and pS (Figure 5a).[17] However, since the composition of the exposed segments was mainly pS, the morphology resembled that of both spin-coated pS films, or pS brushes. Treatment of the copolymer brush with either methanol (Figure 5b), or annealing at 130 °C yielded similar features. Alternatively, treatment of the wafer in a mixture of toluene and hexane resulted in the selective presentation of pDMS segments to the brush surface. AFM of the hexane treated brush, using similar tapping forces as for the toluene and methanol treated brushes, imaged a featureless surface topography due to dissipation of energy from applied tapping forces by the pDMS layer (Figure 5c). However, upon increase of applied tapping forces, granular features were again discernable in AFM height images indicating that after sufficient deformation of the pDMS layer, the glassy features of the underlying pS phase were observed.

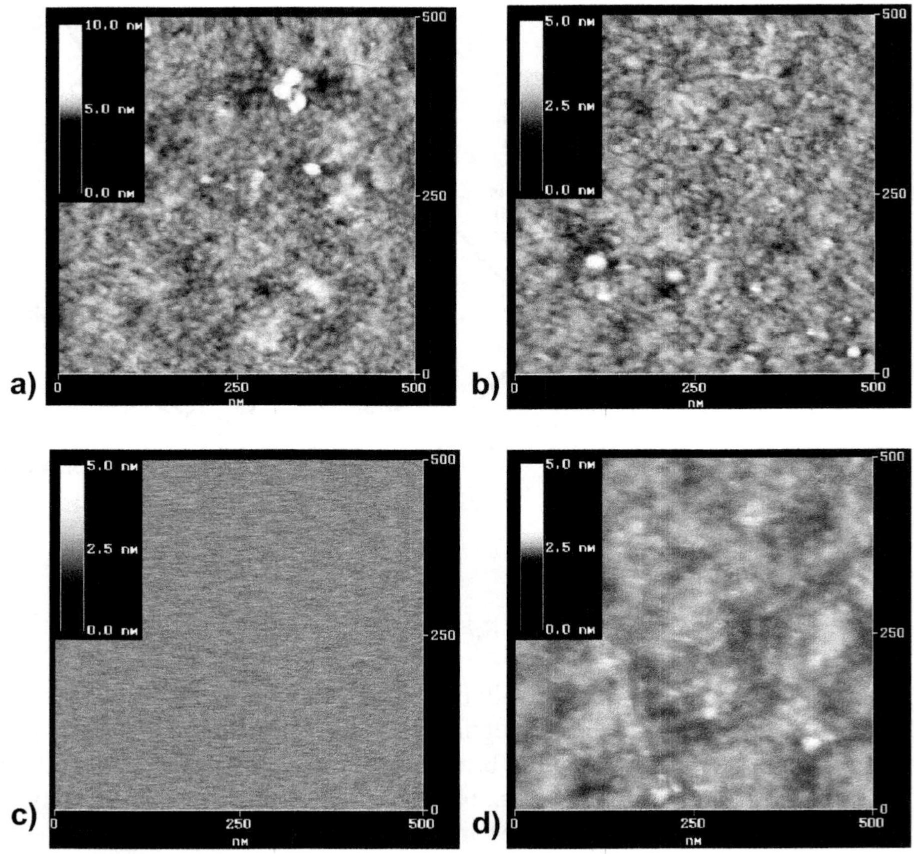

Figure 5. AFM height images (500 nm scale) of pDMS-*b*-pS-*b*-pDMSA brushes after the following treatments; a) after immersion in toluene and drying with nitrogen, b) after immersion in toluene, gradual addition of methanol and drying with nitrogen c) immersion in toluene, gradual addition of hexane and drying with nitrogen d) AFM image with an increase in tapping forces after immersion in toluene, gradual addition of hexane and drying with nitrogen. Ref. [17].

5. BRUSHES FROM FLAT SURFACES

Well-defined copolymers can be not only attached to flat surfaces but also grown from flat surfaces by ATRP.[18-20] This can be done with or without sacrificial initiator. Typically, Cl-Si or MeO-Si terminated ATRP initiators are attached to the oxidized silicon wafer via condensation with silanol groups on the surface. The alternative approach includes light initiated hydrosilation of Si-H groups on the surface with functional alkenes. Modification of tethered polymer composition has been demonstrated to effect both surface properties and wettability of various substrates. The incorporation of well-defined random copolymers had a profound effect on the wetting behavior of the polymer brushes. The surface properties of Si wafers has also been modified by the incorporation of well-defined hydrophobic and hydrophilic polymer segments using ATRP. By the homopolymerization of a perfluorinated monomer (3,3,4,4,5,5,6,6,7,7,8,8,9,9,10,10,10-heptadecafluorodecyl acrylate), a polymer brush with a hydrophobic surface was prepared, as evidenced by water contact angle measurements (119 $\pm$ 2 °). Hydrophilic Si surfaces were made by the controlled growth of a poly(styrene-*b*-(*t*-butyl acrylate)) (p(S-*b*-*t*BA) followed by hydrolysis to yield the corresponding poly(styrene-*b*-acrylic acid)(p(S-*b*-AA)) polymer brush. Contact angle measurements from water droplets indicated a decrease in the contact angle going from p(S-*b*-*t*BA) (86 $\pm$ 4°) to p(S-*b*-AA) (18 $\pm$ 2°) surfaces.[19]

6. BLOCK AND GRAFT COPOLYMERS WITH POLY(DIMETHYLSILOXANE)

The synthesis of copolymers containing poly(dimethylsiloxane) (PDMS) was also conducted using various approaches. To prepare these structures, PDMS prepared from both living anionic polymerization and condensation processes were functionalized with either ATRP initiating groups, or moieties polymerizable using CRP, i.e. methacrylate or acrylate functionalities.[21] Subsequently, AB and ABC block copolymers were prepared by ATRP. Correspondingly, macromonomers were incorporated using ATRP, other CRP methods and also conventional free radical polymerization. The properties of the resulting polymers depend very strongly on the molecular structure of the block/graft copolymers.[22,23]

7. CONCLUSIONS

The synthesis of nanostructured materials from discrete colloids and flat surfaces has been demonstrated. Using ATRP, a variety of inorganic substrates can be functionalized with well-defined block copolymers possessing both glassy and rubbery segments. The versatility of this approach is significant, as block copolymers can either be initiated from a functionalized surface, or preformed and then grafted to an inorganic substrate.

ACKNOWLEDGMENTS

Financial support from the National Science Foundation (DMR-00-90409) is appreciated.

REFERENCES

1. Webster, O. W. (1991) *Science* **251**, 887.
2. Matyjaszewski, K. (2001) *Macromol. Symp.* **174**, 51.
3. Hawker, C. J. (1997) *Acc. Chem. Res.* **30**, 373.
4. Patten, T. E.; Matyjaszewski, K. (1998) *Adv. Mater.* **10**, 901.
5. Wang, J.-S.; Matyjaszewski, K. (1995) *J. Am. Chem. Soc.* **117**, 5614.
6. Matyjaszewski, K.; Xia, J. (2001) *Chem. Rev.* **101**, 2921.
7. Kamigaito, M.; Ando, T.; Sawamoto, M. (2001) *Chem. Rev.* **101**, 3689.
8. Pyun, J.; Matyjaszewski, K. (2001) *Chem. Mater.* **13**, 3436.
9. Von Werne, T.; Patten, T. E. (1999) *J. Am. Chem. Soc.* **121**, 7409.
10. Pyun, J.; Matyjaszewski, K.; Kowalewski, T.; Savin, D.; Patterson, G.; Kickelbick, G.; Huesing, N. (2001) *J. Am. Chem. Soc.* **123**, 9445.
11. Savin, D. A.; Pyun, J.; Patterson, G. D.; Kowalewski, T.; Matyjaszewski, K. (2002) *J. Polym. Sci., Part B: Polym. Phys.* **40**, 2667.
12. Pyun, J.; Jia, S.; Kowalewski, T.; Patterson, G. D.; Matyjaszewski, K. (2003) *Macromolecules* **36**, 5094.
13. Sheiko, S. S.; Prokhorova, S. A.; Beers, K. L.; Matyjaszewski, K.; Potemkin, I. I.; Khokhlov, A. R.; Moeller, M. (2001) *Macromolecules* **34**, 8354.
14. Pyun, J.; Kowalewski, T.; Matyjaszewski, K. (2003) *Macromol. Rapid Comm.* **24**, 1043.
15. Liu, T.; Jia, S.; Kowalewski, T.; Matyjaszewski, K.; Casado-Portilla, R.; Belmont, J. (2003) *Langmuir* **19**, 6342.
16. Pyun, J.; Jia, S.; Kowalewski, T.; Savin, D. A.; Patterson, G. D.; Liu, T.; Matyjaszewski, K. (2002) *Polym. Prepr. (Am. Chem. Soc., Div. Polym. Chem.)* **43**, 328.
17. Pyun, J.; Jia, S.; Kowalewski, T.; Matyjaszewski, K. (2004) *Macromol. Chem. Phys.* **205**, 411.
18. Ejaz, M.; Yamamoto, S.; Ohno, K.; Tsujii, Y.; Fukuda, T. (1998) *Macromolecules* **31**, 5934.
19. Matyjaszewski, K.; Miller, P. J.; Shukla, N.; Immaraporn, B.; Gelman, A.; Luokala, B. B.; Siclovan, T. M.; Kickelbick, G.; Vallant, T.; Hoffmann, H.; Pakula, T. (1999) *Macromolecules* **32**, 8716.
20. Zhao, B.; Brittain, W. J. (2000) *Prog. Polym. Sci.* **25**, 677.
21. Nakagawa, Y.; Miller, P. J.; Matyjaszewski, K. (1998) *Polymer* **39**, 5163.
22. Shinoda, H.; Miller, P. J.; Matyjaszewski, K. (2001) *Macromolecules* **34**, 3186.
23. Shinoda, H.; Matyjaszewski, K. (2001) *Macromol. Rapid Commun.* **22**, 1176.

ROLE OF INTERFACES IN MULTICOMPONENT POLYMER SYSTEMS AND BIOCOMPOSITES

GYORGY J. MAROSI, and GYORGY BERTALAN
Department of Organic Chemical Technology, Budapest University of Technology and Economics 1111 Budapest, Műegyetem rkp. 3, phone:+36 1 463 3654, email:gmarosi@mail.bme.hu

Abstract: Wide consideration and classification of the application of surface and interface phenomena in various areas of material science is presented. Interface-related subjects of engineering polymer systems and biomaterials are discussed on a common basis. After a short description of the fundamentals of interphase formation in artificial and biological multicomponent systems, the role of interphases in engineering and biomaterials is illustrated with a series of examples. Beyond the adhesion and mechanical properties in polymeric systems, several other properties, in which the role of the interfacial phenomena has not yet been clarified adequately (such as flame retardancy, photostability, transport characteristics etc.), are discussed. New or less known terms, such as adaptive, smart interphases and reactive surfactants are defined and their application introduced.

Key words: multicomponent systems, biocomposites, interphases, interfacial phenomena

1. INTRODUCTION

Surface-interface phenomena play an essential role in the whole area of material science, including heterogeneous catalysis, ceramics, multicomponent polymer systems and biomaterials, however, the used terms are not uniformly defined and the discussion is segmented to the different application fields. In this paper we try to give a wider consideration and

F. Ciardelli and S. Penczek (eds.),
Modification and Blending of Synthetic and Natural Macromolecules, 135–154.

classification of interfacial layers characteristic to various engineering and biomaterials and define the relevant terms.

The term *surface* is generally used for the border between different kinds of phases (such as condensed-gas, liquid-vapour), while the term *interface* is applied to a sharp boundary between two similar (condensed-condensed, liquid-liquid) phases. In most cases, however, a layer can be found between two phases with properties differing from those of the bulk material on either side. This layer at the surface or interface can hardly be described as a two-dimensional system but rather as a three-dimensional *interphase, interlayer,* or *interfacial zone*. The interphases, including surfaces or interfaces, form boundary layers of different width between two phases, usually in the micro- or nanometer range.[1] The use of terms "interphase" or "interface" depends in some cases on the point of view (macroscopic or microscopic view) but in this work focusing to the region between different phases we prefer to use the interphase (IP).

IP is a zone spontaneously formed or consciously engineered if phases of different character (such as hard-soft, charged-neutral, polar-apolar, living-lifeless) are attached to each other. The rules that govern the phenomena and the design of IP-s in different materials are basically the same. Apart from certain differences various similarities can be found in the surface related aspects of packaging, blending, formation of (nano-)composites, fire protection, metal-coating, microelectronics, (nano-)tribology, biocompatibility, catalysis, etc.

The materials developed by the nature are composed of structure with optimal interfacial layers (e.g. wood, skin etc.). Compared to these the structure of man-made multicomponent systems is much simpler, but their interfaces should also exhibit complex morphology based on complex physical and chemical interactions.

In engineering polymer systems the main goal of interphase design is obtaining maximum adhesion and mechanical properties. The molecular engineering provides further possibilities for tailoring interphases in order to improve other properties such as fire retardancy, thermal- and photostability, migration to the surface and other transport processes, but there are only very few investigations to utilize these possibilities. We extended our scope from polymer composites and blends to the flame retardancy, transport processes of different stabilisers and biological systems designing/modifying some of the relevant IP-s.

Based on these experiences it was a challenge for us to discuss all the interface-related subjects (artificial and bio-composites, transport properties, etc.) on a common basis. The different types of IP-s and their main characteristics are discussed in this work accordingly following the categories summarized in Table 1.

Table 1. Types, features and utilization of interfacial layers in polymeric systems and biomaterials

Interphase	**Characteristics**	**Effect**	**Application**	**Main processes**
Spontaneous IP	easy debonding, separation from the matrix	adsorption, nucleation	transparent films	adsorption, stress concentration
Modified IP of low molecular mass	prevail polarity, charge, reactivity differences	self assembly oriented, mobility change	controlled transport, (bio)compatibility	transport, adsorption, reaction
Modified IP of high molecular mass	thick, hierarchic, prevail elasticity differences	local stresses, incompatibility, (bio)compatibility	composites, blends, biomaterials	adhesion, stress distribution
Adaptive IP*	thermal, light, pH, biological sensitivity	barrier effect, decomposition, mineralization	(fire)stabilization, biomimetics, (bio)composites	induced transformation, kinetics
Smart IP*	recognition system, patterned	dynamic response, targeting	drug delivery, artificial photosynthesis hybrid composites	recognition followed by transformation

* the new terms will be defined under the relevant headings.

Before describing the systems/processes belonging to each type of interphases, the common background is briefly summarized.

2. BACKGROUND

This chapter gives an introduction to the theory of interaction between adjacent solid-solid, solid-molten phases leading to formation IP-s. Effects changing the IP width are also discussed.

The formation and structure of the IP-s are governed by thermodynamics and kinetics of multicomponent systems. The free energy of interfaces at boundary layers of phases can be lower or higher than the surface energy of individual phases. The first case means a thermodynamic driving force for the formation of stable IP, while in the second case the thermodynamic incompatibility can be overcome by the energy of mixing at the formation of a dispersion, which would, however, segregate if kinetic control does not hinder the disintegration. In case of strong physical or chemical bonding at the interfaces the contact between the phases is thermodynamically stabile.

In multicomponent polymer systems the properties are affected by the interactions between the following phases:

- inorganic - polymer

- polymer - polymer, and
- inorganic – inorganic.

The first establishes adhesion between the polymer matrix and inorganic component (such as reinforcement), the second determines the IP between different polymers, while the third is characteristic for the contact of inorganic inclusions with each other.

The characteristics of the interactions are

- the effective contact area,
- the wetting,
- the adhesion,
- the thickness,
- the structure and the mechanical properties of the IP.

The ***effective contact area*** is determined by the amount, size and form of phases, geometry, and porosity of the surfaces and kinetics of wetting. Interlocking, achieved by good wetting of pores of solid surface by the polymer, and interdiffusion, between neighbouring polymer phases, increase the contact substantially.

Wetting is determined by the ratio between the surface tension of the solid and liquid phases and their interfacial tension. Acid-base interactions promote the wetting substantially. The wetting process can not be described by these terms alone because the viscosity of liquid phase, the geometry of the surface to be wetted and the outer forces promoting the wetting should be considered as well.

The ***adhesion force*** is the highest in the case of covalent bonding at the interface, while the acid-base and van der Waals interactions may also lead to considerable adhesion interaction between the phases. (In order to achieve covalent bonding the reaction of functional groups with other components or contaminants, before the contact of the phases, should be avoided). The macroscopic deformation of multicomponent polymer systems depends on micro-mechanics determined mainly by the value of adhesion force between the components. The micro-mechanics of multicomponent systems must, however, involve several phenomena not accounted for in early models and concept of adhesion. For example interpenetration of macromolecules at the IP multiplies the sites for intermolecular contact. On the other hand poor fracture performance within the boundary phase decreases the adhesion. The interface reinforcement achieved by chain pull-out and chain scission of connecting macromolecular chains, as well as craze formation around the interfacial zone may adsorb considerable energy, when the adhesion contact is tried to be released.

Adhesive forces become more important as the stiffness and size of the contacting materials are reduced. The ratio of the adhesion energy and the modulus of elastic layer (G_c/E) is quite small for metals and ceramics, and adhesive forces are only sufficient to exceed the yield strain for contacts of nanometer dimensions.[2] The ratio is still quite small for engineering thermoplastics, but becomes relatively large for elastomers and other highly compliant materials, including living cells[3,4] and soft tissues.

The ***thickness of the interfacial layer*** depends on the compatibility of the phases, temperature and on the external forces. The interfacial layer between incompatible polymer phases is typically in the range of 2 to 50 nm, depending on the value of interpenetration.[5] Macromolecular IP formed around reinforcing additives (fibers or particles), in order to avoid local stress concentration, may reach the μm size, while the layer formed by living organism around implanted biomaterials can be even thicker. The width of the IP zone, above an indifferent range (<6 nm) is one of the main parameter controlling the fracture toughness of the interface. Craze formation, plastic deformation, due to effective entanglements of macromolecules at IP results in steep increase in fracture toughness. Above a limit thickness value the adhesion becomes independent of the IP width.

The ***structure and the mechanical properties*** of the IP are in close correlation. Interfaces, having low interaction between the neighbouring phases, do not bear the stresses originating from external loading or differences of shrinkage. The stress concentration at the interface, due to these effects, leads to separation of the phases from each other and void formation around the inclusions. The mechanical properties of such IP-s are obviously poor.

A general rule for IP-s of disperse systems is defined in the Oswald-Buzágh continuity principle.[6] According to this principle the inclusions in an optimal disperse system are embedded continuously into the surrounding matrix due to their adsorbed interlayer. At an optimal IP the transition between the different phases should be harmonic must instead of sharp change. This principle was utilized when a multilayer IP structure was proposed for polymer composites. The method developed for thermoplastic polymer systems, based on an early patent, forms multilayer interfacial structure around filler particles by means of surfactants and elastomers.[7] A model was proposed for the arrangement of the surfactant and elastomer molecules in the modified polymer composites.[8, 9] According to this model the IP around the inclusion consists of a thin layer of oriented surfactant molecules and a relatively thick elastomer layer (Fig. 1).

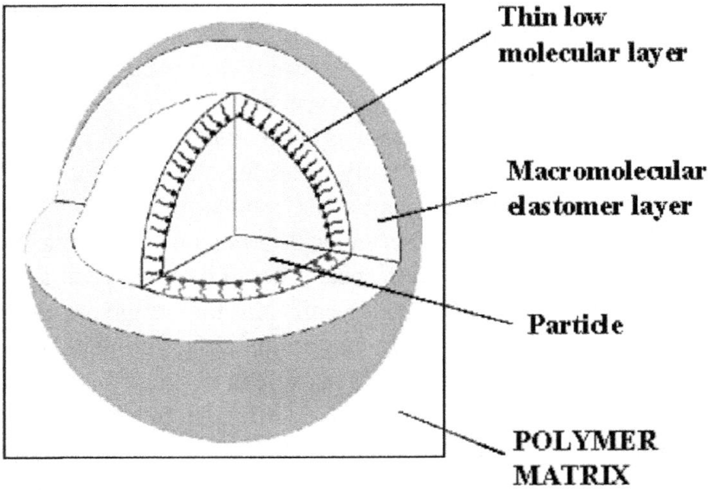

Figure 1. Scheme of multilayer interphase

It was found that such IP promotes the homogeneous distribution of the components and the maintenance of interaction with the polymer matrix and thus substantially affect the final mechanical properties of the polymer composites. External loading initiates stress concentration debonding and void formation around inclusions in case of thin unmodified IP. In contrast the thick multilayer IP built up in a hierarchic way terminates the cracks reaching the IP, absorbs and distribute the local stresses through crazing mechanism.[10] Further advantages of elastomer interlayer on the stress distribution were described earlier.[11,12]

3. SPONTANAEOUS INTERPHASES

As shown, various kinds of differences may occur at the border of phases. These include nucleation activity, polarity, charge, chemical potential etc. The differences generate formation of spontaneous interphases or allow the design of optimal interfacial structures.

Adsorption is the common way of spontaneous IP formation due to the surplus energy of the surface of materials. Pure surfaces occur at normal environment rarely as molecules (water, hydrocarbons) from the surrounding medium adsorb readily especially on inorganic surfaces of high surface energy. The strongly absorbed molecules such as water (already at <1% rel. humidity) decrease the surface energy substantially. The hardly removable adsorbed surface layers may cause problems at processing or application (e.g. partial desorption, decrease of the homogeneity and adhesion etc.). Such adsorbed layers hinder the interfacial interaction in composites. Many types of polymer-metal IP-s would be better described by a polymer-contaminant-metal oxide/metal system, where the adsorbed contaminant layer hinders the adsorption considerably.[13, 14]

Similar, but much more complex cascade process occurs at the interfaces of biomaterials when implanted into biological systems. This process, referred to as the foreign-body reaction, ends in an "encapsulation" of the material, which means the development of a scar tissue surrounding the inclusion.[15] The first step is the adsorption of proteins at the surface of material in a "non-specific" way, which means a physical adsorption (by means of electrostatic, van der Waals forces). The subsequent events are the adherence of disk-shape cells to the interface, which release proteins that direct the formation of a coating fibrin matrix. Finally huge aggregates (thrombus) of 50–200 μm size are formed around the original nuclei. The thrombus can cause vascular obstruction or even more severe consequence: embolism (block of an artery). Thus the spontaneous adsorption process is mostly disadvantageous both in composite and biomedical technology.

Nucleation of crystallites induced by certain particles in crystalline polymers is also spontaneous processes in connection with adsorption. The ordered adsorption of segments of macromolecules on the active sites of nucleating agents initiate transcrystallisation at the interfacial zone much earlier than the crystallization in the bulk starts. Such a transcrystallisation process was demonstrated around talc particles embedded in polypropylene matrix [16]. Nucleation improves the optical properties of crystalline polymers, but uncontrolled development of transcrystalline interfacial layer may lead to a rigid structure, which is not advantageous.

4. MODIFIED INTERPHASE OF LOW MOLECULAR MASS

Designed interfacial layers are constructed in order to make incompatible phases compatible by bridging the differences at IP. Optimally selected interfacial additives, instead of spontaneous IP-s, may offer numerous possibilities for improving the characteristics of multicomponent systems and for ensuring a desired set of properties. The wetting and adhesion of inclusions, processability and mechanical properties, i.e. strength, fracture toughness, elongation, etc. can be optimized by means of IP modification.

Surfactants **and** ***coupling agents*** are the most widely used additives for interface modification in multicomponent polymer systems. The main role of surfactants is to promote the wetting and the homogeneity of dispersions (thus the effective loading level of the dispersed phase can be substantially increased), while the coupling agents improve the adhesion between phases of different polarity through chemical bonding (resulting in increased stiffness and strength).[17] In the course of the IP modification these molecules must replace the spontaneously adsorbed weak boundary layers. In multicomponent systems, however, it is not easy for the interface modifiers to reach the right interface. It is feasible for the surfactants (due to their surface-active character) but in case of coupling agents the number of molecules located at the interface is insufficient unless a separate complicated and costly surface treatment step is included into the technology.

Reactive surfactants have been developed for combining the benefits of surfactants and coupling agents.[18,19] These interfacial additives can be defined as follows: reactive surfactants (RS) are amphiphile molecules that tend to absorb at interfaces preferably and following the adsorption react chemically with both phases. RS surfactants, like other surfactants form self assembly oriented layer at the interface in which the polar and alpolar sites of the molecules are oriented towards the relevant phases. After the reaction with the phases the oriented IP structure becomes permanent. Various types of RS have been synthesised by means of Diels-Alder reaction or by esterification.[19] Dienophile compounds e.g. maleic anhydride was used as reaction partners to the Diels-Alder reaction. One example is shown in reaction scheme (1). These additives contain carboxylic or carboxylic anhydride groups in the polar region and reactive double bonds in a non-polar hydrocarbon chain. Such chemical structure proved to be appropriate for combining the function of dispersing and coupling agents, thus they are capable of separating the agglomerates and bond the phases chemically to each other. Several advantages of such type of reactive interface modification were found in multicomponent polymer systems.

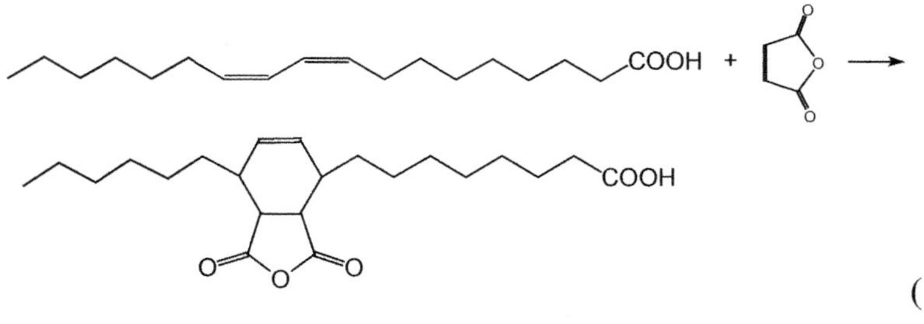

(1)

Reactive surfactants may perform advantageous effect on the following parameters simultaneously:

- wettability,
- area of effective interaction between the phases,
- adhesion,
- adsorption of molecules in the IP-s,
- rheology.

In line reactive interface modification of multicomponent polyolefin systems in a compounding machine require the use of RS-s that are able to find their right place during a short residence time. The interface modification in filled/reinforced polymer systems and polymer blends resulted in improved mechanical properties, while in pigmented PE it contributed to higher photostability.[20-22] Furthermore, the efficiency of intumescent flame retardant stabilizer additives in polyolefins could be enhanced.[23]

Controlled transport achieved by the IP modification is the background behind both kind of improved stability. The schematic drawing in Fig. 2 shows how the modification of IP contributes to controlled transport ad thus improved stability. The first case is the adsorption of thermal/photo-stabilisers on the surface of filler or pigment particles that decreases the stability substantially (see in Figure 2a). The second is the migration of flame retardant (FR) additive components to the surface of products causing processing and aesthetic problems (see in Figure 2b). The concept of solving the mentioned stability problems by applying RS coating layer around the particles (and thus controlling the undesirable transport processes) is shown in Figure 2c.

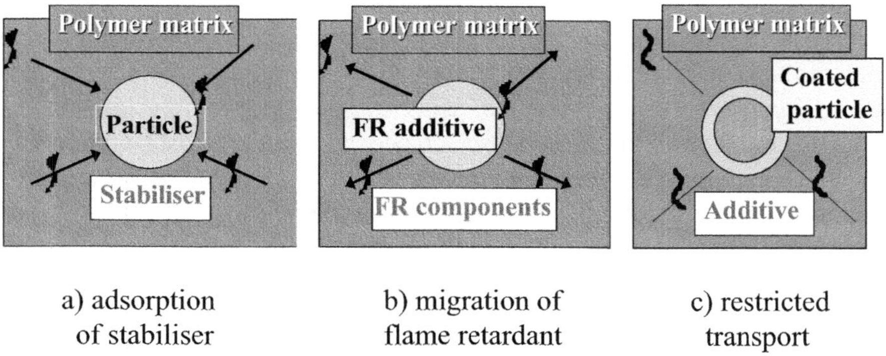

Figure 2. Scheme of the undesirable transport processes and their control through modification of interface

5. MODIFIED INTERPHASE OF HIGH MOLECULAR MASS

Materials having macromolecular IP-s are the most widely used multicomponent systems. Polymer-layered metals, reinforced polymer systems, polymer blends and various kinds of biomaterial implants can be found among these types. Compatibilisation by means of bridging the polarity differences with surface active and reactive interface modifiers (as described in the previous part) satisfies the requirement described in the Oswald-Buzágh principle only partially. The transition between the phases is harmonic only if the differences in elasticity were bridged as well. For this purpose the introduction of macromolecular IP-s have to be considered.

Polymer composites clearly show if their IP is imperfect because pull out or delamination is the dominant destruction mechanism, limiting the reinforcing effect, in that case. In fiber-reinforced composites the transversal strength is especially affected by the characteristics of IP (adhesion, thickness, structure). Filled and reinforced materials (composites) need to satisfy a range of diverse structural and mechanical requirements, e.g. strength, modulus of elasticity, fracture toughness, impact performance, damage tolerance, elongation, etc. These properties depend on mechanical/rheological characteristics of the matrix and reinforcing additives, volume fraction of the components, interfacial adhesion and micro-mechanical properties of the interphase zone. Among the listed factors the IP is the most critical in respect of composite's performance. *Acid-base interactions* play an important role in interphase formation of composites. Stronger interaction leads to a thicker interphase and decreased mobility.[24] *Orientation of macromolecules at the interface* improves the strength of composites.[25] *IP modifications* are divided into four arbitrary groups, i.e. non-reactive and reactive treatment, introduction of a soft interlayer and application of functionalized polymers around the particles.[26] *Macromolecular IP-s* are formed using various technologies in order to achieve uniformly and continuously coated particles. Aerosol and solvent-based procedures allowed forming uniform, well-adhering coatings of polydivinylbenzene on silica beads.[27] Relatively thick (100-500 nm), soft elastomer interlayers could be formed "in situ" during a single step compounding process. *Hierarchic IP-s* containing elastomer and low molecular (such as reactive surfactant) internal layer around inclusions could be produced the same way. The principles affecting the arrangement of the elastomer phase in the composites, i.e. its share between the interphase and disperse phase, have been published from the beginning of eighties.[28,29] It was found that the polarity, viscosity and melting regime of the IP-forming elastomer determine the structure-property relationship of the composite through the IP.[30,31]. The elastomers, similarly to the catalysts, are more effective in a coating layer around filler particles.[31] *Functionalized, grafted polymer brushes* have been developed in wide variety as compatibilization/biocompatibilization methods. Principles of this kind of IP design are applicable to either composite materials (with synthetic/natural fillers/fibres), or blends or for biomaterials. A review on the effect of surface-grafted molecular brushes on the adhesion performance of polymer composite interfaces has been published recently.[32]

Polyelectrolyte-coated particles contain special IP. Electric field originating from the accumulation of surface charge at the solid/liquid interface extends over a thickness of several molecular layers. The force of interaction between two or more adsorbed layers can be modified by neutral and charged polymer brushes of grafted polymers.[33]

Polymer blends are compatibilized also by addition of block or graft copolymers.[34-36] Owing to the resulting reduced interfacial tension the dispersed domain sizes are found to be decreased and their size distribution is made more uniform. However, a problem with introducing a pre-made copolymer to the blend is that it is difficult for the copolymer to reach the interface. Further, the copolymers can form micelles.[37,38] *Reactive Processing,* in which the copolymer is formed at the interface during mixing of the blend components containing reactive sites for formation of the necessary covalent bonds, is more attractive from a commercial standpoint.[39,40] The reactive additives, dispersed into the components of the blend, should have a surface active character for accumulating at the surfaces during the mixing process. This is ensured if *reactive surfactants* (RS) are used.[41] As described above, these additives find the interfaces, emulsify the phases and react with them during the time of compounding.

Improved interfacial interaction and decreased domain sizes achieved by RS during reactive extrusion of incompatible PP-PA blend is shown in Figure 3.

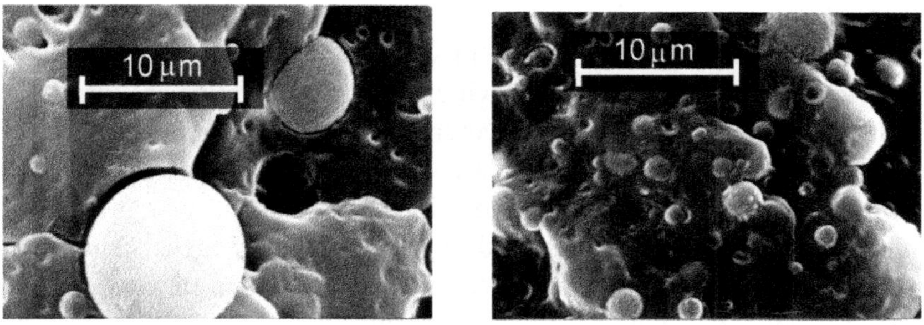

a b

Figure 3. SEM micrographs of the fractured surfaces of the PP-PA blends (broken under liquid nitrogen) (a) without RS, (b) in presence of 1% RS.

The interfacial thickness is an important feature of the polymer–polymer interface. Fracture toughness of blends is considered as resulting from the energy necessary to pull the diffused chains out of the other polymer phase along a chain tube. Thus, the energy can be expressed as a function of the interfacial thickness.[42, 43] In case of some reactive interfaces the interfacial thickness was found to be much larger than in case of non-reactive IP-s. Furthermore it was larger than the calculated coil size of the copolymer formed *in situ* at the interface and the value estimated for the equilibrium interface in a ternary system composed of the immiscible polymer pairs, A, B, and its copolymer A–B.[44,45] The knowledge of the chemical interactions and molecular connectivity at interfaces and their influence on processability and mechanical properties of the polymers and polymer composites is especially essential in the case of thermoset polymer-polymer IP-s, which play a vital role in recycling, repair, welding, and joining of polymer composites.[46]

Biomaterials have to be surrounded also with soft, hydrophilic macromolecular IP in order to inhibit the spontaneous adsorption of non-specific proteins (fouling) described above. Several methods, developed for forming non-fouling surfaces, have been reviewed in.[47] Grafting of hydrophilic layer on the surface of biomaterials is the most frequently applied method. The IP is formed generally using one of the following molecules:

- amphiphile polymer molecules containing ethylene oxide units of min. 500-1,000 molecular mass,
- methacrylate- polyethyleneglycol (PEG) oligomer,
- liposomes immobilized trhough functionalized coupling molecules (PEG, biotin, avidin),
- hydrogel of functionalized polylactic acid and PEG,
- phosphatidyl choline containing methacrylate-polymer,
- polysulfone.

Soft, hydrophilic surfaces are formed also by grafting poly (N-vinyl pyrrolidon) and polyacrylamide in presence of riboflavin. Selective functionalization is the process by which functional groups are introduced at a pre-designated location in a polymer molecule.[48] Controlled anionic and cationic living polymerizations are widely used for preparation of functionalized polymers of predetermined lengths and narrow molecular weight distributions. Carboxyls, hydroxyls, amines, halogens, double bonds, and many other functional groups can be placed at one or two ends in the centre or evenly spaced along polymeric chains.[49,50]

The chemical reactions are generally initiated by the following physical treatments:

- UV irradiation in presence of photo-sensitizer (e.g. benzophenon),
- ozone treatment,
- ion beam assisted deposition (IBAD),
- plasma treatment.

Plasma processes, using microwave- and radio-frequency (RF) excited plasma, have been successfully applied to various polymer types in order to generate multifunctional surfaces. As an example, surfaces made by the RF-plasma deposition of tetraethyleneglycol dimethylether shown to have extremely low protein pickup due to the formed crosslinked PEG-like structure.[47] Plasma polymerization of hexametyl disiloxane on PUR implants improved the biocompatibility as well. The density of grafting, the length and architect of grafted chains have to be designed accurately as these determine the intensity of protein and cell adsorption. High density and highly selective amino functionalization could be achieved by hydrogen admixture to discharges in nitrogen and ammonia.[51] The largest surface density (0.4 polymer chain/nm^2) was achieved by forming the Langmouir-Blodget film of chains on the surface which was fastened then by controlled radical polymerization.

6. ADAPTIVE INTERPHASES

The new term '*adaptive IP*' is suggested for a new type of interlayer. In contrast to the conventional static IP-s, an adaptive IP is able to *react to the change of its environment*, thus a desirable transformation of its structure and behaviour can be induced by changing the characteristics of the neighbouring phase. Heat, light and pH induced transformation facilitates stabilization (i.e. fire retardancy, and photostability) and drug delivery systems while the change of and biological environment can be utilized in advanced biomaterials, (bio)composites.

Stabilization against fire protects the flammable polymers from the severe thermal and oxidative surface attack of flame. The largest part of fire retardant additives is introduced in the form of particles. Modification of the surface of material and the interfaces around the flame retardants strongly affects the fire resistance of polymer systems. In normal environment, similarly to other multicomponent systems, the main role of IP in fire retarded systems is the compatibilization for facilitating the processability and the mechanical properties.[52] At elevated temperature caused by fire attack, however, the IP-s around the particles should promote their transport to the surface where a protective barrier IP should be formed. Hierarchic IP, containing an organosilicon layer, around ammonium polyphosphate (APP) flame retardant particles, according to the scheme 2, was found to meet the described requirements in polyolefins.[53]

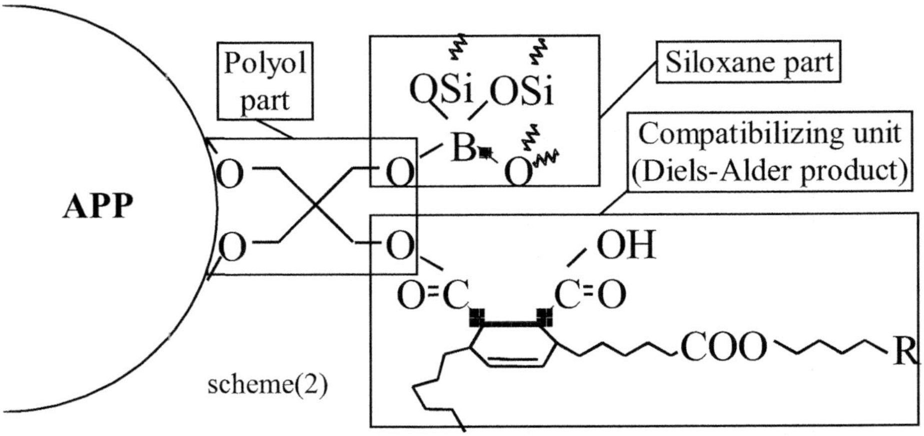

scheme(2)

As shown in the scheme, the compatibilization (at normal temperature) is solved by a reactive surfactant of long hydrocarbon chain, which acts as a leaving group at the temperature of fire. After the decomposition of the compatibilizing unit, the uncovered incompatible polysiloxane coating, as a carrier layer, delivers the flame retardants to the surface.[54] Furthermore at high temperature the boron containing polysiloxane transforms to a durable ceramic IP protecting the material against the heat and oxidative effect of fire.[55] Silcone modified layered nanoparticles in polymer nanocomposites may contribute to the fire retardancy through similar mechanism.[56] Thus the surface phenomena and especially the adaptive interface play an important part in the fire retardancy performance of polymeric systems. The hydrothermal stability of flame retardant additives was also improved by IP modification.[57]

The *photostability* of polymers depends on surface effects as well because the light interacts with the materials at their surface. Modification of the surface layer through controlled transport of stabilizer molecules, induced by heat- and/or light, can be a strategy for enhancing the efficiency of the current stabilizer molecules.

Drug delivery systems mostly contain thick IP around enclosed drugs. Thermally and chemically adaptive IP-s can be utilized in various new types of drug delivery systems. Enhanced temperature and acidosis in the neighbourhood of pathologic processes allows nanoparticles covered by heat- and pH-sensitive polymers (such as copolymer of poly(*N*-isopropyl acrylamide) and polylactic acid) to release drugs at that areas. Drug delivery systems coated by acrylic polymer and certain polysaccharides results in pH dependent release of drugs.

Biomimetics, biocomposites require sophisticated IP modification. Change of adaptive IP-s according to biological environment can be utilized for formation of biomimetic materials to be used in bone repair and regeneration, coatings, wound dressing, artificial skin etc. The major surface properties that should be modified for these purposes include two kinds of biocompatibility. One is the surface property that elicits the least foreign-body reactions and the other is the cell- and tissue-bonding capability. In addition, physiologically active surfaces with, for instance, selective adsorbability may be required.[58] Biomimetic IP modification means generally conjugation with peptides, proteins, and/or nucleotides. The nature's use of proteins as signalling agents comes from a few specific proteins in fixed conformations and orientations so they optimally deliver signals. The body views the other type non-physiologic proteinaceous layer as an unrecognized foreign invader that must be walled off.[59] Therefore biomimetic modification of IP-s must control the conformation and orientation of proteins and thus form topographic, chemical and visco-elastic patterns on surfaces (matching proteins at the nanometer scale and cells at the micrometer scale) so that the body will specifically recognize them. The non-specific adsorption of proteins and other biomolecules must be inhibited. Promising approach is the attachment of peptides containing domains of proteins of extracellular matrix (ECM, natural environment of cells) to the surface of biomaterials. Hierarchic IP-s including these active peptides can induce tissue formation conforming to the cell type seeded on the material. Biomimetic surface engineering for bone regeneration can affect bone cell adhesion and preferentially induce mineralization on the modified surfaces. Presence of hydroxyapatite on the modified surface appears to have the ability to induce bone formation. Bones and other biominerals are biocomposites containing unique IP-s between mineral and organic units. Molluscs for example synthesize nacre of 95% calcium carbonate and 5% protein, which is mechanically tough due to optimal IP between the phases.[60]

7. SMART INTERPHASES

Dynamic interphases that "communicate" with the environment due to a *recognition mechanism* and undergo *rapid shifts* in surface properties with small external changes are smart IP-s. Development in this area tries emulating nature's methods to make more functional, controllable surfaces. The structure and behaviour of such IP can be controlled through external signal or can be programmed for a certain purpose. Molecular recognition surfaces are formed by coupling to enzymes or other specific biological receptors.[47] Receptors are the ultimate "smart" materials that can adjust their function in response to environmental changes or biochemical modification. A variety of approaches have been developed for forming controllable surfaces, many of these address pharmacological strategies.

Targeted drug delivery, gene therapy and controlled drug release are the areas where the advantages of smart IP-s can be clearly utilized. Targeted systems are able to find the selected part of a substratum. The targets are generally the brain, arteries, lymphatic system, lung, liver etc. Polymer nanoparticles of 20-60 nm size coated with a hydrophilic layer are used for forming targeted drug delivery system. Biochemical control is developed through immobilization of special peptides, glycoproteins, enzymes, etc. in the IP of nanoparticles. This requires the functionalization of the hydrophilic polymer coating by means of the described methods. Inadequate structure of the grafted layer inhibits the interaction with the target area. For achieving controlled branching in the layer various dendrimers and star polymers has been developeded. Magnetic and electric field can be also used in the targeting mechanism if sensitive elements are built in the delivery system.[61]

The formation of ***pH-sensitive membrane*** is another area where smart IP can be applied. The surfaces of pores of such membranes are grafted with polyacrylic acid. These chains change their form dynamically according to the pH modifying the transport through the membrane in this way (Fig. 4).

Similar controllable IP can be formed on the surface of particles. As an example by postgrafting of hydrophilic polymer to grafted hydrophobic chain on carbon black, amphiphilic carbon black was obtained, the dispensability in solvent and the surface wettability of which were readily controlled by pH and temperature.[62]

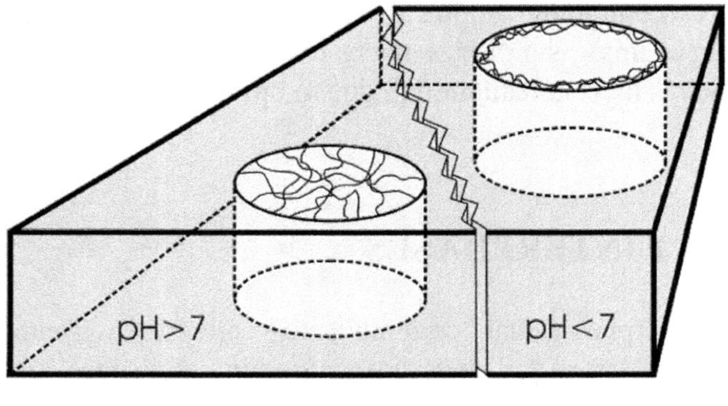

Figure 4 Scheme of pH-dependent behaviour of grafted membrane.

In the near future very active research is expected for application of smart IP-s in the area of *bioelectronics* and *artificial photosynthesis*. The challenge at both areas is to form chemically and topographically patterned surfaces that adsorb and retain biomolecular building blocks of the circuit elements (responsible for transferring the charge initiated by electric field or electromagnetic field of light) at right location, conformation and orientation without loss of functionality. Sophisticated programming of the surfaces is an important aim for designers of *tissue engineering* scaffolds. The development of smart IP-s for *hybrid polymer composites* and *biodegradable polymer systems* is also a challenge for the future.

8. CONCLUSIONS

Fundamentals of interfacial phenomena are general for whole field of materials science and these are especially important for multicomponent materials, in which both the surface and interfacial layers has to be designed. Common consideration of the interface related characteristics of multicomponent engineering polymer systems and biomaterials is advantageous because analogies can be found that initiate development of both areas. Based on the examples presented in this work one can conclude that an optimal IP encompasses a harmonic transition between the neighbouring phases by means of multicomponent, multilayer, hierarchic, structure. A complex interfacial layer contains oriented or specially arranged, functionalized and sometimes patterned molecules.

In situ methods for modification of the interfacial zone can be performed using reactive modifiers, such as coupling agents/connector molecules or reactive surfactants (recently proposed by the authors). These methods are more economic than the separate treatments. Soft macromolecular interphases are essential for overcoming the local stresses at the phase borders.

There are many areas, where the potential of modified interphases are not entirely discovered. For example further research is needed in the field of the composites, transport processes, flame retardancy, rheology, nano-and biotechnology.

New type of dynamic interphases, called adaptive and smart ones, has been proposed and defined in this work for advanced material systems. As the structure of materials is getting more and more complex, it seems that the adaptive and/or smart interphases will be more preferred in the currently developed multicomponent systems. These are common in the nature, giving many examples for optimal design of artificial interfacial structures.

ACKNOWLEDGMENT

This work has been financially supported by the IKMA foundation and Hungarian Research Fund through projects NKFP 00169/2001, NKFP 3A /0036/2002 and OTKA T32941.

REFERENCES

1. Geckeler K. E., Rupp F., Geis-Gerstorfer J., Adv. Mater. **9** (6) 513-518 (1997).
2. Landman U., Luedtke W. D., J. Vacuum Sci. Technol. **B9** 414-423 (1991).
3. Zhu C., Bao G., Wang N., Ann. Rev. Biomed. Eng. **2** 189-226 (2000).
4. Shull K.R., Mater. Sci. Eng. R. **36** (1): 1-45 (2002).
5. Stamm M., Schubert D.W., Ann. Rev. Mater. Sci. **25**: 325-356 (1995).
6. Buzágh A., Kolloid Z. 125 (1952) 14-21.
7. Hungarian Pat. 167063 (1975); US Pat. 4116897 (1978); German Pat. 2453491 (1986).
8. Bánhegyi Gy., Marosi Gy., Bertalan Gy., Karasz F.E., Colloid Polym. Sci., **270** 113 (1992).
9. Marosi Gy., Bertalan Gy., Rusznák I., Anna P., Molnár I., in Polymer Composites (Ed. B. Sedlacek) Walter de Gruyter , Berlin- New York, 1986, 457-464.
10. Rusznák I., Bertalan Gy., Anna P., Marosi Gy., Molnár I., Plaste Kautsch. **32**, 254-258 (1985).
11. Matonis V. A., Small N. C., Polym. Eng. Sci. **9**, 90 (1969).
12. Kardos J. L., in " Molecular Characterisation of Composites Interfaces" (Ed. Ishida H. Kumar G.) Plenum Publ. Corp. 1985.
13. Chakraborty et al A. K. Chakraborty, H. T. Davis, M. Tirrell, J. Polym. Sci. A: Polym. Chem., **28**, 3185 (1990).
14. Van Oss C. J., in Polymer Surfaces and Interfaces II, (Eds: Ferit W. J., Munro H. S., Richards R. W.), Wiley, Chichester, UK 1993, p.267.
15. Anderson J. M. in: Ratner B.D. (Ed.), Biomaterials – an Introduction into Materials in Medicine, Academic Press, San Diego, 1996, p. 165.
16. Marosi Gy., Lágner R., Bertalan Gy., Anna P., Tohl A., J. Therm. Anal. **47**, 1163 (1996)
17. Chodak I., Novak I., in: "Polypropylene, An A-Z Reference", Karger-Kocsis J., Ed., Kluwer Academic Publisher, Dordrecht/Boston/London, 1999, p.790.
18. Hungarian Patent 218 016 (2000).
19. Marosi Gy., Anna P., Csontos I., Márton A., Bertalan Gy., Macromol. Symp. **176**, 189-198 (2001).
20. Bertalan Gy., Marosi Gy., Anna P., Ravadits I., Csontos I., Tóth A., Solid State Ionics **141-142**, 211-215 (2001).
21. Maatoug M. A., Anna P., Bertalan Gy., Ravadits I., Marosi Gy., Csontos I., Márton A., Tóth A., Macromol. Mater. Eng. **282**, 30-36 (2000).

22. Anna P., Bertalan Gy., Marosi Gy, Ravadits I., Maatoug M. A., Polym. Degrad. Stab. 73, 463-466 (2001).
23. Marosi Gy., Bertalan Gy., Balogh I., Tohl A., Anna P., Szentirmay K., in Flame Retardants (ed. S. J. Grayson, British Plastics Federation) Interscience Commun. LTD Publ. London 1996, 115.
24. Moczo J, Fekete E, Pukanszky B, J., Adhesion 78, (10) 861-875 (2002).
25. Marosi Gy., Tóhl A., Bertalan Gy., Anna P., Maatoug A. M., Ravadics I., Bertóti I., Tóth A., Composites Part A 29A, 1305 (1998).
26. Pukanszky B, Fekete E., Adv. Polym. Sci. 139, 109-153 (1999).
27. Partch R, Brown S J., of Adhesion 67 (1-4) 259-276 (1998).
28. Rusznák I., Bertalan Gy., Trézl L., Horváth V., Huszár A., Székely G., Molnár I., Muanyag Gumi, 16 257 (1979).
29. Rusznák I., Bertalan Gy., Anna P., Marosi Gy., Molnár I., Plaste Kautsch., 32, 254 (1985).
30. Marosi Gy., Bertalan Gy., Rusznák I., Anna P., Colloids and Surfaces, 23 (3), 185 (1986).
31. Marosi Gy., Bertalan Gy., Anna P., Rusznák I., J. Polymer Eng., 12, 34 (1993).
32. Gutowski WV, Li S, Russell L, Filippou C, Hoobin P, Petinakis S, Composite Interfaces, 9 (1): 89-133 (2002).
33. Netz R.R., Andelman D., Phys. Rep.-Rev. Phys. Lett. 380 (1-2) 1-95 (2003).
34. Paul DR, Newman S, (eds.) Polymer blends, 1-2. New York: Academic Press, 1978.
35. Fayt R, Jerome R, Teyssie P., J Polym Sci Polym Lett Ed; 19, 79 (1981).
36. Utracki LA., Polymer alloys and blends. Munich: Hanser, 1989.
37. Nakayama A, Inoue T, Gue´gan P, Macosko CW., Polym Prepr. **34** (2) 840 (1993).
38. Dai HK, Kramer EJ, Shull KR., Macromolecules; **25**, 220 1(992).
39. Ide F, Hasegawa A., J. Appl. Polym. Sci., **18**, 963 (1974).
40. MacKnight WJ, Lenz RW, Musto PV, Somani R., J. Polym. Eng. Sci. **25**, 1124 (1985).
41. Marosi Gy., Anna P., Márton A., Csontos I., Matkó Sz., Szép A., Kiss É., Progr. Coll. Polym. Sci. (accepted).
42. Foster KL, Wool RP. Macromolecules; **24**, 1397 (1991).
43. Wool RP. Polymer interfaces. 9. Hanser, Munich, Vienna, New York, 1995.
44. Yukioka S, Inoue T., Polymer **35**, 1182 (1994).
45. Koriyama H, Oyama HT, Ougizawa T, Inoue T, Weber M, Koch E, Polymer **40,** (23) 6381-6393 (1999).
46. Raghavan J, Wool RP, J. Appl. Polym. Sci. **71**, (5) 775-785 (1999).
47. Castner D.G., Ratner B.D., Surface Science **500,** (1-3) 28-60 (2002).
48. Hulubei C., Mater. Plast. **39,** (4) 209-212 (2002).
49. Jagur-Grodzinski J., Reactive Functional Polym. **49,** (1) 1-54 (2001).
50. Jagur-Grodzinski J., J., Polym. Sci. Part A-Polym. Chem. **40,** (13) 2116-2133 (2002).
51. Meyer-Plath A.A., Schroder K., Finke B., Ohl A., Vacuum **71,** (3) 391-406 (2003).
52. Marosi Gy., Anna P., Bertalan Gy., Szabó Sz., Ravadits I., Papp J., in Fire and Polymers (eds. G. Nelson, C. Wilkie) ACS Ser., 797 Washington 2001 161-171.

53. Ravadits I., Toth A., Marosi G., Marton A., Szep A., Polym. Degrad. Stab. **74,** (3) 419-422 (2001).
54. Marosi Gy., Ravadits I., Bertalan Gy., Anna P., Maatoug M. A., Tóth A., Tran M. D., in Fire Retardancy of Polymers: the Use of Intumescence, (ed. G. Camino, Le. Bras, S. Bourbigot, R. Delobel), Royal Soc. of Chem. Cambridge, 1998, p. 325.
55. Marosi G, Marton A, Anna P, Bertalan G, Marosfoi B, Szep A, Polym. Degrad. Stab. **77,** (2) 259-265 (2002).
56. Marosi G, Anna P, Marton A, Bertalan G, Bota A, Toth A, Mohai M, Racz I Polym. Adv. Technol. **13,** (10-12) 1103-1111 (2002).
57. Marosi Gy., Csontos I., Ravadits I., Tohl A., Anna P., Sommer F., Botreau M., Tran M. D., J. Therm. Anal. Calorim., **56**, 1071-1080 (1999).
58. Ikada Y, Biomaterials **15,** (10) 725-736 (1994).
59. Ratner B., D., Macromol. Sci., **130**, 327-335 (1998).
60. Sellinger A., Weis P.M., Nguyen A., Lu Y., Assink R.A., Gong W., Brinker C.J., Nature **394,** 256-260 (1998)
61. Zrinyi M, Szilagyi A, Filipcsei G, Feher J, Szalma J, Moczar G, Polym. Adv. Technol. **12** (9) 501- 505 (2001)
62. Tsubokawa N, Bull. Chem. Soc. Jap. **75,** (10): 2115-2136 (2002).

REACTIVE PROCESSING OF POLYMER BLEND USING REACTIVE COMPATIBILIZATION AND DYNAMIC CROSSLINKING: PHASE MORPHOLOGY CONTROL AND MICROSTRUCTURE – PROPERTY RELATIONS

CHAREF HARRATS, and GABRIEL GROENINCKX

Catholic University of Leuven, Department of Chemistry, Division of Polymer Chemistry, Laboratory of Macromolecular Structural Chemistry, Celestijnenlaan, 200 F, B-3001 Heverlee-Leuven, Belgium

Abstract: This chapter covers the area of polymer blends produced via melt-processing. The blends are first introduced based on their miscible or immiscible nature. Advantages and disadvantages of one class over the other are briefly highlighted. The need to compatibilize immiscible polymer blends is emphasized. The two main strategies of compatibilization, i.e., physical blending which consists of adding a pre-formed copolymer to compatibilize the blend and reactive blending which is based on the in-situ generation of the compatibilizing agent, are compared. The key-factors such as viscosity ratio, blend composition, interfacial tension and mixing conditions that control the phase morphology in immiscible blends are discussed in detail. A large section is devoted to the discussion of compatibilization with extensive consideration of all aspects of the reactive compatibilization approach. Moreover, in-situ copolymer formation and compatibilization activity are discussed in terms of the effect of : - the miscibility of the compatibilizer precursor, - the configuration of reactive end groups and, - the content of reactive groups. In the last section, the thermoplastic vulcanizates which are a particular case of polymer blends are described in terms of their composition, their phase morphology development and their specific elasticity and recovery behaviour.

Key words: Polymer blends, reactive processing, compatibilization, phase morphology, dynamic vulcanization

F. Ciardelli and S. Penczek (eds.), Modification and Blending of Synthetic and Natural Macromolecules, 155–199.

1. INTRODUCTION TO POLYMER BLENDS

Polymer blending is an economical strategy to design new materials from existing polymers as old as polymer industry itself. After the great progress recorded in monomer synthesis and new polymer or copolymer design, the innovation was slowing down by a cost criteria. Designing new random, block, or graft copolymers is more expensive than developing novel polymer mixtures. Indeed, driven by a tough competition, the development of new materials via blending of existing homopolymers and copolymers has been initiated during the seventies. The major advantages of blending can be summarized as follows:

- opportunity to develop new properties or improve the existing ones to meet specific customer needs,
- significant material cost reduction without substantial loss in properties,
- improvement of the processability of some interesting polymers that are difficult to shape as neat,
- ability of recycling two, three or more polymers of different nature,

The development of commercial polymer blends has been well documented by Utracki.[1]

1.1 Miscibility versus immiscibility in polymer mixtures

Even though more than 400 miscible polymer pairs have been reported,[2,3] most mixed polymer systems are not miscible on a molecular scale and tend to phase separate during mixing into individual components domains. This does not mean that immiscible polymer blends are not interesting for pratical use. Some properties such as impact strength and barrier properties can only be reached through immiscible, two or three phase polymer blends. The properties of miscible polymer systems are intermediate between those of the individual components whereas synergies are obtained in immiscible polymer blends provided that the phase morphology and the interfacial adhesion are judiciously controlled. For most polymer pairs the interfacial tension $\gamma_{1,2}$ is significantly high; comprised within 0.5 to 15 $mN{\cdot}m^{-1}$, resulting in phase separation upon mixing[4]. Phase separated mixtures require improved interfacial adhesion to promote the necessary load bearing properties acceptable for commercial applications. Unfortunately in immiscible polymers the Flory parameter, χ . is large within 0.05-0.5. Consequently the interfacial contact is narrow (1-5 Å) which means that there is a little interpenetration of molecules at the interface.[5] As a result, failure of the interface between two glassy immiscible polymers thus requires only to overcome the Van der Waals bonds as in the fracture of low-molecular weight polymers. In uncompatibized polymer blends the interfaces are the most vulnerable locations to mechanical fracture. When subjected to mechanical stresses, they most likely fail well before the individual components of the blend. The third critical aspect in immiscible polymer blends is their phase morphology unstability. The state of dispersion of one phase in another is controlled by both thermodynamic energy (interfacial tension) and thermo-mechanical energy (shearing action exerted on the phases during mixing). The stability of the phase morphology depends on the balance between these two energetical parameters. In a melt-blending process, the phase morphology is developed as a result of the shearing forces overcoming the interfacial tension which tends to resist the deformation and break-up of particles. Upon cessation of the shearing action, the interfacial tension remains the unique driving force for the system to evolve. Indeed, the particles are driven into close proximity and coalesce in a larger phases. These deficiencies encountered in immiscible polymer blends can be approached via a compatibilization strategy. The concept of compatibilization resulted from the need of controlling and stabilizing the phase morphology and controlling the interfacial adhesion between the phases of the immiscible blend. The strategy was inspired from colloid science where the idea of adding a molecule having hydrophobic and hydrophilic entities called surfactant or emulsifier was able to disperse and stabilize emulsions of water-in-oil or oil-in-water. Addition of a block or graft copolymer to its parent immiscible homopolymer blends plays the same role. It has also an additional role of enhancing the interfacial adhesion between the phases. The copolymer or the additive used is known under various names such as "compatibilizer", "interfacial agent", "compatibilizing agent", "adhesion promotor" and "emulsifier". Basically, there are two main strategies of compatibilization ; physical (non-reactive) compatibilization and reactive compatibilization or reactive blending.

1.2 Physical versus reactive blending

In reactive blending, the compatibilization of immiscible polymers is ensured by a chemical reaction initiated during the process of melt-mixing. The *in situ* formed compatibilizing agent (block or graft copolymer, crosslinked species, ionic associations, etc.) reduces the interfacial tension between the immiscible blend components, enhances the adhesion between the phases and, as a consequence, imparts to the blend acceptable mechanical properties.[6-9] In physical blending the compatibilizing agent is chemically synthesized prior to the blending operation, and subsequently added to the blend components as a non-reactive component. Owing to its chemical and molecular characteristics, the added agent is able to locate at the interface, reduces the interfacial tension between the blend components, and promotes adhesion between the phases.[10-14]

In table 1 is summarized each action of the compatibilizer and its effect on the immiscible blend.

Table 1. Action of the compatibilizer and its consequence on the blend

Action of the compatibilizer	Consequence of the action
- reduces the interfacial tension	- reduction of the particle size
- locates at the interface	- suppression or limitation of coalescence process
- stays at the interface	- stabilization of the phase morphology
- own sequences anchor into and /or develop chemical interactions with the respective homopolymers	- promotion of interfacial adhesion between the phases

The main similarities between reactive and physical blending can be described as follows:

- The compatibilizing agent is expected to be located at the interface between the phases.
- Compatibilization in both types of blending results in particle size reduction of the dispersed phase, enhanced interfacial adhesion between the phases, and a thermally stable phase morphology during post-processing steps such as annealing, re-extrusion and injection molding.
- Both methods of blending lead to the design of compatibilized blends with attractive properties.
- For industrial purposes melt-extrusion is used as the main compounding operation for both types of blending.

Reactive blending is a very cost-effective process that allows the formulation of new multiphase polymeric materials. The copolymer responsible for compatibilization of the blend components is formed *in-situ* during the melt-process. It is expected that the reactive precursors generate the compatibilizing block or graft copolymer at the interface of the immiscible polymer blend. As a consequence, micelle formation in one or the other phase is expected to be minor compared to when the pre-made block or graft copolymers are used, which can easily self-organize in the phase where they are the most energetically stable. This situation is expected to be more probable in extrusion processes where the residence time is short so that the mixing equilibrium is often not completely attained.

The reactive molecules used in reactive blending are usually prepared using less sophisticated synthesis routes. In contrast physical blending requires an additional step for the synthesis and the design of the compatibilizing agent; this approach is mainly used when reactive blending is technically not feasible. There are a few cases where the compatibilization can only be ensured by using a copolymer prepared separately under one or more of the following conditions: in solution media (organic or aqueous), via anionic, cationic or co-ordination polymerization, or using a succession of chemical reactions carried out under different operating conditions. These reactive conditions, which are not exclusive, are not easily transferable to extrusion facilities traditionally employed in reactive blending. In contrast, a large amount of semi-crystalline polymers has a limited solubility in a common solvent. As consequence, melt reactive processing remains the only alternative for the synthesis of block copolymers based on polymers suffering from solubility limitations. Typical examples of this situation are copolymers based on polyethersulphones prepared via reactive extrusion.[15]

Physical blending remains, however, more convenient for fundamental investigation since it involves well defined molecular characteristics of the added compatibilizing agent. Judicious correlation between the compatibilization efficiency of the copolymer and its molecular characteristics is merely established. In contrast, the copolymer formed *in-situ* in the reactive blending is more difficult to separate and characterize. Several attempts have been made and are still being made to identify and characterize the *in-situ* formed copolymers. NMR as well as FTIR spectroscopy investigations carried out on selectively extracted polymers have largely been used for the characterization of the chemically formed species at the end of the blending process. Practical difficulties associated with the swelling and the partial solubility of the blend components make this method less efficient. When the concentration of the reactive groups is low, the spectroscopic techniques are also found to be inefficient, particularly when the groups under characterization do not exhibit distinct and intense absorption bands.

2. PHASE MORPHOLOGY DEVELOPMENT IN IMMISCIBLE POLYMER BLENDS

In practice, the blending operation generally starts with the components in pellets or powder form having an average particle size of approximately 3 and 0.2 mm, respectively. Target morphologies for the final product often require domain sizes in the range of 0.1 to 10 μm in order to achieve superior performance characteristics. It means that a size reduction of about three orders of magnitude is imposed to the initially introduced components in the mixing equipment.[16] Phase morphology development in immiscible polymer blends can be considered as the big change in dimension and shape the components undergo during their transfomation and compounding operations. The minor phase in an immiscible polymer blend in the melt is deformable, as opposed to composite materials, which contain a rigid minor phase. A wide range of sizes and shapes (spherical, ellipsoidal, cylindrical, ribbonlike, co-continuous, etc.) can thus be obtained for the dispersed phase during processing.[3,17] The final phase morphology in a blend is the result of a balance between deformation-disintegration phenomena and coalescence.

2.1 Deformation, break-up and coalescence phenomena during melt-blending

2.1.1 Basic principles

Historically, Einstein was the first to derive an expression for the viscosity of hard spheres in a dilute suspension.[18] Later on Oldroyd[19] considered the case where the spheres are themselves liquid. Taylor extended the case to a system where the suspending medium as well as the dispersed spheres are Newtonian liquids.[20,21] It was observed that when the radius of the drop is great enough or the rate of distortion is high, the drop breaks-up. Taylor derived the following two equations:

$$Ca = \eta_m R\dot{\gamma} / \sigma \qquad \text{(Eq. 1)}$$

$$E = Ca\left[(19p + 16)/(16p + 16)\right] \qquad \text{(Eq.2)}$$

From the above equations it is possible to calculate the size of the largest drop that exists in a fluid undergoing distortion at any shear rate. In these equations the governing parameters for droplet beak-up are the viscosity ratio p (viscosity of the dispersed phase to that of the matrix), the type of flow (elongational, shear, combined, etc.), the capillary number Ca which is the ratio between the deforming stress $\eta_m\dot{\gamma}$ (matrix viscosity x shear rate) imposed by the flow on the droplet and the interfacial forces σ / R, where σ is the interfacial tension and R is the radius of the drop. If Ca is small, the interfacial forces dominate and a steady drop shape develops. Upon exceeding a critical value Ca crit., the drop becomes unstable and finally breaks-up. A dimensionless parameter, E, (Eq.2) was derived which allows to describe dispersed-particle disintegration for Newtonian systems in shear-flow fields. According to Taylor, the apparent deformation D (Eq.3) of the droplet has a value of 0.5, where L is the length of the particle and B is the breadth:

$$D = (L - B)/(L + B) \qquad \text{(Eq.3)}$$

It has been demonstrated experimentally by Taylor that for values of p from 0.1 to 1, droplet break-up occurred at D values between 0.5 and 0.6. The expression of E in equation 2 indicates that the viscosity ratio, the shear stress, the droplet diameter, and the interfacial tension are critical variables to consider in controlling particle deformation and break-up in Newtonian fluids. In that equation, however, the coalescence, which has been later found to be critical in a break-up process, has not been considered. Grace[20] has constructed a plot of the critical capillary number as a function of the viscosity ratio, p, under two types of flow ; a simple shear flow and a hyperbolic (elongational) flow field (Fig. 1). It is shown that droplets are stable when their Ca number is below a critical value, the deformation and break-up is easier at p within 0.25-1 range for shear flow and the elongational flow field is more effective for break-up and dispersion than the shear flow. It can also be seen that at a viscosity ratio $p > 4 - 5$, it is not possible to break-up the drop in simple shear flow.

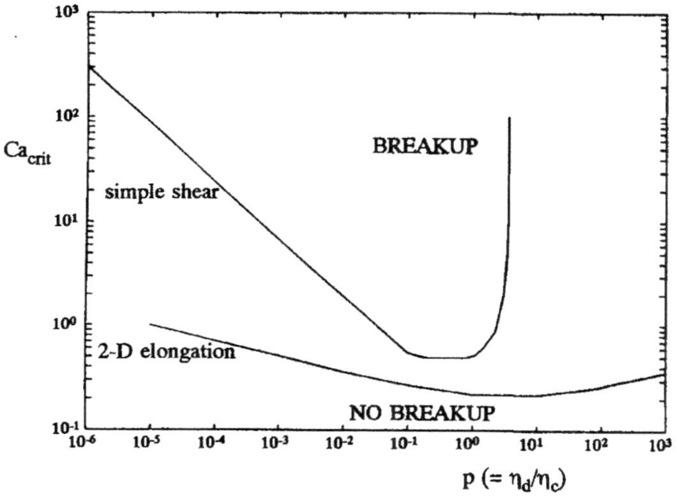

Figure 1. Critical capillary number versus the viscosity ratio in simple shear and plane hyperbolic (two-dimensional elongational) flow. Ref [20].

2.1.2 Effect of component characteristics on phase morphology

Favis[21] has summarized in a well structured chapter the effect of the viscosity ratio, blend composition, elasticity, shear stress and interfacial modification in immiscible blends on the final phase morphology.

2.1.2.1 Viscosity ratio

In a binary blend, the ratio of the viscosity of the dispersed phase to that of the matrix is one of the most critical variables for controlling the phase morphology. In many situations if the minor phase has a lower viscosity than the major one, the minor phase will be finely dispersed. In contrast, if the major phase exhibits a lower viscosity than the minor phase, a coarse dispersed phase will be developed. In the table 2 some examples of results reported in literature on the effect of the viscosity ratio, on the particle size of the dispersed phase in a binary blend are given.

Table 2. Examples of reported results on the effect of viscosity ratio on phase morphology of various blends

Affecting variable	Reported results	Authors and references
Viscosity ratio	A linear relationship between the particle size and the viscosity ratio in a blend of EPDM dispersed in PP matrix	Karger-Kocsis et al.[22]
	Particle size in EPDM/polybutadiene blend increases with the ratio of the mixing viscosities (torque ratio) of the components (phase morphology characterized via TEM)	Avgeropolous et al.[23]
	Drops do not break-up at a viscosity ratio $p > 3.7$ (see fig. 1)	Mason[24], Grace[20]
	- Dispersed rubber phase in polyamide matrix was shown to break-up during twin screw extrusion (i.e. complex flow field) even when $p > 4$. - as the viscosity moves away from a ratio of 1, the particles become larger.	Wu[25]
	Polycarbonate / Polypropylene blends in internal mixer - Significant particle disintegration occured even at a torque ratio (torque of dispersed phase/torque of the matrix) equal to 13. - Particle size increases by a factor of 3 to 4 as the torque ratio changes from 2 to 13. - the minimum particle size was reached at a torque ratio of 0.25.	Favis and Chalifoux[26]

2.1.2.2 Blend composition

In a binary blend A / B, the phase morphology can be of an A-droplet-in-B-matrix when A is minor, a B-droplet-in-A-matrix when B is minor, and a third particular situation in between where the two phases A and B are co-continuous.

As the output phase morphology in a blending operation is a balance between particle break-up and coalescence, increasing the concentration of the A component in an A/B blend will favor coalescence because of the increased particle to particle collision probability resulting in an increased particle size. Coalescence has been reported to start at concentration of the dispersed phase as low as 1 wt %[27]. In Fig 2 the particle size dependence on blend composition is illustrated over the whole composition range for a PC-PP blend.[28] At low PC content, a gradual increase in particle size is observed as a function of increasing PC content. Around 40 wt % PC, the dimensions of the particles become very sensitive to the concentration. Whereas within a concentration range of 45 - 55 wt % PC, a phase inversion region is delimited where phase co-continuity occurs at about 50 wt %.

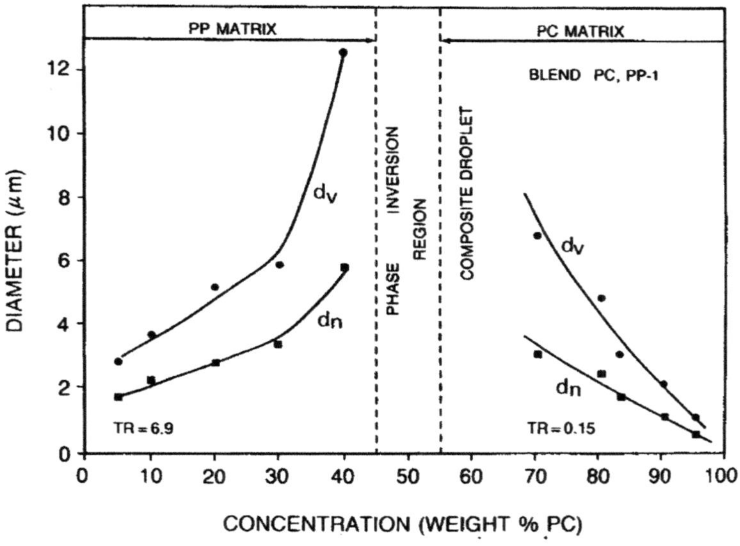

Figure 2..Dependence of phase size - d_n (■) and d_v (●) – on composition for a blend of PC and PP. Ref. [28].

Several authors have developed empirical equations to describe the effect of coalescence in the development of phase morphology. Equation (4) relates the particle size of the dispersed phase at equilibrium to its composition in the blend.[29] This equilibrium equation was derived from a more complex expression where a rate constant for breaking the drops and one for their coalescence have been defined to account for the continuous process of phase morphology development resulting from a competition between break-up and coalescence.

$$d_e \approx \frac{24P_r\sigma}{\pi\tau_{12}}\left(\Phi_d + \frac{4P_rE_{DK}}{\pi\tau_{12}}\Phi_d^2\right) \qquad \text{(Eq 4)}$$

τ_{12} is the shear stress, σ is the interfacial tension, Φ_d is the volume fraction of the dispersed phase, E_{DK} is the bulk breaking energy, and P_r is the probability that a collision between two close particles will result in coalescence and thus a particle size increase. It is clear from the expression that the particle size at equilibrium diminishes as the shear stress increases, the interfacial tension decreases, and the volume fraction of the dispersed phase decreases. This theory has been verified experimentally for several immiscible blends where a master curve of particle size as a function of composition was found to follow a Φ_d +k Φ_d^2 dependence.[30] Fortelny et al.[31] have proposed an expression (Eq 5) which accounts for drop break-up (first term in Eq. 5) and coalescence process (second term in Eq. 5):

$$R = (\sigma_{12}(We)_c/(\eta_m\gamma)) \times (\sigma_{12}\alpha\phi/(\eta_m f_1)) \qquad \text{(Eq 5)}$$

where R is the particle radius, $(We)_c$ is the critical Weber (or capillary) number, η_m the viscosity of the matrix, σ_{12} the interfacial tension, γ the shear rate, α the probability of coalescence after collision, ϕ the volume fraction of the dispersed phase, and f_1 the slope of the function F (We) which describes the frequency of phase break-up at the critical Weber number.

Elmendorp et al. also developed an expression to describe shear-induced coalescence of spherical particles.[27] A critical coalescence time has been defined which accounts for the time that passes between the arrival of a droplet at a liquid-liquid interface and the rupture of the intervening film. The critical coalescence time t_c has been defined as:

$$t_c = (3\eta_m R/2\sigma)\ln(R/2h_c) \qquad \text{(Eq 6)}$$

The above discussion clearly reveals that particle size versus composition dependence which is affected by coalescence is controlled to a large extent by the interfacial tension between the two phases of a binary blend. Several reports showed that the coalescence can be suppressed or at least its role minimized over a wide composition range. e.g. Favis et al. have demonstrated that in a PVC/PE blend of a low interfacial tension of 3.4 mN/m, the particle size reduction (suppressed coalescence) is much more significant than in a PA/PE blend having an interfacial tension as high as 14-18 mN/m).[32-34]

The interfacial tension in immiscible polymer blends is reduced by the addition of suitable interfacial agents such as block or graft copolymers, pre-formed or generated in-situ during the blending or the processing operation. The action of adding these agents to immiscible blends in an attempt to decrease their interfacial tension and reduce the coalescence phenomena during phase morphology development is called compatibilization. Literature reporting on the compatibilization of polymer blends is too enormous to be exclusively listed in this section. Addition of hydrogenated polybutadiene-b-polystyrene block copolymers as compatibilizing agents to polyethylene/polystyrene immiscible blends has been extensively investigated by Teyssié et al.[35-39] Diblock copolymers were found to be more efficient than graft, triblock or star-shaped copolymers. In these studies the most efficient interfacial agents were reported to be diblock copolymers of a well balanced composition. Stable dispersions were produced by adding as low as 1-2 wt % of an efficient copolymer.

In addition to interfacial tension, composition and viscosity, the elasticity of the components has also been considered in the developement of the phase morphology. The role of this parameter is the least understood among all parameters considered. Van Oene[40] is the first who pointed out that in capillary flow the dispersion of particles undergoes stratification and droplet-fiber formation. In addition to the role of the particle size and the interfacial tension, these two morphologies were also controlled by the differences in viscoelastic properties between the two phases. e.g. in a PS /PMMA blend, the PMMA phase, having the largest normal stress function (more elastic than PS) exhibits a droplet morphology type in the PS matrix. After the addition of a low molecular weight PMMA to the blend, a stratified PMMA phase was obtained. It was also noted that when the droplet size was smaller than 1 μm, the difference in morphologies (stratified versus droplet type) disappeared. It was concluded that the elastic contribution to the interfacial tension was not anymore dominant. The statement of Van Oene is that the phase of higher elasticity has the tendency to encapsulate the one with lower elasticity. He showed that it is difficult to deform a highly elastic material. He could then develop an expression where the contribution of material elasticity to the interfacial tension is considered (Eq 7).

$$\sigma_{eff} = \sigma + \frac{d}{12}(N_{2d} - N_{2m}) \qquad \text{(Eq 7)}$$

where σ_{eff} is the effective interfacial tension under dynamic conditions, σ is the static interfacial tension, d is the droplet diameter and N_{2d} and N_{2m}, are the second normal stress functions for the dispersed phase and for the matrix, respectively. Furthermore, it has been also shown that in an extensional flow field when a Newtonian droplet is deformed by a viscoelastic matrix, no lower limiting droplet size exists beyond which disruption of particles becomes impossible.[41,42] Elmendorp[43] demonstrated experimentally that the normal stress exhibited by droplet in a viscoelastic fluid stabilizes it, as predicted by Van Oene. Levitt et al.[44] have observed that polypropylene droplets were elongated perpendicular to the flow direction in a polystyrene matrix which was highly elastic. The extent of particle stretching in the perpendicular direction of flow was found to be proportional to the normal stress differences between the phases. Particle contraction was observed upon cessation of the shearing action which confirms the role of the elasticity on deformation.

The shear stress, $\tau_{12} = \eta_m \dot{\gamma}$ ($\tau_{12} = \eta_m \dot{\gamma}$), which is a machine parameter has a direct effect on particle deformation. The particle size is inversely proportional to the applied shear stress. Increasing the shear stress results in a particle size reduction. This interrelation has been verified unambiguously in a polystyrene / polyethylene immiscible blend.[45] Its has been shown that increasing the shear stress resulted in a much finer particle size, and also that the viscosity ratio was less predominant compared to the shear stress. The morphology was not influenced by variations in molecular weight of polyethylene or by increase in the mixing temperature. Other authors reported that a variation of the shear stress by a factor of two or three does not influence significantly the particle size.[26,46,47] Apparently, the variation of shear stress should be much higher than that in order to be effective in affecting predominantly the phase morphology size.

The shear stress was reported to have an effect on the shape of the dispersed phase. In a polyamide/polyethylene blend extruded through a slit die, an increase of shear stress from 17 kPa to 29 kPa resulted in a change of the nylon phase from a spherical to a fiber-like particle shape.[48]

2.2 Main types of phase morphologies : effect of material characteristics and mixing conditions

There exists in polymer blends three major types of phase morphologies; a particle (droplet, fiber, platelet, etc..)-in-matrix, a droplet-in-droplet-in-matrix (composite or subinclusion morphology) and a co-continuous phase morphology where both phases are mutually inter-connected throughout the whole volume of the blend.

The mechanism and the parameters that control the droplet formation in a binary immiscible blend has been discussed in the section above. The particle size of the dispersed phase in a polymer blend is measured on pictures observed by microscopy. For well mixed blends, the dispersed phase exhibits log-normal behavior.[25] That is, the observed frequency of size versus the log of the diameter of the particles results in a normal distribution. This distribution does not appear to be substantially altered by changes in interfacial energy or compatibilization (copolymer formation), although there may be an effect on the mean size of the particles.

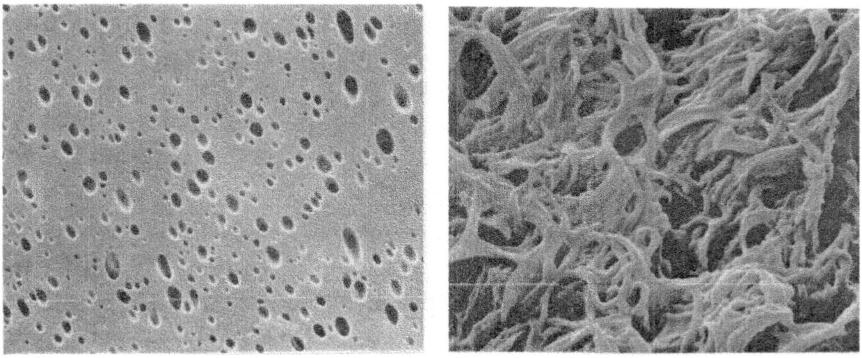

Figure 3. SEM pictures of (a) left: 80 LDPE / 20 PS dispersed phase morphology; cryo-smoohed surface and solvent etched minor PS phase; (b) right: 20 LDPE / 80 PS co-continuous phase morphology; polystyrene major phase was extracted using THF solvent.

In many melt-mixing devices not all of the resin may experience the same shear history, and bimodal or multimodal distributions may be generated. Deviation from log-normal behavior may be provoked by re-processing of instabilized blends or by segregation effects. The most complete description of the dispersed phase size is contained in the frequency-distribution histogram or the cumulative frequency histogram. It is often advantageous to construct independent plots based on both the number fractions and the weight fractions of particles in each group. The number fraction provides the most important information about the particle size on the small end of the distribution curve, while the weight fraction reveals the importance of larger particles. To characterize a droplet-in-matrix phase morphology, it is more judicious to smoothen the surface at a temperature below the glass transition temperature of the matrix, then, etch the particles using a selective solvent. In figure 3 SEM pictures of polystyrene / polyethylene blends are presented where polystyrene particles have been removed by etching using tetrahydrofuran, a selective solvent for polystyrene. For more details on this topic, a good literature survey on microscopic characterization of multiphase polymer blends has been reported by Hobbs.[49]

Another phenomenon which has to be considered in the phase-morphology development is the transition from a droplet to a fiber-like dispersion. This transformation depends on the extent of the deforming forces, the capillary instabilities, the coalescence as well as on the interfacial tension between the phases. The type of flow to which a particle is subjected, whether it is a shear flow or an elongational flow field is a crucial parameter to consider when targeting a fiber-like dispersion. Elongational flow fields are more effective in inducing a droplet-fiber transition. Fibers were reported to be produced at the entrance of a capillary where elongational flow is generated.[50] Uniaxial melt drawing which generates an elongational flow field was also reported to generate fiber dispersions.[48,51] The coalescence of particles in a concentrated dispersed phase system where particle-particle interactions are intense can also generate fibers. This phenomenon was demonstrated mainly in the entrance zone of a capillary rheometer.[52,53]

The droplet-in-droplet-in-matrix or composite droplet morphology is a structure well known in high impact polystyrene (HIPS). It is composed of polystyrene matrix and minor rubber phase which contains sub-inclusions of polystyrene. This morphology results from the polymerization of a mixture of 5 to 10 % of polybutadiene in styrene monomer.[54] Polystyrene sub-inclusions are entrapped within polybutadiene particles after polymerization. Polyamide 6 was reported to be toughened by a composite dispersed phase of styrene-ethylene-butylene-styrene (SEBS) block copolymer and polyphenylene oxide (PPO) as a thermoplastic.[55]

Hobbs and coworkers reported about the spontaneous development of a composite droplet morphology of a series of ternary blends.[56] This investigation highlighted the phase morphology when three components, having by pairs different interfacial tension, were melt-blended together. The encapsulation effect responsible for the generation of composite droplets was explained in terms of the differences in interfacial tensions between each two components. It was deduced that the subinclusion formation could be predicted using Harkin's equation (Eq 8),

$$\lambda_{31} = \sigma_{12} + \sigma_{32} + \sigma_{13} \tag{Eq 8}$$

where λ_{31} is the spreading coefficient for component 3 to encapsulate component 1 and the three terms to the right are the interfacial tensions between the respective 1,2; 3,2 and 1,3 polymers. When λ_{31} is positive, encapsulation of component 3 about component 1 will take place. Compatibilization of one pair has been shown to result in a composite droplet morphology, for which the third component forms subinclusions in the dispersed phase.[57,58]

The mixing conditions play a crucial role in the development of phase morphology. These include the mixing equipment and the mixing parameters such as the mixing temperature, the mixing time and the rotation speed of the mixing elements. Three major types of equipment are employed in the processing of polymer blends, internal mixers and extruders (single and twin screw extruders), and injection molding which is used to shape an already compounded blend. These devices are equipped with screws or rotor blades revolving in a closed chamber or a barrel, respectively. The design of the screws or the rotor blades conditions the type of flow fields which, as has been mentioned above, are key parameters in phase morphology development.

Internal mixers are mainly used for rubber-based blends but they are also useful on a laboratory scale for blending a wide range of thermoplastics. They are practical for blending small amounts of material and allow a monitoring of torque (a viscosity dependent parameter). Favis has reported about the influence of the mixing time on the size of the dispersed phase in a Brabender mixer[46]. It was found that for highly immiscible polycarbonate/ polypropylene blends, the most significant particle size reduction or phase morphology generation took place within the first two minutes of the mixing operation. Between 2 and 20 minutes of mixing time, very little particle size reduction of the dispersed phase was noticed. An increase in rotor speed had only a minor effect on futher size reduction. Similar observations were also made by Schreiber and Olguin in polyolefin / elastomer blends prepared using internal mixer.[59]

Single-screw extruders, although limited in developing high shear rates and good distributive mixing, remain, nevertheless a widespread tool for blending polymers. Lindt et al. have studied phase morphology evolution in a single-screw extruder. It was revealed that the blend undergoes a continuous change in its developing morphology during the passage through the melting zone. Within a fraction of a second, the length scale of the dispersed phase drops by several orders of magnitude.[60]

The most appropriate mixing tool for polymer blends remains the twin-screw extruder. Corotating and counter-rotating twin-screw extruders are high-intensity mixing devices consisting of two screws with a kneading section for intensive mixing. An entire screw can be built using a set of kneading elements to meet specific mixing requirements. Many studies were devoted to the comparison of the mixing efficiency between a static mixer and a twin-screw extruder[61,62]; the static mixer produced a coarser dispersion than the twin-screw extruder (factor of two has been reported). The dependence of phase size on the viscosity ratio for blends prepared in a twin-screw extruder was found to resemble that observed for Newtonian fluids in an elongational flow field. Detailed aspects on phase morphology development in a twin-screw extruder is well documented elsewhere.[21]

3. COMPATIBILIZATION OF TWO-PHASE POLYMER BLENDS

There exists two categories of polymer blends : miscible and immiscible. The miscible blends are thermodynamically soluble molecular mixtures and are composed of one phase having a single glass transion temperature. The properties of miscible blends can often be predicted from the composition weighted average of the properties of the individual components. In contrast, immiscible blends are phase separated, with each phase recovering the own glass-transition temperature of the corresponding component. The overall performance of immiscible blends depends on the properties of the individual components, the interfacial adhesion between the phases and also significantly on the phase morphology developed. The properties of an immiscible blend cannot be easily predicted from the properties of the individual component as it is the case for miscible blends. To control the phase morphology, the interfacial adhesion between the phases and promote a stable (against coalescence during reprocessing) blend, immiscible blends have to be compatibilized. A compatible blend is characterized by a finely dispersed phase, good adhesion between the blend phases and strong resistance to phase coalescence, and as a result, a blend of technologically desirable properties is produced.

There are several methods of compatibilization of immiscible polymer blends. These can be grouped under two main strategies, physical compatibilization and reactive compatibilization. The first one consists of the addition of a pre-formed suitable substance such as a copolymer or a coupling agent to the immiscible blend, whereas, the second strategy involves the generation of the compatibilizer *in-situ* during the melt-processing of the blend. Excellent literature reviews are available on the compatibilization of multiphase polymer blends.[63-72]

3.1 Physical compatibilization via addition of preformed block or graft copolymers

The preformed block or graft copolymer used as compatibilizing agent contains a minimum of two different segments. Each segment is miscible or at least has sufficient chemical interactions with one blend component. The copolymer segments are not necessarily identical to the blend components. Copolymer A-C can compatibilize an A/B immiscible blend, provided that segment C is miscible or exhibits sufficient interactions with B component. An A/B immiscible blend can be compatibilized using a copolymer C-D provided that C and D segments are miscible or have sufficient chemical interactions with components A and B, respectively.

A sufficient amount of the added copolymer is expected to locate at the interface between the immiscible blend phases, reduces the interfacial tension between the blend components and, as a consequence, facilitates the particle break-up and is also able to stabilize the dispersion against coalescence. If mechanical performance is targeted, the copolymer should be carefully selected to promote adhesion between the blend phases. The molecular weight of the segments of the copolymer should be larger or equal to a critical molecular weight which induces molecular entanglements. Fayt and co-workers have compared the compatibilization efficiency of preformed compatibilizers in improving the ultimate mechanical properties of immiscible blends.[73-80] The reported conclusions state that a block copolymer is more efficient than a graft copolymer, a diblock copolymer is more effective than a triblock or star-shaped copolymer and, a tapered diblock copolymer is superior to a pure diblock of the same composition and molecular weight. A preformed block or graft copolymer used as compatibilizing agent in immiscible polymer blends suffers from a lack of quantitative location at the interface between the phases. Because of the high viscosity of the melt-mixed medium, the copolymer forms micelles in one of the pure phases where it is the more stable.[81-83] In table 3 are listed a series of blends compatibilized using preformed copolymers.

Table 3. Blends compatibilized by pre-made copolymers

Blend components	Major component	Minor component	Pre-made compatibilizer	References
A/B/A-B	PE or PS	PS or PE	HPB-b-PS, HPB-b-PI-b-PS, SBS, SEBS	73-80
	PE or HIPS	PE or HIPS	HPB-b-PS	84, 85
	PBT or PS	PBT or PS	PS-b-PET or PS-b-PBT	86
A/B/A-C block	PPO or PBT	PPO or PBT	PS-b-PET or PS-b-PBT	86
	SAN	SBS (Solprene)	PS-b-PMMA or PB-b-	80, 87
	PVF_2 or PMS	PMS or PVF_2	PMMA	88
	PVF_2	HIPS	PMS-b-PMMA	80, 89
	PVF_2	PE or PP	PS-b-PMMA	90
	PET or PS	PS or PET	HPB-b-PMMA, HPI-b-	91
	PS/PPO or PBT	PBT or PS/PPO	PMMA	86
	PVC	PS	PS-b-PCL	80, 92, 93
	ABS	PE	PS-b-PET or PS-b-PBT	94
	PVF_2	PS	PCL-b-PS or PMMA-	80
	PVC	SEBS	b-PS	80
	PET	HDPE	HPB-b-PMMA	95, 96
	PS or PC	PC or PS	PS-b-PMMA	97
A/B/C-D	Phenoxy	PPO	HPB-b-PMMA	98
	Surlyn	PPO	S-EB-S	99
	PPO	SAN	PS-b-PCL	100
	SAN	PPO	PS-b-PMMA	101
A/B/A-B graft	EPDM	PMMA	PS-b-PVP	102
	PF	PMMA or PS	PS-b-PMMA	103
	PBT	PS	PS-b-PMMA	104
	PS or PE	PE or PS	EPDM-g-MMA	105-107
			PF-g-MMA or PF-g-PS	
			PBT-g-PS	
			PS-g-PE	

3.2 Reactive compatibilization

The reactive compatibilization process is a sub-category of the broader class of interchain copolymer formation reactions performed by reactive extrusion[69]. In this case compatibilization is ensured via the *in-situ* formation of a block or a graft copolymer during the process of melt-mixing, through an interfacial chemical reaction between functional groups available on the polymer chains in the blend system.[108-118] The functional groups may be those naturally occurring in polycondensation polymers such as polyesters or polyamides or added via a grafting reaction along the chain for addition polymers such as polyolefins and vinylics. The functional groups should be selected so that the interfacial reaction occurs within the time frame tolerated for the processing operation (in the order of a minute in extruders). The generated interchain bonding must be stable enough to survive the thermal and shearing treatment during the process of blending. Because of the limited yield of the interfacial reaction due to the short extrusion time and low molecular diffusion (high viscosity of the reaction medium), highly reactive groups are required. In some situations higher content of functional species are added in order to ensure the minimum yield required for blend performance. In this case, the reactive groups which are not converted during the reactive compatibilization can further react during subsequent processing (re-extrusion or injection molding), which might result in deleterious consequences. This clearly means that the kinetics and yield aspects of the interfacial reaction in reactive compatibilization are of utmost importance, which explains why each blend system has its own 'pack' of experimental conditions such as mixing time, mixing temperature, screw design, molecular weight and reactive group content of the precursors. Although the experimental tuning is severe, the control of the molecular structure of the compatibilizer generated *in-situ* remains qualitative. Reactive compatibilization has several advantages, mostly economical, over the physical compatibilization :

a. The copolymer is made as needed during the melt-blending process and a separate commercialization of a copolymer is not required.
b. The copolymer is formed directly at the interface between the immiscible polymers where it is needed to stabilize the phase morphology developed. In contrast, when a compatibilizing copolymer is added as a separate entity to a blend, it must overcome the viscous forces and diffuse to its expected location at polymer-polymer interface. It may, however, form micelles as a separate phase that is useless for compatibilization.
c. A second fundamental advantage of *in-situ* copolymer formation is that the molecular weight of each of the two distinct polymeric segments in the copolymer is usually the same as that of the individual bulk polymer phase in which the segment must dissolve allowing, thus, for a maximum interfacial adhesion.

The main disadvantage of reactive blending resides in the need to have reactive functional groups on the polymers to be compatibilized.

3.2.1 Functionalization of polymers

There exists a variety of ways to incorporate useful functional groups into polymers for reactive compatibilization. Maleic anhydride has been introduced into various polymers by grafting[119-123], by copolymerization via radical polymerization[124], by end-capping[125] or by more specific techniques.[126] The living anionic polymerization is appropriate for the end-functionalization of anionically polymerizable polymers. Macosko et al. described a synthesis procedure to prepare polymers end-capped by NH_2,

OH, SH, COOH, CHO, and epoxy groups.[127] This synthesis control allowed to compare the reactivity in melt-processing of various pairs of reactive groups such as epoxide/COOH and, aromatic and aliphatic NH_2/ anhydride.[128] As a result of that the pair of primary aliphatic amine with cyclic anhydride was found to be the fastest and the most efficient reaction.[129,130] The most commonly used pairs of reactive groups in reactive compatibilization are given in figure 4.

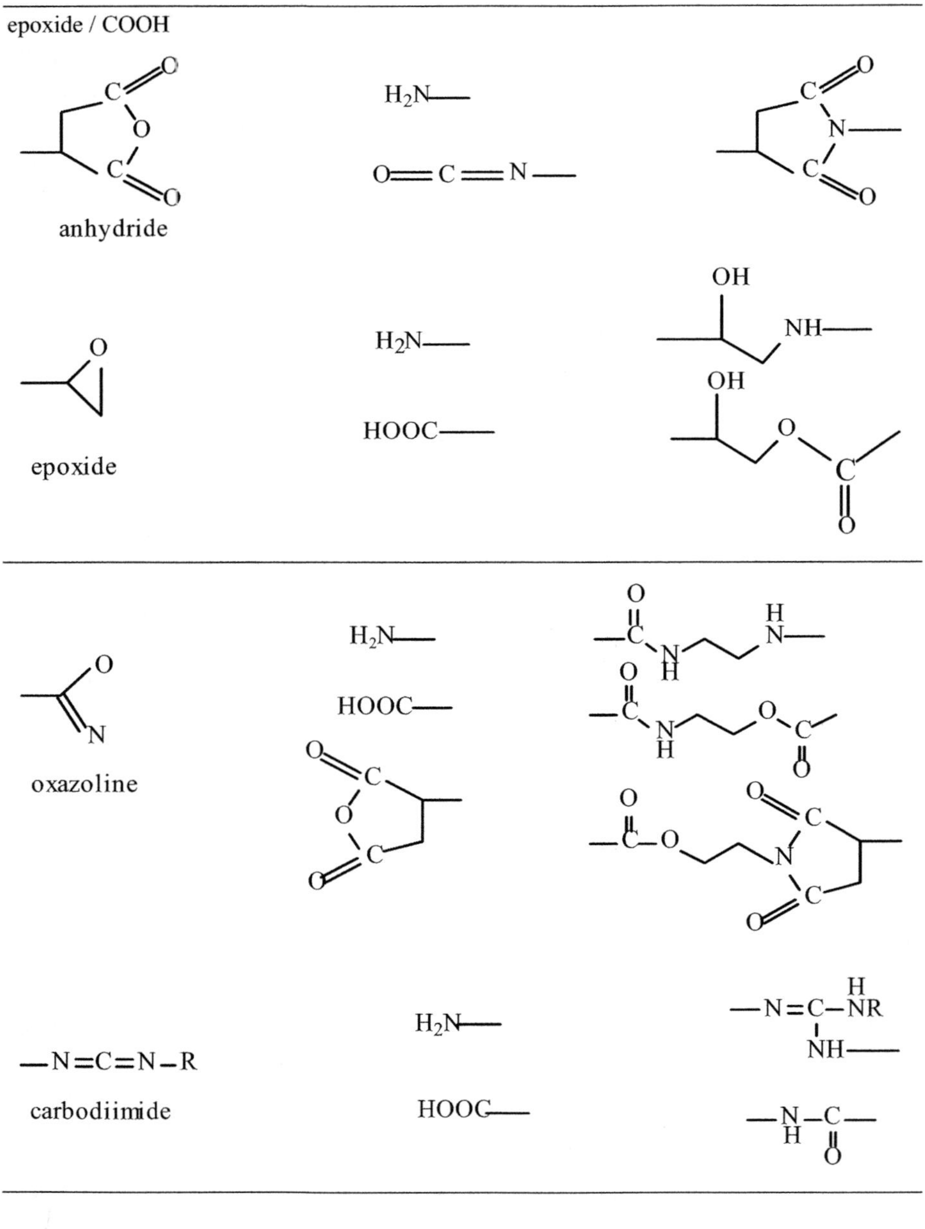

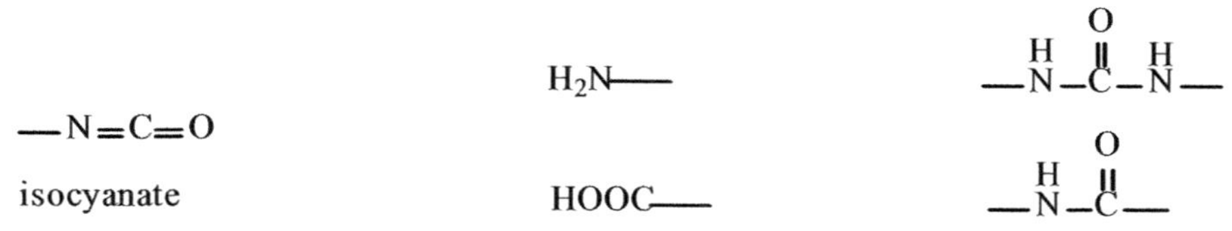

Figure 4. Reactive groups commonly encountered in reactive compatibilization

3.2.2 Types of interfacial chemical reactions involved

Depending on the functionalization of the reactive precursor, a variety of situations can be distinguished with respect to the in-situ formed copolymer. Brown[131] has detailed in a very structured and a complete table the various types of copolymers formed in-situ during reactive compatibilization. It is noteworthy to reproduce below in table 4 a summarized version of the content of that classification.

Table 4. Copolymer formation processes in extruder reactors

Process	Type of Copolymer Obtained
Redistribution "Transreaction"	
a. By reactive end-groups of one polymer A attacking main chain of a second polymer B	Block + Random AAAAABBBBB + AABBBBAAA+ AABABBA, etc
b. By chain cleavage / recombination involving each polymer	Same as above
Graft Copolymer	
c. By direct reaction of end-group of polymer A with pendent group of polymer B.	BBBBBBBB A A A A A A
d. By reaction of end-group of polymer A with pendent group of polymer B in the presence of a condensing agent.	

e.	By reaction of end-group of polymer A with pendent group of polymer B in the presence of a coupling agent 'c'.	BBBBBBBBBB c c A A A A A A
f.	By reaction of pendent groups of polymer A with main chain of polymer B in a degradative process.	
Block Copolymer Formation		
g.	By direct reaction of end-group of polymer A with end-group of polymer B	AAAAA-BBBBB
h.	By reaction of end-group of polymer A with end-group of polymer B in the presence of a condensing agent	AAAAA-BBBBB
i.	By reaction of end-group of polymer A with end-group of polymer B in the presence of a coupling agent 'c'.	AAAAA-c-BBBBB
j.	By reaction of end-group of polymer A with main chain of polymer B in degradative process.	AAAAA-BBBB + BBB
Crosslinked Copolymer Formation		
k.	Covalent crosslinking by direct reaction of pendent functionality of polymer A with pendent fuctionality of polymer B	crosslinked
l.	Covalent crosslinking by reaction of pendent functionality of polymer A with pendent functionality of polymer B in the presence of a condensing agent.	crosslinked

m.	Main chain of polymer A reacts with main chain of polymer B in the presence of a radical initiator	crosslinked
n.	Covalent crosslinking by reaction of pendent functionality of polymer A with pendent functinality of polymer B in the presence of a coupling agent 'c'.	crosslinked
Ionic Bond Formation		
o.	Ion-ion association mediated by metal cations as linking agents 'c'	Block, graft or crosslinked
p.	Ion-neutral donor group association mediated by metal cations	
q.	Ion-ion association mediated by interchain protonation of a basic polymer by an acidic polymer	

In the same report a series of reactive blends, corresponding to the various types of reactions, redistribution reaction, graft copolymer formation, block copolymer formation, covalent crosslinking reaction and ionic interactions have been listed together with the processing conditions used for their preparation and the characterization methods used to probe the formation of the compatibilizing entity.

In contrast to physical compatibilization, the phase morphology development in reactive compatibilization is a dynamic process which depends on a very broad spectrum of input parameters; most of these are molecular, directly involved in the copolymer generation. As more surface area is created by breaking down the dispersed phase through interfacial reactions, the probability becomes higher for functional groups to meet each other and react to form more copolymer, which reduces further the phase size via interfacial tension reduction and barrier action against coalescence.

3.2.3 Copolymer formation and compatibilization activity

3.2.3.1 Effect of the miscibility of the compatibilizer precursor with the phases

The level of miscibility of the precursor with the phases will certainly affect the final location of the *in-situ* formed copolymer during the reactive compatibilization. As shown in the schematic representation of figure 5 at least three situations can be envisionned:

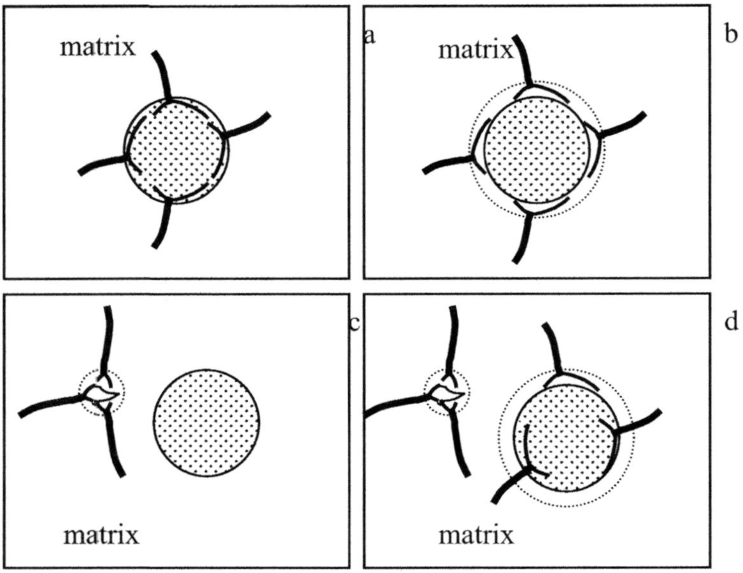

Figure 5. Schematic representation of the possibilities of copolymer location in a reactively compatibilized blend depending on the level of its interaction with the dispersed phase.

1. In figure 5a the precursor is miscible with the dispersed phase where it is initially incorporated. In this case the chemical links formed with the matrix of *in-situ* formed copolymer resides at the interface whereas the segment of the precursor are totally located in the dispersed phase. This location is expected to result in the most efficient compatibilizing activity (reduced size of the dispersed phase and increased barrier against coalescence).
2. The second possibility illustrated in figure 5b shows the case where the precursor resides at the interface between the phases with more affinity to the dispersed phase. Here the formed copolymer will certainly perform

its compatibilizing activity but is less efficient in preventing coalescence; copolymer mobility and reorganization or rejection from the interface is expected.

3. The third case illustrates the miscibility of the precursor with matrix phase. The formed copolymer will be located preferentially in the matrix away from the interface. In this blend the interface will be weaker than in the above two cases. The formation of the highly grafted copolymer dispersed in the matrix will increase its viscosity and hence can affect the phase morphology via a rheological modification of the conditions of phase break-up. In reality, the three above senarios can co-exist in a real reactively compatibilized blend as illustrated in figure 5d.

The copolymer location in relation to the level of interaction between the compatibilizer precursor and the phases has been reported for SAN dispersed phase in polyamide matrix using as compatibilizer precursor an imidized polymer.[132]

Another series of blends where the miscibility or immiscibility of the compatibilizer precursor with the phases to compatibilize is of a crucial importance is styrene-co-maleic anhydride copolymer used in PMMA / PA6, PPO/PA6 or in PS / PA6 blends.[133-137] SMA compatibilizers having 14, 17, 20, 25, 28 and 33 weight % of maleic anhydride have been used in PA6/ PMMA blends. With respect to their compatibilizing efficiency, it is very important to take into account the miscibility behaviour between SMA and PMMA. PMMA/ SMA blends always show an LCST behaviour. At the extrusion temperature of 240 °C, SMA is miscible with PMMA for MA contents between 10 and 35 wt %. As the cloud point of the PMMA/SMA blends depends on the molecular weight of PMMA and SMA and on the blend composition, the miscibility behaviour was checked for the components used.

The compatibilized blends PA6/(PMMA/SMA) under consideration have the composition 75/(20/5). The miscibility behaviour was thus analyzed for the PMMA/SMA blends in a weight ratio of 80 / 20; they were prepared at 240 °C in a mini-extruder. The transparency of the blends was considered as a first indication of miscibility. The blends of PMMA with SMA14, SMA17, SMA20, SMA25 and SMA28 were transparent after 3 minutes of extrusion. The blend PMMA/SMA33 (80/20) was opaque at short extrusion times but became transparent after longer extrusion times (10 min). The blend PMMA/SMA8 remained opaque, even after longer extrusion times (10min) indicating that this blend is not miscible at 240 °C. The miscibility of PMMA/SMA was evaluated using DSC on blend samples quenched in liquid nitrogen. The results of the DSC measurements are given in figure 6 which clearly shows that SMA is more miscible with PMMA at the midconcentration of about 20 wt % of maleic anhydride. The more one deviates from this concentration in either directions, low or high MA content, the less the system is miscible.

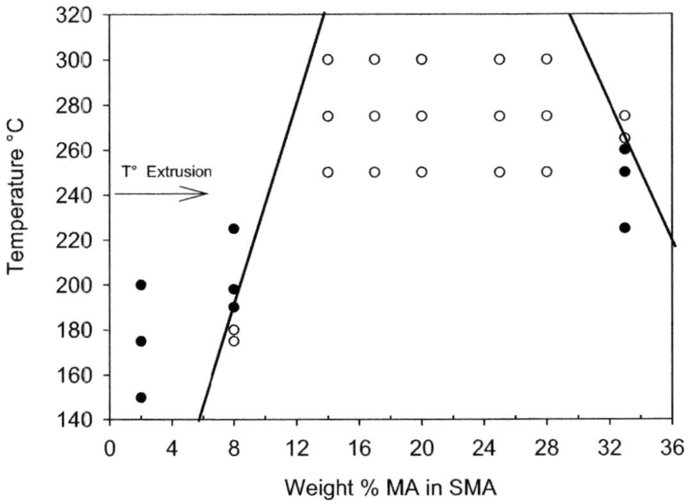

Figure 6. Miscibility of PMMA / SMA 80/20 blends as a function of the MA content in SMA (○) miscible; (●) immiscible

The weight average particle diameter of PMMA dispersed phase in SMA compatibilized PA6 / PMMA blends is plotted in figure 7 as a function of the extrusion time when SMA20, SMA28 and SMA33 were used. Compared to the case of SMA20, a completely different behaviour is observed with SMA28 and SMA33 as compatibilizers. The dispersed phase particle size can still decrease after 3 minutes of extrusion for the blends containing SMA28 or SMA33, while this is not the case for SMA20. This behaviour can be directly correlated with the extent of miscibility of SMA and the minor phase PMMA. Indeed, in the beginning of the melt-extrusion process, SMA-g-PA6 graft copolymer is formed at the interface causing a significant particle size decrease. As the compounding process continues further, the SMA copolymer reacts more quantitatively so that an unbalanced graft copolymer, richer in PA6, is built-up. It has a strong tendency to be driven out from the interface into the PA6 phase where it is expected to be the most stable. This is made easier when the extent of miscibility of the SMA (28 and 33 wt % MA) with PMMA does not 'counter-balance' the unfavorable compositional effect (a graft copolymer richer in PA6 exhibits more interaction with the PA6 matrix). This correlation is also illustrated in figure 8 with SMA copolymers covering a broader range of MA concentrations.

The weight average particle size of PMMA / PA6 blends is plotted as a function of the weight % of maleic anhydride of SMA copolymer after 20 minutes of extrusion time. The figure clearly reveals that for the compatibilizers closer to the boundaries of the miscibility region of PMMA and SMA (refer to fig. 6), a larger PMMA particle size is obtained.

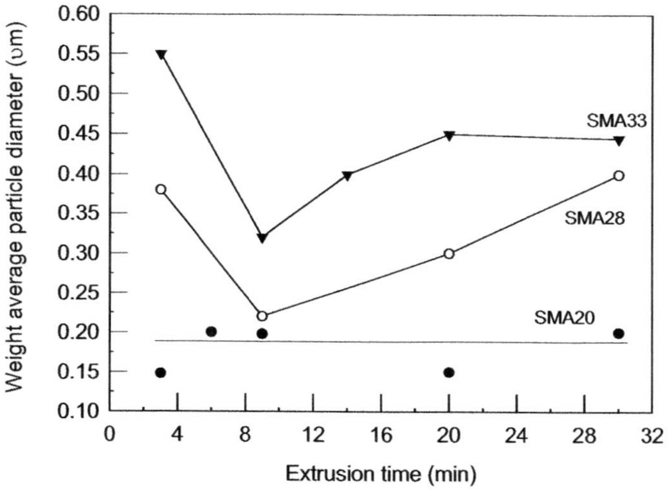

Figure 7. Weight average particle diameter as a function of the extrusion time for the blend PA6/(PMMA/SMA) 75/(20/5) with different MA contents for SMA

The unstability of the graft copolymer formed at the interface is also illustrated for the melt-blended and SMA compatibilized PA6/PPO blend system where the particle size of the PPO dispersed phase is plotted as a function of the extrusion time (Fig. 9). In this case SMA8 is less miscible with PPO than SMA2. As can be seen, the particle size of SMA8 modified blend is smaller than that of SMA2 modified blend at short extrusion times. The kinetics of interfacial reaction is faster with SMA8 containing four times more MA reactive groups than SMA2. Beyond 15 minutes of extrusion time, the *in-situ* formed SMA-g-PA6 copolymer becomes reacher in PA6 grafts causing an unbalanced structural design which drives the copolymer out off the interface into the PA6 matrix, forming micelles. The moderate miscibility of SMA8 with PPO could no more 'counter-balance' the strong interactions the copolymer develops with the PA6 matrix. As a result of this 'decompatibilization' process, the PPO dispersed phase coarsens via a coalescence process.

A third example of copolymer expelling or pull-out from the interface between the individual phases of the blend is illustrated for the SMA reactively compatibilized PA6/PS blend system. In this system the same SMA2 which is at the limit of miscibility with polystyrene homopolymer results in the *in-situ* formation of SMA-g-PA 6 compatibilizer which is moderately stable at the interface. Indeed, we could show by doubling the PA6 matrix molecular weight and therefore its melt-viscosity that the formed SMA-g-PA6 copolymer was really driven out of the interface by the larger shearing forces exerted on it. The shearing forces developed by the PA6 matrix of which the molecular weight is 44,000 were superior to the interactions the SMA has with the polystyrene minor phase. Contrary to the blend where the PA6 matrix has a molecular weight of 18,000, the blend where the molecular weight of PA6 is 44,000 exhibited a considerable particle coalescence process as a result of compatibilizer inefficiency. The graft copolymer is expelled from the interface forming micelles in the PA6 matrix, as can be seen from figure 10a. One would expect a dispersed phase size reduction (drop break-up process) when the matrix viscosity is higher (large MW) compared to a blend where the matrix viscosity is low, as is the case in figure 10b. The opposite effect is observed which clearly demonstrated the 'just stable' location of the reactively formed SMA-g-PA6 copolymer at the interface. A larger PA6 matrix viscosity completely alleviated the compatibilization efficiency and caused the opposite effect.

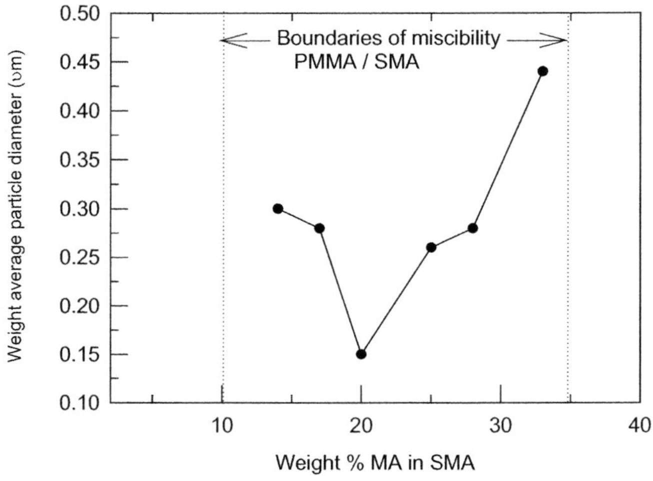

Figure 8. Weight average particle diameter of the blend PA6/(PMMA/SMA) 75/(20/5) as a function of the MA content in SMA. All blends contain 5 wt % of SMA and were compounded at 240 °C for 20 min.

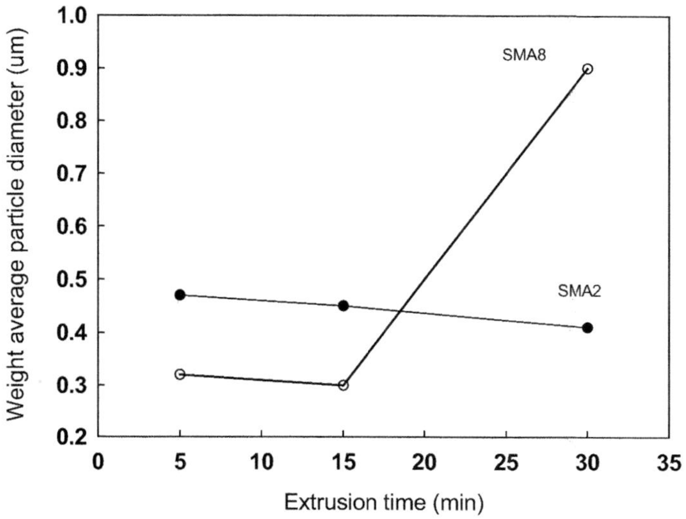

Figure 9. Weight average particle diameter of PA6/(PPO/SMA) blends in a weight ratio 75/(20/5) as a function of the extrusion time. (●) SMA2 ; (O) SMA8.

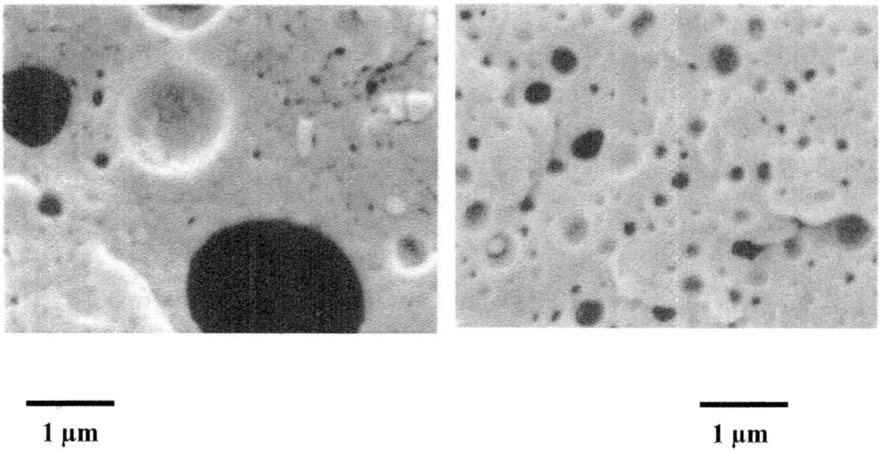

Figure 10. SEM of cryo-smoothed and THF etched PA6/(PS/SMA2) blend : 80/(16/4); left a) MW of PA6 = 44,000, right b) MW of PA6 = 18,000

3.2.3.2 Effect of the configuration of reactive end-groups

The most illustrative case of the effect of the chain end functionality on the phase morphology development are the significant differences observed by Keskkula and co-workers[138,139] in reactively compatibilized blends of nylon 6 having only one amine end-group or nylon 6,6 which is amino difunctional with a maleated SEBS copolymer. It was reported that in the nylon 6 matrix, SEBS particles were extremely small and essentially spherical, while in nylon 6,6, the rubber domains were much larger and more complex in shape. Indeed, the monofunctional chains of nylon 6 simply graft onto the maleated SEBS copolymer with one attachment per chain, while the difunctional chains of nylon 6,6 can form bridges between rubber particles or loops within a particle i.e. crosslinking effects, that make it difficult to break the rubber into fine particles.

When the reactive groups are incorporated along the chain, their distribution is also of an utmost importance in affecting the reactive compatibilization efficiency. Baker et al. have investigated the compatibilization activity of PP-g-GMA and PP-g-(S-ran-GMA) precursors in PP/Acrylonitrile-ran-butadiene-ran-acrylic acid) binary blend.[140] At a fixed content of GMA, the PP-g-GMA was more efficient in toughening PP. The observed efficiency was ascribed to the less restricted access of the acrylic acid groups to the GMA units in the grafted precursors than in the randomly copolymerized version with the bulky styrene comonomer PP-g-(S-ran-GMA).

In another example PS-g-GMA precursor was used to reactively compatiblize a PS with two types of reactive polyethylenes, PE-ran-acrylic acid and PE-graft-acrylic acid.[141] PE-ran-acrylic acid was found to efficienlty reduce the particle size compared to the grafted version , PE-graft-acrylic acid. The superiority of the randomly copolymerized version in terms of compatibilization activity was ascribed to the accessibility of the epoxy groups of the GMA units of PS-g-GMA to a maximum of acrylic acid units in the PE-ran-AA copolymer than in the PE-graft-AA one because of its AA phase separation.

3.2.3.3 Effect of the reactive group content of reactive polymers

The reactive group content will affect the extent of the interfacial compatibilization reaction and the molecular architecture of the *in-situ* formed compatibilizer. By increasing the number of the reactive groups on a reactive polymer from one to multiple groups, the structure of the graft changes from a single to a multiple-graft comb-like structure. The extent of interfacial reaction which is function of the content of the reactive groups available will affect the stability of the formed copolymer at the interface as well as its physical entanglement with the compatibilized phases.

De Roover reported that the particle size of polypropylene dispersed in polyamide 6 decreased drastically with increasing maleic anhydride content in PP-g-MA reactive precursor. A direct relationship between the particle size of dispersed polypropylene and the extent of the grafting reaction was proposed.[142]

Paul et al. also reported about the effect of maleic anhydride content on the phase morphology (mainly particle size and its distribution) of polypropylene dispersed in polyamide 6, as compatibilized by maleic anhydride grafted polypropylene.[143] It was clearly established that the size and size distribution of polypropylene phase depend on both the content of maleic anhydride in PP-g-MA and the extent of miscibility between polypropylene and PP-g-MA.

Pagnoulle et al. made an interesting review on the effect of the type and the content of reactive groups, amine or carbamate attached to SAN, on particle size and impact strength of reactively compatibilized SAN / EPR, where EPR is composed of 50:50 wt to wt of EPDM and EP-g-MA. When amine groups were used, the size of EPR rubber particles was independent of the amine group content. Amine groups were found to be 3.3 times more reactive towards maleic anhydride groups than carbamate groups. The impact strength was unaffected by a change in carbamate content, whereas it is significantly deteriorated upon increasing amine groups content in SAN. The differences between amine and carbamate reactives groups were explained by the molecular architecture of the *in-situ* formed SAN-g-EP copolymer resulting from the reaction of each of the two groups with maleic anhydride. The thermolysis of carbamate gives rise to an amine group which in turn reacts with maleic anhydride of EPR creating a new graft on SAN chain. It has been claimed that SAN-NH_2 precursor results in a copolymer having a more flat conformation at the interface and less tendency to be entangled with SAN matrix when the amine group content is increased. More entanglements were expected when the carbamate groups are used instead.[144,145]

4. DYNAMIC VULCANIZATION OF RUBBER / THERMOPLASTIC BLENDS WITH HIGH RUBBER CONTENT (TPVS)

4.1 Concept of dynamic vulcanization and particularities

Thermoplastic vulcanizates "TPVs" have been introduced in the beginning of the eighties as a special class of thermoplastic elastomers. Thermoplastic vulcanizates can be developed at much lower cost compared to the pure elastomeric block copolymers. They are binary blends of an elastomer, usually crosslinked, dispersed at a high concentration, usually more than 50 wt %, in a thermoplastic matrix. They can be considered as a new class of materials which combines the excellent processing characteristics of thermoplastics and the elastic properties of rubbers. These materials flow at elevated temperature and behave like an elastomer at room temperature. The performance of the thermoplastic vulcanizate is controlled via the selective vulcanization of the rubber phase under dynamic shear during the compounding process. This way a fine phase morphology of the rubber phase can be generated. Interfacial adhesion between the rubber phase and the thermoplastic matrix can be strengthened via the addition of a suitable compatibilizer.

Thermoplastic vulcanisates exhibit a full strain recovery as compared to the pure thermoplastics although their matrices are thermoplastic. They also exhibit higher ultimate tensile properties than pure rubbers. This characteristic stress-strain mechanical behaviour is illustrated in figure 14 in comparison to a pure rubber or a pure thermoplastic.

Dynamic vulcanization consists of the crosslinking of the rubber phase during its melt-mixing with the thermoplastic phase. It was first discovered by Gessler et al.[146] during the development of impact resistant polypropylene containing halo butyl rubber as a dispersed phase dynamically vulcanized using zinc oxide. Owing to the dynamic vulcanization, it is possible to disperse up to 70 wt % of the rubber phase in a very fine way into the thermoplastic matrix. Fischer was the first to patent an application of vulcanized EPDM dispersed in polypropylene matrix.[147] The more extended contributions in improving the properties of thermoplastic vulcanizates came from Coran et al.[148-156] A full control of the vulcanization process was achieved without affecting the processability of the thermoplastic matrix

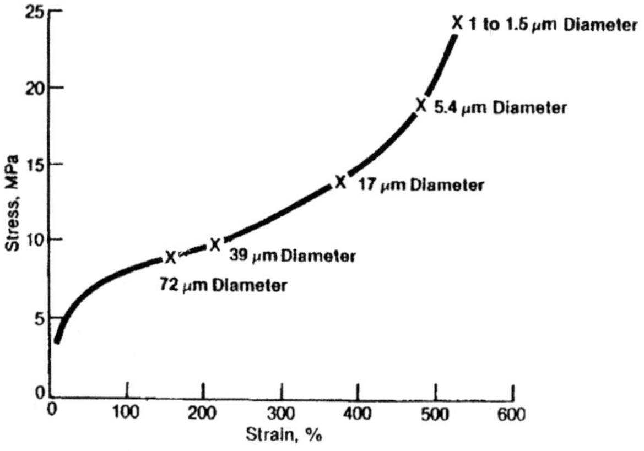

Figure 11. Effect of vulcanized-rubber particle size on the mechanical properties of EPDM-PP TPVs. Ref.[158].

The particle size of the rubber dispersed phase should be comprised within 1 to 2 μm range for a TPV to exhibit the best tensile properties.[157] In figure 11 a composite curve is shown where the tensile strength to break is plotted as a function of the strain to break for TPVs of various particle sizes.

4.2 Phase morphology development in TPVs

The targeted phase morphology for a TPV is that of an elastomeric dispersed phase and a thermoplastic continuous matrix. The dynamic vulcanization process which is ensured by the addition of a selective crosslinking agent to the rubber phase, is responsible for the control of the viscoelastic properties differences between the vulcanizing rubber phase and the thermoplastic phase. When a critical viscoelasticity ratio is reached, phase inversion occurs; i.e. the rubber phase becomes dispersed although it is far major in content and the thermoplastic becomes continuous.

In the initial stages of the dynamic curing process, the degree of crosslinking increases during mixing and the continuous rubber phase breaks up into droplets. The crosslinking of the rubber particles suppresses their coalescence preventing a co-continuous phase morphology to persist. In figure 14 the morphology transformation due to the dynamic vulcanization at the end of the mixing process is schematically illustrated.

The development of a co-continuous phase morphology prior to dynamic vulcanization was reported to be necessary for the preparation of PP/EPDM and PE/NBR TPVs.[159,160] Co-continuous phase morphologies can be developed in binary blends even at concentrations as low as 20 wt % of the minor phase[161].

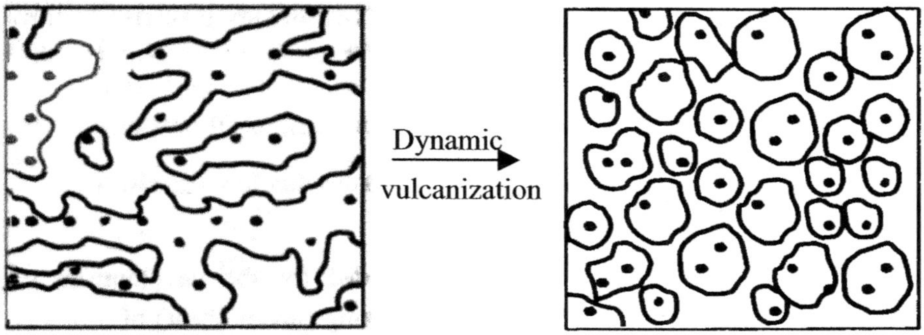

Figure 12. Morphological change from a co-continuous phase morphology to a rubber dispersed in a thermoplastic matrix, as a result of dynamic vulcanization.

4.3 Role of compatibilization

Design of TPVs from incompatible pairs of rubber and thermoplastic results in poor mechanical properties because of a lack in interfacial adhesion between the phases. As in immiscible polymer blends, the resulting particle size is large as a consequence of a high interfacial tension. Therefore a compatibilizer and a compatibilization approach is required to compatibilize TPVs obtained from dissimilar components. A successful design of a TPV from nitrile rubber in polypropylene thermoplastic was obtained after reactive compatibilization using maleic anhydride modified polypropylene.[162] A graft copolymer is generated during the mixing and the dynamic vulcanization of the compound from the reaction of anhydride groups of the PP-g-MA and the amine end groups of NBR rubber. The mechanical properties of the reactively compatibilized blend were excellent when compared to the non compatibilized system.

The preparation of performant TPVs based on engineering thermoplastics is difficult without compatibilization because most of the engineering polymers such as polyesters and polyamides which are polar are uncompatible with the common rubbers such EPR and EPDM. Moffet et al. sucessfully designed a TPV containing 50 wt % rubber in polybutylene terephthalate matrix by reactive compatibilization using EPDM grafted with 3 % of glycidyl methacrylate.[163]

Nylon 6/rubber TPVs with high amount of dispersed rubber, displaying good strain recovery and mechanical properties were prepared using both reactive compatibilization and dynamic vulcanization.[164] By slightly crosslinking the rubber phase during melt-mixing, it was possible to finely disperse 60 wt % of EPDM in nylon 6 matrix. EPDM-g-maleic anhydride has been used as a reactive precursor for the generation of nylon 6-grafted-EPDM compatibilizer. By fine tuning the viscosity ratio between the rubber phase and the nylon 6 phase as well as the concentration of maleic anhydride groups in MA-g-EPDM copolymer it was possible to control the type of phase morphology developed. For example, for a viscosity ratio of rubber/ nylon of 3.8 and a MA-g-EPDM copolymer containing 0.7 wt % MA, the phase morphology developed was co-continuous and the desired TPV was not obtained. Whereas when the viscosity ratio was increased to 8.7 and the MA content to 1 wt %, TPV was obtained where the rubber particle size was within 0.2 - 0.8 μm range. In this system a peroxide (2,5-dimethyl-2,5 bis (t-butylperoxy) hexyne vulcanizing agent was employed. The influence of the amount of maleic anhydride groups in EPDM has been investigated. As shown in figure 13, observation of the phase morphology using transmission electron microscopy has revealed that for a fixed composition of 50 wt % rubber, phase inversion occurred at a concentration of MA up to 0.4 wt %, and small rubber particles were developed when the MA concentration was increased to a range of 0.4 - 1 wt %.

4.4 Role of vulcanizing agents

The effect of complete dynamic vulcanization has been demonstrated by Coran et al.[148-156] The complete vulcanization of the rubber phase leads to a significant improvement of the mechanical properties of the TPV.[165] If the composition of a TPV and the solubility of the different components in solvents are known, then calculations can be made in order to determine whether full cure was achieved during dynamic vulcanization. It has been accepted that if the cross-link density is higher than 7×10^{-5} moles/cm^3 and/or the elastomer is at least 97 % cured, then full cure has been achieved.

In the vulcanization of rubber a very broad choice in vulcanization chemicals, accelerators and retarders exists which also depends on the type of rubber used.[166,167] The sulfur vulcanizing system is the most extensively studied because of its wide use in tyre manufacture. The disadvantage of using sulfur vulcanization resides in the undesirable sulfurous odor. The use of peroxide with polyolefin as the thermoplastics results in side reaction because of the free radicals that are generated. For example in the case of polyethylene, the peroxy radicals cause crosslinking of polyethylene matrix, yielding a very viscous and difficult material to process. When polypropylene is used the radicals abstract a hydrogen from the main chain, a more stable tertiary free radical is generated, which undergoes chain scission. To remedy to that situation additifs such as polyisobutylene, which degrades preferentially, has been added to the compound.[168] Abdou-Sabet and Fath proposed phenolic curatives which demonstrated an efficient vulcanization of EPDM rubber in polypropylene matrix compared to peroxides.[169] Various vulcanizing agents can be used for EPDM / PP TPVs production. In terms of tensile strength, compression set at 100 °C and resistance to oil, dimethyl alkyl phenol is more efficient than the peroxide or the sulfur vulcanizing agent. A lower ultimate elongation of 350 % compared to 530 % and 450 % in the case of sulfur and peroxide, respectively, is a good indication of the larger extent of crosslinking imparted by the phenolic curative.

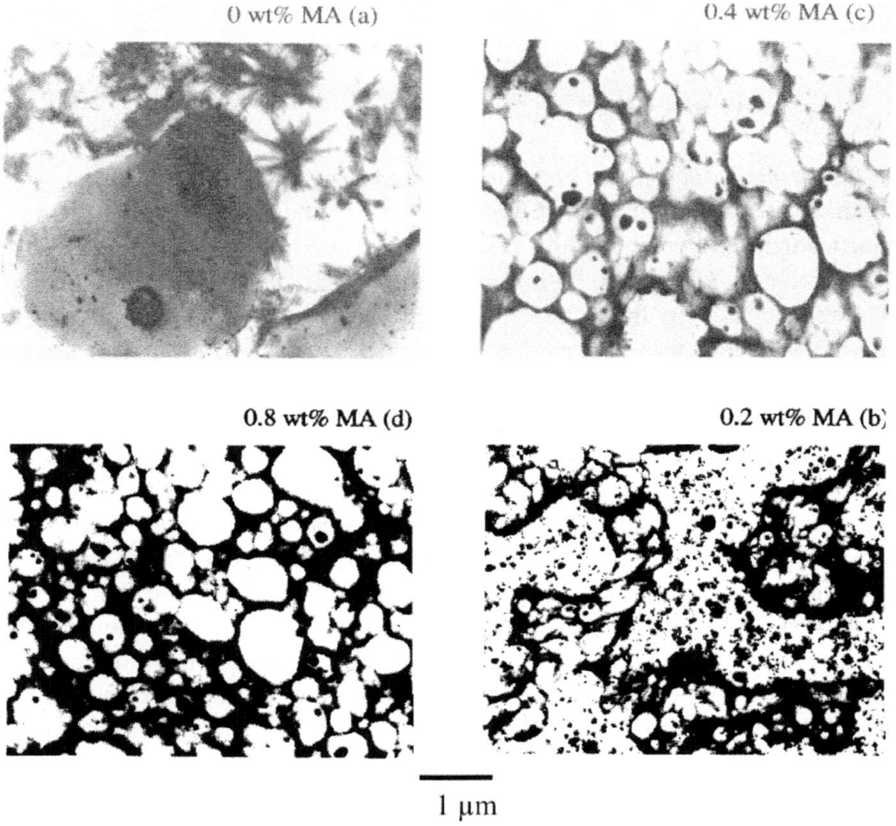

Figure 13. Influence of maleic anhydride (MA) concentration in the EPDM rubber on the phase morphology development of nylon 6/rubber blends with 50 wt % rubber.

4.5 Mechanical behaviour: deformation and recovery

As shown in figure 14, TPVs exhibit a high elasticity and an excellent strain recovery although they are consisting of a thermoplastic matrix which is expected to deform via irreversible plastic deformation. To explain this behaviour several theoretical models were proposed. Finite element analysis was applied on a two dimensional model of particles in a matrix.[170,171] The calculations predicted that in PP/EPDM TPV at least 70 % of the PP matrix is deformed beyond its yield point when the material is stretched. The other fraction of PP could act as a glue between the rubber particles. It was thus assumed that the elastic behaviour of the TPVs is provided by these adhesion points. In reality the elastic forces of the rubber phase are not strong enough to pull back the large amount of plastically deformed PP matrix.

Kawabata[172] suggested a three dimensional model of rubber particles in a thermoplastic matrix. The adhesion between the rubber particles and the matrix was assumed to be very weak apart from a few points at the interface, so that a large amount of voids is formed when the TPV is deformed. As in the previous model it was proposed that these adhesion points link the rubber particles together in a continuous network leading to a macroscopic elastic behaviour. The model does not provide explanation for TPV systems where the interfacial adhesion between the rubber and thermoplastic matrix is very good, as is the case for PP/EPDM TPVs

Soliman et al.[173] used infrared spectroscopy in combination with stress-strain measurements to characterize real-time during deformation and recovery the orientation of the phases. They reported that EPDM is streched during deformation while only a small fraction of PP is deformed. The observed behaviour was ascribed to inhomogenous deformation exhibited by the PP matrix. This hypothesis of mechanism of deformation and recovery has also been verified from IR-strain recovery experiments for a TPV based on EPDM rubber dispersed in nylon 6 thermoplastic matrix.[164]

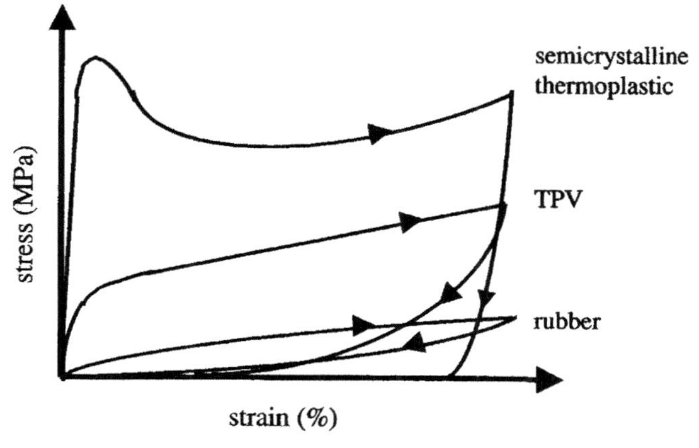

Figure 14. Typical stress-strain and recovery curves of a thermoplastic vulanizate compared to those of a pure rubber and a pure thermoplastic polymer.

Plastic deformation will be initiated and concentrated at locations where the matrix ligaments are the thinnest. The ligaments localized transverse to the applied stress direction will bend or buckle, initially elastically and subsequently with increasing plasticity. Those ligaments which are laying in the stress direction will yield and draw. The thicker regions of the matrix will remain almost unyielded and act like rigid blocks at relatively lower strains. These rigid blocks will hold the rubber particles together, as suggested by Kikuchi et al.[170,171] and Kawabata et al.[172] The continuous substructure of the rubber particles and the only slightly deformed thicker regions of the matrix interconnect the rubber phase allowing for a high degree of elasticity. Therefore one can deduce that the thinner matrix ligaments will be highly plastically deformed but embedded in an elastic matrix. If the elastic restoring force in the rubber phase is high enough, the highly stretched nylon parts could undergo reverse plastic deformation recovering their initial state. However it is more likely that the elastic forces of the rubber phase will cause the highly plastically deformed nylon parts to bend and or to buckle. This hypothetical micro-structural model of inhomogeneous deformation and recovery behaviour of TPVs is schematically illustrated in figure 15. Aoyama et al.[174] proved using finite element methods that in an ideal PBT / rubber TPV (50 / 50 wt%), in which the dispersed phase consists of perfect spheres, that the plastic deformation will be localized at the boundaries between these rubber spheres. The yielding of the matrix is thus induced at the equatorial zone of the rubber particles.

By performing real-time atomic force microscopy (AFM) on thin films and transmission electron microscopy (TEM) on ultra-microtome sections of deformed and stress-released bars of Nylon 6 / EPDM TPV, it was possible to obtain profound insight into the deformation and recovery mechanisms as illustrated in figure 15. Inhomogeneous deformation of the matrix phase has been observed and proved to be the key mechanism of the elastic behaviour of the TPVs. During stretching, the matrix of nylon 6 / rubber TPVs deforms inhomogeneously, mainly the plastic deformation is initiated in those regions where the nylon matrix between the rubber particles is the thinnest. Even at high strains, the thick ligaments of the nylon matrix remain almost undeformed and act as adhesion points holding the rubber particles together. When the external force is released from the sample, the elastic force of the dispersed rubber phase pulls back the highly plastically deformed nylon parts by either buckling or bending. As shown in figure 16 from the phase contrast images, the elastic restoring forces in the rubber phase are high enough to pull back the highly plastically deformed thin nylon ligaments. These ligaments bend and in most cases buckle due to the high amount of plastic deformation.[175]

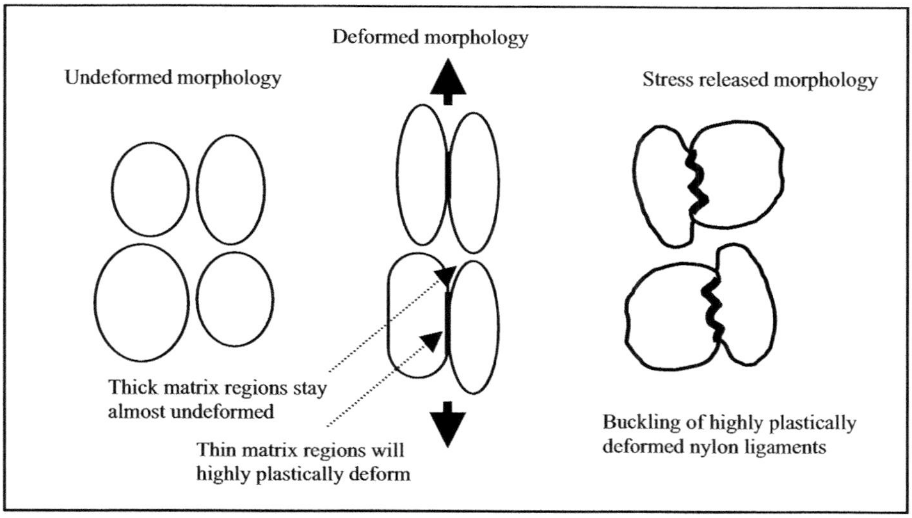

Figure 15. Schematic illustration of an inhomogeneous deformation model of a thermoplastic vulcanizate. Ref. [164]

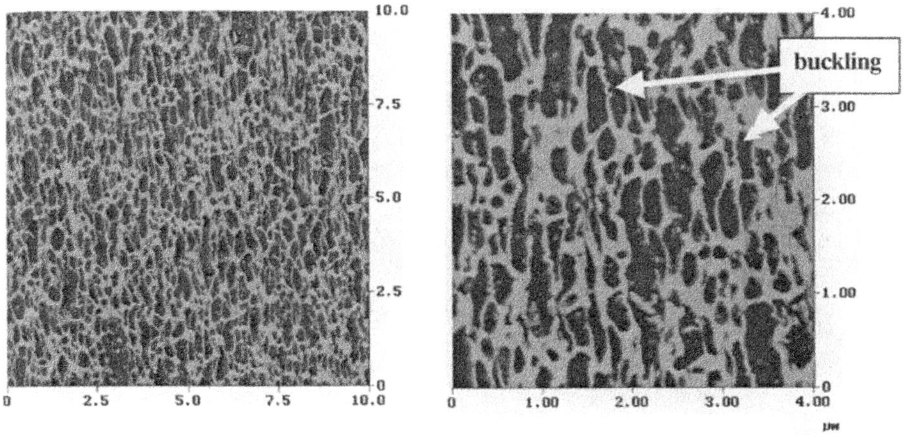

Figure 16. AFM phase contrast images (different magnifications) of a stress releazed 60 EPDM / 40 PA6 TPV film, after an applied strain of 100 % (recovery of 50 %). Ref.[175].

ACKNOWLEDGEMENTS

The authors are indebted to KULeuven Research Council (GOA 98 / 06 Project)

REFERENCES

1. L. A. Utracki , Commercial Polymer Blends (1998) Chapman & Hall, London.
2. D. R. Paul and J. W. Barlow, Polymer blends (or alloys), J. Macromol. Sci.-Rev. Macromol. Chem., C18, 109 (1980).
3. L. A. Utracki and B. D. Favis, Polymer Alloys and Blends, in Handbook of Polym. Sci. and Technology, 4, 121 (1989).
4. W. E. Baker and G. –H. Hu, Chap. 1 (pp. 1-11) in Reactive Polymer Blending, W. Baker, C. Scott, G.-H. Hu, Ed. Hanser (2001).
5. E. J. Kramer, L. J. Norton, C. –A Dai, Y. Sha and C. –Y. Hui, Faraday Discuss 31, 98 (1994).
6. F. Ide and A. Hasegawa, J. Appl. Polym. Sci., 18, 963 (1974).
7. M. Xanthos, Polym. Eng. Sci., 28, 1392 (1988).
8. N. Dharmarajan, S. Datta, G. Ver Strate and L. Ban, Polymer, 36, 3849 (1995).
9. V. J. Tricca, S. Ziaee, J. W. Barlow, H. Keskkula and D. R. Paul, Polymer, 32, 1401 (1991).
10. W. J. Coumans, D. Heikens and S. D. Sjoerdsma, Polymer, 21, 103 (1980).
11. W. M. Barentsen and D. Heikens, Polymer, 14, 579 (1973)
12. D. Heikens, N. Hoen, W. Barentsen, P. Piet and H. Ladan, J. polym. Sci.: Polym. Symp., 62, 309 (1978.
13. D. R. Paul and S. Newman (Eds) " Polymer Blends", Chap. 12, Academic Press, New york (1978).
14. R. Fayt, R. Jérôme and Ph. Teyssié, J. Polym. Sci., Polym. Lett. Ed., 24, 25 (1981).
15. M. J. Mullins and S. Woo, European Patent 0353 478 A1 (1990).
16. C. E. Scott and N. D. B. Lazo, Chap. 5 (pp. 113-139) in Reactive Polymer Blending, W. Baker, C. Scott, G.-H. Hu, ed. Hanser (2001).
17. B. D. Favis, Canadian J. Chem. Eng., 69, 619 (1991).
18. A. Einstein, Ann. Physik, 19, 289 (1906).
19. J. G. Oldroyd, Proc. Roy. Soc., London, A232, 567 (1955).
20. H. P. Grace, Chem. Eng. Commun., 14, 225 (1982).
21. Basil D. Favis, Chap. 16 (PP 501-537) in Polymer Blends Vol. 1: Formulation, D. R. Paul and C.B. Bucknall, Ed. John Wiley, 2000.
22. J. Karger-Kocsis, A. Kallo, and V. N. Kuleznev, Polymer, 25, 279 (1984).
23. G. N. Avgeropoulos, F. C. Weissert, P. H. Biddison and G. C. A. Böhm, Rubb. Chem. Technol., 49, 93 (1976).
24. F. D. Rumscheidt and S. G. Mason, J. Colloid. Sci., 16, 238 (1961).

25. S. Wu, Polym. Eng. Sci., 27, 335 (1987).
26. B. D. Favis and J. P. Chalifoux, Polym. Eng. Sci., 27, 1591 (1987).
27. J. J. Elmendorp and A. K. Van der Vergt, Polym. Eng; Sci., 26, 1332 (1986).
28. B. D. Favis and J. P. Chalifoux, Polymer, 29, 1761 (1988).
29. N. Tokita, Rubber Chem. Technol., 50, 292 (1977).
30. B. D. Favis and J. M. Willis, J. Polym. Sci., Polym. Phys., 28, 2259 (1990).
31. I. Fortelny, Z. Cerna, J. Binko and J. Kovar, J. Appl. Polym. Sci. 48, 1731 (1993).
32. H. Liang, B. D. Favis,Y. S. Yu and A. Eisenberg, Macromolecules, 32, 1637 (1999).
33. N. Chapleau, B. D. Favis and P. J. Carreau, J. Polym. Phys., 36, 1947 (1998).
34. J. M. Willis, V. Caldas and B. D. Favis, J. Mater. Sci., 26, 4742 (1991).
35. R. Fayt, R. Jérôme and Ph. Teyssié, Makromol. Chem., 187, 837 (1986).
36. R. Fayt, R. Jérôme and Ph. Teyssié, j. Polym. Sci., Polym. Lett., 24, 25 (1986).
37. R. Fayt, R. Jérôme and Ph. Teyssié, Polym. Eng. Sci., 27, 328 (1987).
38. R. Fayt, R. Jérôme and Ph. Teyssié, J. Polym. Sci., Polym. Phys., 27, 775 (1989).
39. R. Fayt, C. Harrats, S. Blacher, R. Jérôme and Ph. Teyssié, J. Polym. Sci., Polym. Phys., 69, 178 (1993).
40. H. Van Oene, J. Colloid. Inter. Sci., 40, 448 (1972).
41. R. W. Flumerfelt, Ind. Eng. Chem. Fundam., 11, 312 (1972).
42. J. M. H. Janssen and H. E. H. Meijer, J. Rheol., 37, 597 (1993).
43. J. J. Elmendorp and R. J. Maalcke, Polym. Eng. Sci., 25, 1041 (1985).
44. L. Levitt, C. W. Macosko, and S. D. Pearson, Polym. Eng. Sci., 36, 1647 (1996).
45. K. Min, J. L. White, and J. F. Fellers, J. Appl. Polym. Sci., 29, 2117 (1984).
46. B. D. Favis, J. Appl. Polym. Sci., 39, 285 (1990).
47. U. Sandararaj and C. W. Macosko, Macromolecules, 28, 2647 (1995)
48. R. Conzalez-Nunez, D. Dekee, and B. D. Favis, Polymer, 37, 4689 (1996).
49. S. Y. Hobbs and V. H. Watkins, Chapter 9 (PP 239-284) in " Polymer Blends Volume 1: Formulation, D. R. Paul and C. B. Bucknall, Ed. John Wiley, 2000.
50. V. E. Dreval, G. V. Vinogradov, E. P. Plotnikova, M. P. Zabugina, N. P. Plotnikova, E. V. Kotova and Z. Pelzbauer, Rheol. Acta, 22, 102 (1983).
51. R. Gonzalez-Nunez, C.F. Chan Man Fong, B. D. Favis and D. DeKee, J. Appl. Polym. Sci., 62, 1627 (1996).
52. M. V. Tsebrenko, A. V. Yudin, T. I. Ablazova and G. V. Vinogradov, Polymer, 17, 831 (1976).
53. M. V. Tsebrenko, G. P. Danilova and A. Y. Malkin, J. Non-Newt. Fl. Mech., 31, 1 (1989).
54. J. L. Amos, O. R. McIntyre and J. L. McCurdy, U.S. Patent 2,694,692 (1954).
55. J. R. Campbell, S. Y. Hobbs, T. J. Shea and V. H. Watkins, Polym. Eng. Sci., 30, 1056 (1990).
56. S. H. Hobbs, M. E. Dekkers and V. H. Watkins, Polymer, 29, 1598 (1988).
57. A. Legros, P. J. Carreau, B. D. Favis and A. Michel, Polymer, 38, 5085 (1997).
58. H. F. Guo, S. Packirisamy, N. V. Gvozdic and D. J. Meier, Polymer, 38, 785 (1997).
59. H. P. Schreiber and A. Olguin, Polym. Eng. Sci., 23, 129 (1983).

60. J. T. Lindt and A. K. Ghosh, Polym. Eng. Sci., 32, 1802 (1992).
61. B. D. Favis and D. Therrien, Polymer, 32, 1474 (1991).
62. A. P. Plochocki, S. S. Dagli, J. Starita and J. E. Curry, J. Elastomers. Plast., 18, 256 (1986).
63. L. A. Utracki, In Polymer Alloys and Blends (1989) Hanser, Munich New York.
64. N. G. Gaylord, in Compatibilization Concepts in Polymer Applications, in Copolymers, Polyblends and Composites. N. A. J. Platzer (Ed.) (1975) ACS, Wachington DC, 142, pp. 76-84.
65. M. Xanthos, Polym. Eng. Sci., 28, 1392 (1988).
66. G. Groeninckx, C. Harrats and S. Thomas, Chapter 3 ' In "Reactive Polymer Blending", W. Baker, C. Scott and G. –H. Hu (Eds), Hanser Publishers, Munich, 2001.
67. M. Saleem and W. E. Baker, J. Appl. Polym. Sci., 655, 39 (1990).
68. M. Xanthos and S. S. Dagli, Polym. Eng. Sci., 31, 929 (1991).
69. S. B. Brown, In Survey of Chemical Reactions of Monomers and Polymers during Extrusion Processing in Reactive Extrusion, Principles and Practice. M. Xanthos (Ed.) (1992) Hanser Publishers, Munich, Vienna, New York, pp. 75-199.
70. N. C. Lu and W. E. Baker, Adv. Polym. Technol., 249, 11 (1992).
71. D. R. Paul In Interfacial Agents ("compatibilizers") for Polymer Blends, in Polymer Blends. D. R. Paul and S. Newman (Eds.) (1978) Academic Press, new york, Vol. II, pp. 35-62.
72. M. Van Duin, C. E. Koning, C. Pagnoulle and R. Jérôme, Prog. Polym. Sci., 707, 23 (1998).
73. R. Fayt, R. Jérôme and Ph. Teyssié, J. Polym. Sci., Polym. Lett. Ed., 79, 19 (1981).
74. R. Fayt, R. Jérôme and Ph. Teyssié, J. Polym. Sci., Polym. Lett. Ed., 2209, 20 (1982).
75. R. Fayt, R. Jérôme and Ph. Teyssié, J. Polym. Sci., Polym. Phys. Ed., 775, 27 (1989).
76. R. Fayt, R. Jérôme and Ph. Teyssié, J. Polym. Sci., Polym. Phys. Ed., 1269, 27 (1981).
77. R. Fayt, P. Hadjiandreou and Ph. Teyssié, J. Polym. Sci., Polym. Chem. Ed., 337, 23 (1985).
78. R. Fayt, R. Jérôme and Ph. Teyssié, J. Polym. Sci., Polym. Lett. Ed., 25, 24 (1986).
79. R. Fayt, R. Jérôme and Ph. Teyssié, Makromol. Chem., 837, 187 (1986).
80. R. Fayt, R. Jérôme and Ph. Teyssié, Polym. Eng. Sci., 328, 27 (1987).
81. K. Shull, Macromolecules, 2346, 26 (1993).
82. M. D. Whitmore and J. Noolandi, Macromolecules, 657, 18 (1985).
83. R. J. Roe, Macromolecules, 371, 19 (1986).
84. R. Fayt and Ph. Teyssié, Polym. Eng. Sci., 937, 30 (1990).
85. B. Brahimi, A. Ait-Kadi, A. Ajji, R. Fayt, J. Polym. Sci., Polym. Phys. Ed., 945, 29 (1991).
86. M. Yoshida, J. J. Ha, K. Hin, J. L. White and R. P. Quirk, Polym. Eng. Sci., 30, 30 (1990).
87. R. Fayt and Ph. Teyssié, 2077, 19 (1986).
88. T. Ouhadi, R. Fayt, R. Jérôme and Ph. Teyssié, Polymer Communications, 212, 27 (1986).

89. T. Ouhadi, R. Fayt, R. Jérôme and Ph. Teyssié, J. Polym. Sci., Polym. Phys. Ed., 973, 24 (1986).
90. T. Ouhadi, R. Fayt, R. Jérôme and Ph. Teyssié, J. Appl. Polym. Sci., 5647, 32 (1986).
91. I. D. Mckay, J. Appl. Polym. Sci., 1593, 43 (1991).
92. J. Heuschen, J. M. Vion, R. Jérôme and Ph. Teyssié, Polymer, 1473, 31 (1990).
93. R. Fayt, R. Jérôme and Ph. Teyssié, J. Polym. Sci., Polym. Chem. Ed., 2823, 27 (1989).
94. R. Fayt, R. Jérôme and Ph. Teyssié, Polym. Eng. Sci., 538, 29 (1989).
95. J. W. Barlow and D. R. Paul, Polym. Eng. Sci., 525, 24 (1984).
96. T. D. Traugott, J. W. Barlow and D. R. Paul, J. Appl. Polym. Sci., 2947, 28 (1983).
97. I. D. Mckay, J. Appl. Polym. Sci., 281, 42 (1991).
98. H. C. Kim, K. H. Nam and W. H. Jo, Polymer, 4043, 34 (1993).
99. W. H. Jo, B. C. Jo, and J. C. Cho, J. Polym. Sci., Polym. Phys. Ed., 1661, 32 (1994).
100. W. H. Jo, H. C. Kim and D. H. Baik, Macromolecules, 2231, 24 (1991).
101. C. Auschra, R. Stadler, I. G. Voight-Martin, Polymer, 2085, 34 (1993).
102. P. G. Anderson, US Patent No. 4476283, 1984.
103. M. Jayabahan and T. Bahakrishnan, Polym. Eng. Sci., 553, 25 (1985).
104. T. O. Ahn, Y. C. Ha, M. –H. Oh and S. –S Lee, J. Macromol. Sci., Phys., 215, B34 (1995).
105. C. E. Locke and D. R. Paul, J. Appl. Polym. Sci., 2957, 13 (1973).
106. C. E. Locke and D. R. Paul, Polym. Eng. Sci., 308, 13 (1973).
107. D. Heikens and W. M. Barentsen, Polymer, 69, 18 (1971).
108. F. Ide and A. Hasegawa, J. Appl. Polym. Sci., 18, 963 (1974).
109. B. Majumdar, H. Keskkula, D. R. Paul and N. G. Harvey, Polymer, 35, 4263 (1994).
110. B. Majumdar, H. Keskkula and D. R. Paul, J. Appl. Polym. Sci., 54, 339 (1994).
111. R. J. M. Borggreve and R. J. Gaymans, Polymer, **30**, 63, 71 and 78 (1989).
112. A. J. Oshinski, H. Keskkula, and D. R. Paul, Polymer, 33, 268 (1992).
113. Y. Takeda, H. Keskkula and D. R. Paul, Polymer, 33, 3173 (1992).
114. G. Serpe, J. Jarrin and F. Dawans, Polym. Eng. Sci., 30, 553 (1990).
115. M. Lu, H. Keskkula and D. R. Paul, Polym. Eng. Sci., 34, 33 (1993).
116. K. Dijkstra and R. J. Gaymans, Polymer, 35, 323 (1994).
117. E. A. Flexman, Polym. Eng. Sci., 19, 564 (1979).
118. K. Yokioka and T. Inoue, Polymer, 35, 1182 (1994).
119. B. C. Triveli and B. M. Culbertson, Maleic Anhydride, Plenum Press, New York, 1982, P 269.
120. N. G. Gaylord and M. K. Mishra, J. Polym. Sci.: Polym. Lett. Ed., 21, 23 (1986).
121. N. G. Gaylord and R. Mehta, J. Polym. Sci.: Part A : Polym. Chem., 26, 1189 (1988).
122. N. G. Gaylord, M. K. Mishra and R. Mehta, J. Appl. Polym. Sci., 33, 2549 (1987).
123. W. Heinen, C. H. Rosenmoller, C. B. Wenzel, H. J. M. Degroot, J. Lugtenburg and M. van Duin, Macromolecules, 29, 1151 (1996).
124. J. M. Cowie (ed.), Alternating Copolymers, Plenum Press, New York, 1985.
125. I. Park, J. W. Barlow and D. R. Paul, J. Polym. Sci.: Part A: Polym. Chem., 29, 1329 (1991).

126. M. Hallden-Abberton, Polym. Mater. Sci. Eng., 65, 361 (1991).
127. A. Hirao, S. Nakahama, C. W. Macosko, Polym. Prepr. (Am. Chem.Soc., Div. Polym. Chem.), 37, 722 (1996).
128. P. Guégan, C. W. Macosko, T. Ishizone, A. Hirao and S. Nakahama, Macromolecules, 27, 4993 (1994).
129. A. Padwa, K. A. Wolske, Y. Sasaki, C. W. Macosko, J. Polym. Sci., Part A: Polym. Chem., 33, 2165 (1995).
130. C. A. Orr, A. Adedeji, A. Hirao, F. S. Bates, C. M. Macosko, Macromolecules, 30, 1243 (1997).
131. S. B. Brown, Chap. 5, 'Reactive Compatibilization of Polymer Blends' in " Polymer Blends Handbook" L. A. Utracki (Ed.), Kluwer Ac. Publishers, Netherlands, 2003.
132. B. Majumdar, B. H. Keskkula, D. R. Paul, N. G. Harvey, Polymer, 35, 4263 (1994).
133. K. Dedecker and G. Groeninckx, Polymer 39, 4985, (1998)
134. K. Dedecker and G. Groeninckx, Polymer 39, 4993, (1998)
135. K. Dedecker and G. Groeninckx, Pure and Applied Chemistry 70, 1289, (1998)
136. K. Dedecker and G. Groeninckx, J. Appl. Polym. Sci., 73, 889, (1999)
137. K. Dedecker and G. Groeninckx, Macromolecules 32, 2472, (1999)
138. A. J. Oshinski, H. Keskkula, and D. R. Paul, Polymer, 33, 268 (1992).
139. H. Keskkula and D. R. Paul, "Toughened Nylons 6," in Nylon Plastics Handbook, M. I. Kohan (Ed.), Hanser Publishers, Munich, 1995.
140. L-F. Cheng, B. Wong and W. E. Baker, Polym. Eng. Sci., 36, 1594 (1996).
141. S. Kim, J. K. Kim and C. E. Park, Polymer, 38, 1809 (1997).
142. B. De Roover, PhD. Thesis, Université Catholique de Louvain, Belgium (1994).
143. A., Gonzalez-Montiel, H. Keskkula and D. R. Paul, J. Polym. Sci., Part B: Polym. Phys. 33, 1751 (1995).
144. R. Jérôme and C. Pagnoulle, Chapter 4 In "Reactive Polymer Blending", W. Baker, C. Scott and G. –H. Hu (Eds), Hanser Publishers, Munich, 2001.
145. C. Pagnoulle and R. Jérôme, Macromol. Symp. 149, 207 (2000).
146. A. M. Gessler and W.H Haslett, U. S. Patent 3,037,954 (1962).
147. W. K. Fischer, U. S. Patent 3,758,643 (1973); U.S. Patent 3,835,201 (1974) ; U.S. Patent 3,862,106 (1975).
148. A. Y. Coran, B. Das and R. P. Patel, U. S. Patent 4,130,535 (1978).
149. A. Y. Coran and R. P. Patel, Rubber Chem. Technol., 54, 91 (1981).
150. A. Y. Coran and R. P. Patel, Rubber Chem. Technol., 54, 982 (1981).
151. A. Y. Coran and R. P. Patel, Rubber Chem. Technol., 55, 116 (1982).
152. A. Y. Coran, B. Das and R. P. Patel, U. S. Patent 4,355,139 (1982).
153. A. Y. Coran, R. P. Patel and D. Williams, Rubber Chem. Technol., 54, 1063 (1982).
154. A. Y. Coran and R. P. Patel, Rubber Chem. Technol., 55, 210 (1983).
155. A. Y. Coran and R. P. Patel, Rubber Chem. Technol., 56, 1045 (1983).
156. A. Y. Coran, R. P. Patel and D. Williams-Headd, Rubber Chem. Technol., 58, 1014 (1985).
157. A. Y. Coran and R. P. Patel, Rubber Chem. Technol., 53, 781 (1980).

158. A.Y. Coran and R. P. Patel, Rubber Chem. Technol., 53, 141 (1980).
159. H. J. Radusch and T. Pham, Kautschuk Gummi Kunstoffe, 49, 249 (1996).
160. M. D. Ellul, J. Patel and A. J. Tinker, Rubber Chem. Technol., 68, 573 (1995).
161. D. Romanini, E. Garagnani and E. Marchetti, Communication at the International Symposium on New Polymeric Materials, European Physical Society Macromolecular Section, Naples, Italy, June, 1986.
162. A. Y. Coran and R. P. Patel, Rubber Chem. Technol., 58, 1014 (1985).
163. A. J. Moffet and M. E. J. Dekkers, Rubber Chem. Technol., 32, 1, 1992.
164. J. Oderkerk, G. Groeninckx and M. Soliman, Macromolecules, 35, 3946 (2002).
165. S. Abdou-Sabet and R. Patel, Rubber Chem. Technol., 69, 476 (1996).
166. W. Hoffmann, Vulcanization and Vulcanizing Agents, Maclaren and Sons Ltd., London, 1967, P. 306.
167. Kirk-Othmer Encyclopedia of Chemical Technology, Vol. 21, John Wiley & Sons, N.Y., 1997, P. 460.
168. K. Youckura and S. Shimono, U.S. Patent 4,785,045 (1988).
169. S. Abdou-Sabet and M. A. Fath, U.S. Patent 4,311,628 (1982).
170. Y. Kikuchi, T. Fukui, T. Okada and T. Kikuchi, Polym. Eng. Sci., 31, 1029 (1991).
171. Y. Kikuchi, T. Fukui, T. Okada and T. Kikuchi, J. Appl. Polym. Sci.: Appl. Polym. Symposia, 50, 261 (1992).
172. S. Kawabata, S. Kitawaki, H. Arisawa, Y. Yamashita and X. J. Guo, J. Appl. Polym. Sci.: Appl. Polym. Symposia, 50, 245 (1992).
173. M. Soliman, M. Van Es and M. Van Dijk, PMSE, ACS Preprints, 79, 86 (1998).
174. T. Aoyama, A. Juan Carlos, H. Saito, T. Inoue and Y. Niitsu, Polymer, 40, 3657 (1998).
175. J. Oderkerk, G. Schaetzen, B. Goderis, L. Hellemans and G. Groeninckx, Macromolecules, 35, 6623 (2002).

PROCESSING FOR ULTIMATE PROPERTIES

PIET J. LEMSTRA
Eindhoven University of Technology/Dutch Polymer Institute, P.O. Box 513, 5600 MB, Eindhoven, The Netherlands

Abstract: The ultimate properties of synthetic vs. natural polymers are discussed in brief with the emphasis on processibility. The ultimate properties are discussed in various dimensions, 1-D, 2-D and 3-D(imensional). Important lessons from Nature concerning exploiting the polymer chain for engineering purposes are highlighted.

Key words: ultimate properties, stiffness, processibility, fibers, films, composites.

1. INTRODUCTION

Polymer Science & Technology is the paradigm of a multi disciplinary area. The path from monomer via polymer to functional materials and products travels through the various disciplines of macromolecular science, polymer chemistry and physics, rheology, processing and design. Consequently, the successful design of novel and improved polymeric materials and products requires an integrated "chain-of-knowledge" approach[1] and a profound understanding of the disciplines *met en route*.

In the past two decades polymer chemists have synthesized an impressive range of novel polymer structures including intrinsically conducting polymers, polymeric light-emitting diodes (poly-LED's), high T_g polymers, including liquid-crystalline-polymers (LCP's), synthetic polypeptides etc. etc. The success of these novel "speciality polymers" depends to a large extent on their processibility. In this respect, many pitfalls have been and will be encountered. Many newly synthesized polymeric materials proved to

F. Ciardelli and S. Penczek (eds.),
Modification and Blending of Synthetic and Natural Macromolecules, 201–217.

be (highly) intractable usually related to a limited thermal stability during processing.

A possible solution for poor processibility is often the use of solvents (solution-spinning/casting). The use of solvents, however, is limited to products with a high surface/volume ratio such as fibers (1-D) and films/coatings (2-D) in view of full and easy recovery of the solvent. In the case of 3-D(imensional) products, the use of solvents which have to be removed completely after processing/shaping is cumbersome and will be even more in the future due to strict environmental legislation.

Novel creative and environmentally sustainable processing routes and techniques are needed to fully exploit the intrinsic properties of the polymer chain in advanced polymer products. This statement holds equally well for both synthetic and natural (bio)polymers such as starch, cellulose, polypeptides etc. which are usually intractable due to their high molar mass and/or thermal instability.

2. NATURAL VS. SYNTHETIC POLYMERS

Nature without any doubt is the Master polymer chemist at this point in time in comparison with Mankind, viz. the making of synthetic polymers in laboratories or industrial plants. The petrochemical industry uses about 20 monomers to make approximately 100 (co)polymers. The control over polymerisation reactions is increasing all the time using new and/or improved catalysts systems. Stereo-regular polymers such as (isotactic) polypropylene can be produced with high efficiency including copolymers. But in the case of copolymers, the comonomer can be introduced randomly or in blocks but not in a programmed way along the polymer chain like Nature does in the case of Proteins.

Nature, on the other hand, uses also 20 monomers as building blocks in the synthesis of polypeptides, the 20 amino acids, but the way the various amino acids are arranged in the polypeptide chain is fully controlled and encoded by DNA. Consequently, the number of various polypeptides on the earth is an astonishing number, to be compared with Mankind, viz. industry, ca. 100 various (co)polymers.

Even more astonishing is the DNA molecule and its base sequence. In terms in precision and scale, assume that the size of a carbon atom is identical to a tennis ball. At that scale, the length of a human DNA molecule can span the globe (> 40,000 km) whereas a synthetic polymer like UHMW-PE (ultra-high-molecular-weight PE) is at most 30 km in length and not defect-free in contrast with DNA where defects are not allowed in human heritage.

Nature produces polymers with high precision and, if needed, in large amounts. Well know polymers produced in large quantities are:

- Cellulose (plants, trees, bamboo)
- Starch (potatoes, corn)
- Chitosan (animal shells)
- Proteins (tissues, enzymes)
- Natural Rubber (cis-Polyisoprene)

The amount of cellulose produced via photosynthesis in plants is appr. 10^{12} – 10^{14} tonnes/annum (dependent on the season), to be compared with 200 x 10^6 tonnes/annum of synthetic polymers.

The amount of starch is also almost unlimited available all around the world in potatoes, corn (mais), tapioca, cassave etc., viz. also abundantly present in undeveloped countries.

Given the fact that Nature is the Master polymer chemist and natural polymers are available abundantly, why do we need this polluting synthetic plastics, linked to the oil industry?

2.1 The world of synthetic polymers

Synthetic polymers are based on oil and about 5% of the world oil supply is used to make polymers, the building blocks to make polymers, viz. the monomers, are by-products or derivatives of the oil refinery industry (crackers). Polymerisation reactions run usually exothermic and, consequently, polymers are cheap to make in contrast with steel and notably aluminium which are highly endothermic processes.

Plastics (synthetic polymers) were born in the 20th century and especially in the latter decades; their growth has been almost exponential, see Figure 1. Currently, close to 150 million tonnes of plastics are produced per year. In view of their low specific mass (density), 900 - 1500 kg/m^3, plastics are surpassing steel in terms of output volume. Between 1980 and 1990, the production of plastics increased by 62% while that of steel decreased by 20%.

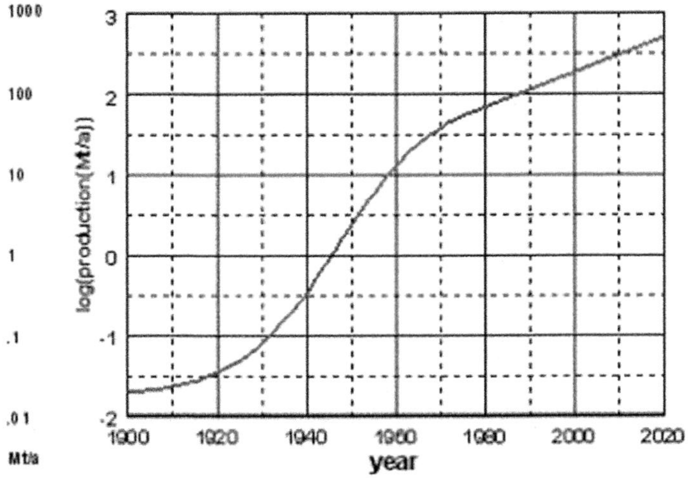

Figure 1. The growth of the world production of plastics in the 20th century

This tremendous growth in the production of synthetic polymers (plastics) correlates very well with the growth of the world population.[2] The reality, however, is that the consumption of plastics is high in Western Europe, the USA and Japan but very low in countries such as Pakistan, Iran etc. The consumption of plastics in Belgium, for example, is the highest in the world with nearly 200 kg/capita per annum (due to large car assembly plants which use plastics for exterior and interior car parts), and the lowest in Pakistan, less than 200 grams/capita per annum. If the production/consumption of plastics in the world would reach the level of Belgium, the world production would increase by a factor of 10, viz. $> 10^9$ tonnes/annum!

The impact of polymers on daily life can be noticed by anyone, notably in cars, electronic equipment (compact disc, TV and computer cases), buildings (houses, exterior and interior, textiles, packaging, sport and leisure articles). For instance, with respect to the automotive industry, plastics are the preferred materials for parts such as bumpers, dashboards, fuel tanks and chairs (synthetic fibres).

2.2 Processing plastics, easy and fast

It is very important to note that synthetic polymers are no homogeneous compounds but heterogeneous in many respects. First, there is no uniform molar mass but a molar mass distribution which can be extremely broad, for instance, in the case of polyolefins (PE + PP) based on Ziegler/Natta catalysis. Second, polymers are processed with additives to enhance processibility and the physical and mechanical product properties. Numerous additives are available ranging form stabilisers, plasticisers, colourants, and reinforcing elements either of particulate (minerals such as clays) or fibrous nature (usually glass fibers). In some cases, the polymer is only used as a binder and more than 50% is made up of additives, e.g. a rubber tyre (reinforced with steel and carbon-black).

The term plastics is used for polymer/additive compounds which are optimised for processibility and product performance. Consequently, synthetic polymers are highly heterogeneous molecular mixtures that typically possess "static averaged" and singular properties requiring additives to tailor the final product performance.

During the past decades a tremendous effort has been made to upgrade plastics. In the German literature, the term 'Polymere Werkstoffe' was coined, viz. polymer + additives to replace the name Kunststoff, with the connotation of "Ersatz", to emphasise the unique character of polymers as construction materials. Unique, because besides substituting classical construction materials such as wood or steel, the specific properties of polymers makes it possible to produce fast and cheap unique constructions which can not be realised easily with steel, wood or glass. Entering the "nano-era", expectations are running high concerning the use of polymer nanocomposites such as (exfoliated) nano-clays and the use of carbon nanotubes as reinforcing elements. The main aim is to obtain superior properties in terms of strength and stiffness at low volume fractions.

The petrochemical industry provides two classes of synthetic polymers, the *thermosets* (about 10%) and the *thermoplasts* (~90%). Thermosets are provided as resins (monomers) and well known examples are urethanes and epoxies (coatings).

The thermoplasts are provided by the industry in the form of pellets and/or powders and are processed via the melt (extrusion, injection-moulding). In the case of thermoplastics, the converters just heat the pellets/powders into the molten state, push the molten polymer through a die via extrusion into a 1-D (fiber) or 2-D (film) shape or in the case of 3-D, via injection-moulding into a mould, and by simply cooling the product is ready. This melt-processing step into complexly shaped parts can be very fast with a high accuracy and reliability. For example, a CD with all its numerous details at sub-micrometer level can be moulded in less than 4 seconds including the metal coating. A plastic (PP) dashboard of a car with its complex shape can be moulded fast and in large and reproducible quantities. To make the very same dashboard out of wood or metal is a quite a job!!

Moreover, the properties of a plastic/polymer product are not only determined by the chemistry of the polymer chain but equally well by the processing step, viz. the orientation of the polymer molecules. A prime example, in this respect, is linear polyethylene. Based on the very same molecules, in terms of molecular structure, flexible containers are made as well as ultra-stiff fibers (Dyneema®; DSM), just a matter of chain orientation.

To summarise, the strong growth of synthetic polymers is based on several factors:

1. Synthetic polymers are **based on oil** (petroleum, viz. petro-polymers) and are consequently **cheap** (as long as the oil price is low), typically 1Euro/kg for commodity polymers such as polyethylene (PE) and polypropylene (PP) and 2-5 Euro/kg for typical engineering plastics such as polycarbonate and polyamides (nylons);
2. The **production costs (energy) are intrinsically low**. Polymerisation processes are often exothermic (heat is generated), compared with the production of steel (melting and refining of ore) and notably aluminum (melting bauxite);
3. **Realisation of complex forms** or moulds is simple, for instance via injection moulding or blow moulding. Well-known examples are the compact disc, an injection moulded polycarbonate disc coated with aluminium, and polyethylene fuel tanks in cars, which possess complex shapes nowadays in order to be fitted in the limited space of the car body;
4. **Insulator for heat** (polymer foam) and **electricity** (cable and wire insulation);
5. **Low specific mass** (density), especially important for the packaging and handling of consumer products (bottles, films and crates). The main advantage is weight reduction;
6. **Limited corrosion**, as opposed to steel;
7. **Large range of properties**, from soft rubber materials to fibres stronger than steel!

2.3 Processing of polymers, Man vs. Nature

Natural (bio) polymers grow in Nature and in comparison with synthetic polymers (heterogeneous molecular mixtures, see above) the control over the molecular structure/architecture is impressive such as a defined length and monomer sequence (polypeptides) and full control on stereo chemistry (cis-polyisoprenes/natural rubber, trans-polyisoprenes/"gutta percha", polyhydroxy butyric acid). Next to the primary structure, viz. the polymer backbone, the secondary and higher order structures are important via self-assembly, i.e. Nature grows its polymers in a complex composite form, for example, cellulose in trees, plants etc.

Processing of natural polymers via the molten state (melt-processing) is often impossible since the polymer will not melt but degrade. Nevertheless, wood is still used extensively but via (often complex) machining.

Another important factor is that natural(bio) polymers usually interact with water, an essential component for their growth. For example, starch is the base (reserve) material in potatoes, corn etc. and is highly hydrophilic and is of high molar mass, $> 10^6$ D., and not processible. Many attempts have been made to make so-called thermoplastic starch by blending starch with lubricants such as glycerol, polymeric additives. Despite the effort, the table below shows the gap between the use of easy to process commodity plastics such as polypropylene (PP) and polyester (PETP) vs. thermoplastic starch.

Production volume (2002)	PP	PETP	Th- Starch
(kilo tonnes)	35,000	30,000	< 15

What can we learn from Nature and how can we improve processibility of natural polymers?

3. ULTIMATE PROPERTIES IN 1-D(IMENSION)

The stiffness (Young's Modulus) of materials is of utmost importance in engineering. Metals and glass are usually considered to be stiff whereas polymers (plastics) are considered to be soft materials.

Table 1 shows that polymeric materials such as rubbers and thermoplasts are indeed quite soft in comparison with notably inorganic materials. Polymers, however, are extremely anisotropic in nature with strong covalent bonds in the main chain and relatively weak interactions between the chains (Van der Waals, Hydrogen bonding). Normally, the polymer chain in plastic products possesses a random coil (amorphous polymers) or a folded-chain conformation (in semi-crystalline polymers).

Table 1 . Stiffness (Young's Modulus) of various materials at ambient temperature

Material	Young's Modulus (GPa)
Rubbers	< 0.1
Amorphous thermoplasts, $T < T_g$	2 – 4
Semi-crystalline thermoplasts	0.1 – 0.3
Wood (fiber direction)	15
Bone	20
Aluminum	70
Glass	70
Steel	200
Ceramics	500
Carbon fibers	500 – 800
Diamond	1200
Polyethylene fibers	100 – 150
Aramid fibers	80 – 130
"M-5"	300

However, when the chain is transferred into an extended conformation, a completely different situation is encountered as we can learn from Nature. For example, cellulose molecules in trees grow in an extended form. The molecules arrange themselves in fibrils, parallel to the fiber-axis. The stiffness of these fibrils is in the range of 15-30 GPa (Young's modulus) and they are responsible for the stiffness of the tree. When a tree should not be too stiff, Nature deliberately imparts disturbances in the crystallization or arrangement of extended cellulose molecules in flora. However, certain types of seaweed, *Algae valonia*, contain fibrils with a stiffness of 70 GPa, which is 50 percent of the theoretical stiffness.

In the early sixties, Treloar[3] already calculated that the stiffness (Young's or E-Modulus) of an extended polyethylene chain is appr. 180 GPa in fact, a fully stretched, extended chain polyethylene molecule is a 1-dimensional diamond, the stiffest material on earth (E-Modulus > 1000 GPa), so it is not too surprising that the question is: "how to reach extended chain structures?"

Polymer chains tend to adopt a *random coil* conformation in solution and in the molten state, except so-called intrinsic *rigid* chains. Polymers that adopt a random coil conformation in solution and in the molten state are called *flexible* chains and the majority of synthetic polymers are flexible chains. To explain the difference in a popular manner: a melt of flexible polymer chains can be compared with cooked spaghetti whereas rigid chains resemble uncooked spaghetti.

A number of different methods to orient synthetic polymer molecules, however, evolved rapidly. A fairly trivial method, which started in the sixties, was the drawing of polyethylene below melting temperature, viz. in the solid state. Deformation in the solid state is essential since in the molten state polymer molecules will slide past each other and the desired (full) orientation cannot be achieved. In the solid state, the relaxation time and viscosity are much higher so partially extended molecules cannot relax and turn back to their original conformation. Pioneer in this field of research hase been Ward at Lees university.[4,5] By optimizing the process, oriented polyethylene structures (mostly tapes) could at last be obtained with a stiffness in the range of 70 GPa, comparable to (fibre)glass.

A breakthrough in the field of polyethylene structures was realized at DSM research at the end of the seventies, where the so-called gel spinning of UHMW-PE, resulted in the development of the DYNEEMA™ fibre. Figure 2 shows the spectacular development in stiffness as a result of orientation processes in this century. The 1980 quantum leap was caused by the discovery of gel spinning at DSM based on ultra-high molecular weight polyethylenes, UHMW-PE. Fibers possessing a stiffness > 100 GPa can be achieved technically, whereas in laboratories stiffnesses above 200 GPa can be realized. This makes polyethylene fibres stiffer than steel! When comparing stiffnesses in relation to the weight of the material, the specific stiffness, the polyethylene fibre is circa seven times stiffer than a steel fibre (at room temperature), see Figure 3.

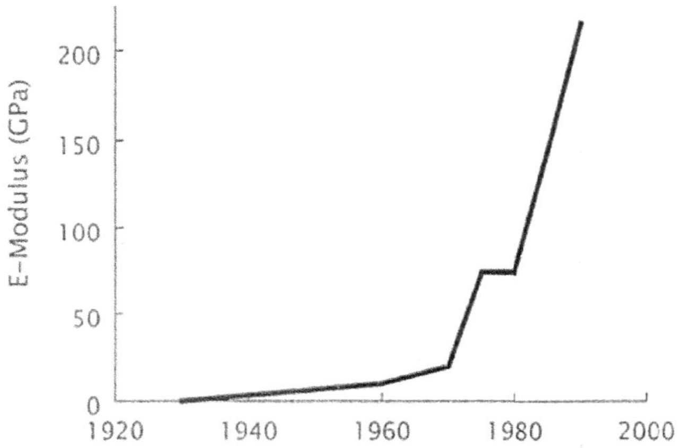

Figure 2. Development of the stiffness, E-Modulus, of polyethylene

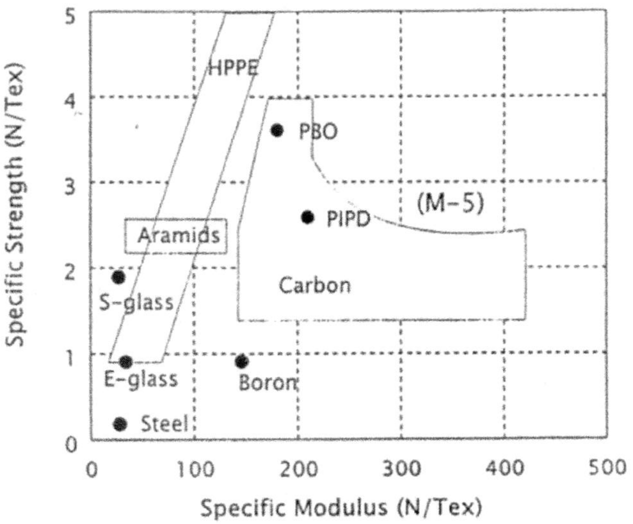

Figure 3. Specific Strength vs. Specific Stiffness of various Fibers; HP-PE is Polyethylene (N/Tex = GPa/density (g/cm^3)

3.1 Rigid chains

A parallel research activity on oriented polymer structures in this century was the development of intrinsic rigid polymers, the so-called rigid chains or rod-like polymers. Already in 1956, Flory postulated a theory on phase-behaviour of rigid polymers. Later on, numerous rigid polymer molecules were synthesized and a new research area was exposed. Well-known examples of rigid polymer molecules are the aromatic polyamides, aramides, especially polyphenyleneterephthalamide or PPTA, which is the building block for aramid fibers KEVLAR™ (DuPont) and TWARON™ (Akzo-Nobel), and the aromatic co-polyesters like VECTRA™ (Hoechst/Celanase).

Intrinsic rigid polymers are intractable systems. For instance, aromatic polyamides cannot be processed through melting as the high melting point decomposes the polymer during the melting process. The aramid fibre is therefore obtained by spinning from a sulphuric acid solution in a 'nematic' or liquid-crystalline phase (PPTA only solves in unpleasant solvents like the mentioned sulphuric acid). Literature often gives the impression that the development of rigid polymer molecules and liquid-crystalline-polymers is simply a result of the application of a theoretical concept, which was already posted by the, earlier discussed, genius Flory in 1956. Though, it's quite a leap to solve PPTA in sulphuric acid and to observe that it is not possible, due to the viscosity, which becomes too high. Even at low concentrations of

PPTA this is a problem. At DuPont, a certain Mrs. Kwolek added more PPTA and noticed that, instead of becoming higher, the viscosity of the PPTA/sulphuric acid decreased. This is known as the discovery of nematic solutions in which the PPTA polymer molecules arrange themselves in domains, comparable to the trunks of trees in a Finnish lake, resulting in a low viscosity, see figure 4.

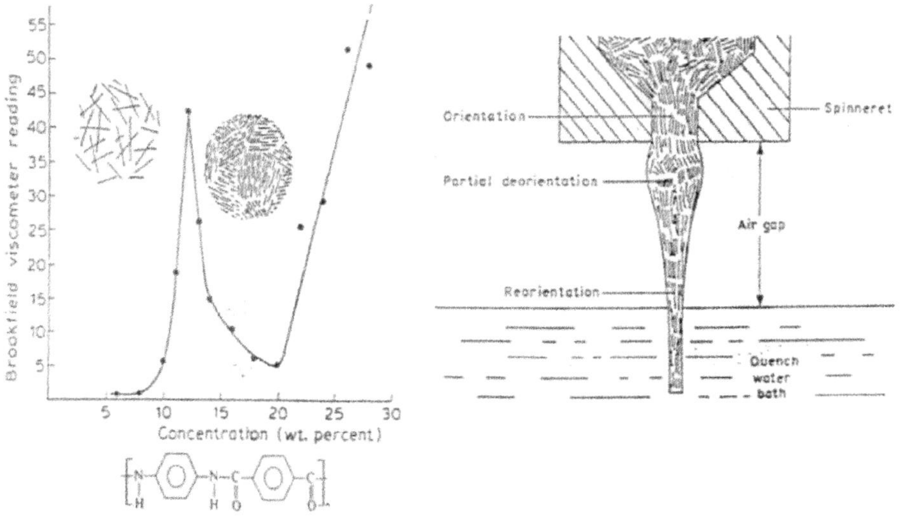

Figure 4. Left: viscosity of PPTA in sulphuric acid; Bottom-left: structure PPTA; Right: spinning of PPTA in sulphuric acid into water

3.2 Epilogue on extended-chains

In the eighties, expectations were running high concerning the application potential of the novel high-strength and high-Modulus organic fibres, respectively the aramids and gel-spun polyethylene. The main dream was to develop lightweight materials based on organic matrices (epoxies, polyesters) and as reinforcing elements these novel fibres. In retrospect, it was surprising that these high-performance fibres could be made on a technological scale.

The process for making high-strength polyethylene fibres looks at first sight quite simple: dissolving PE, spinning and stretching. The technology is not so simple in reality, a lot of solvent has to be recovered and stretching should be done in a proper way and it is far from easy to obtain draw ratios of up to 100x!

The process for making aramid fibres is in fact simpler. Hardly any stretching is involved since the molecules are rigid rods and the molecules align automatically in the fibre direction upon spinning into a water bath. However, sulphuric acid is not so easy to handle and, moreover, the chemistry of making PPTA is much more complex than high molar mass PE.

The various processes of making the fibre, however, is not the real issue. They are made in practice and so, it can be done. To apply these fibres in composite structures is a different issue. The fibres are highly anisotropic: typically 1-dimensional structures of molecules aligned parallel to the fibre direction. In the fibre direction, the mechanical properties are impressive but perpendicular to the fibre or in compression their performance is rather poor. In this respect the cheap glass fibre has a much better performance, it is isotropic by definition (glass), viz. the properties are the same in all directions. Consequently, the aramid and HP-PE fibres failed in their major envisaged application area: structural advanced composites. In due time, however, niche markets have been found such as protective clothing, bullet proof vests, sails etc.

Recently, Akzo Nobel (Sikkema c.s.) developed a novel fibre: PIPD with the code name "M-5". It is uncertain whether this fibre will be developed in future since Akzo Nobel sold their fibre business (Acordis). The structure of "M-5" is shown below. Due to the side chains and hydrogen bonding, this fibre possesses a much enhanced compressive strength, see figure 5.

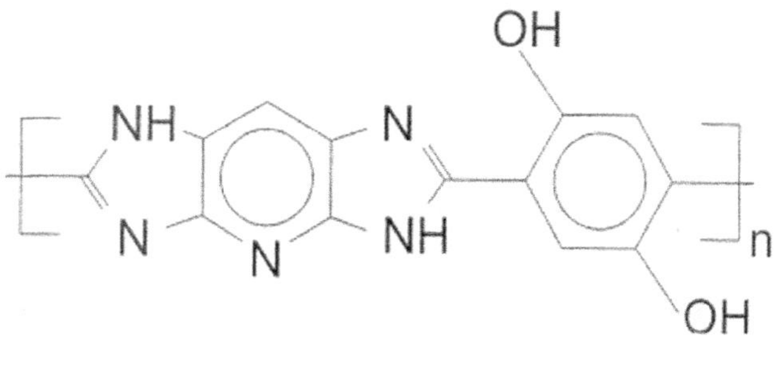

Figure 5. Structure of "M-5"

4. ULTIMATE PROPERTIES IN 2-D

Polymers are used extensively in film applications, notably in the area of packaging. Natural polymers, notably modified cellulose (cellophane) have been used extensively in the past but are replaced more and more by synthetic plastics such as polypropylene and polyester. Stiffness and barrier properties are important properties of packaging materials and become increasingly important for advanced packaging applications such as poly-LED's, to replace glass, aiming at flexible all-plastic displays. Both stiffness and barrier properties are related to the organisation of polymer molecules in the film, viz. the amount of crystallinity and chain arrangement, extended vs. folded conformation. Semi-crystalline polymers with a low T_g , such as PE and PP, are soft and poor barrier materials. In the amorphous, low Tg domains, permanent gases such as O_2 and CO_2 permeate rapidly. Biaxially stretching increases the stiffness and improves the barrier properties, and biaxially stretched PP (BOPP) is a well-known product replacing cellophane. However, compared with uniaxially (1-D) stretching/drawing, both the stiffness and barrier properties remain at a low level. For example, biaxially stretched synthetic polymers such as PE, PP and PETP, can reach stiffness values up to a maximum of 10 GPa and the barrier properties are very poor compared with uniaxially drawn tapes with are virtually impermeable. Consequently, there is room for a lot of improvement.

The low value for the ultimate stiffness is simply related to the fact that oriented polymer structures are only strong and stiff in the (extended/oriented) chain direction upon tensile loading, the 3,3 component in the stiffness matrix below, or more precisely, the E-Modulus in the chain direction is S_{33}^{-1} = 312 GPa. The moduli in all other loading directions are very low. Consequently, high-performance fibers , based on extended organic polymer molecules, are only strong and stiff in the fiber (chain) direction and by any off-axis loading (shear, bending), the low off-axis Moduli will dominate the deformation process. The same applies upon loading films in which there is no preferred direction of the oriented structures in the plane of the film. In this respect we can learn from Nature. It was discovered by Iguchi et al.[7] that dried bacterial cellulose gels, consisting of nearly 100% pure cellulose fibrils, possess an E-Modulus in the plane of the film up to 30 GPa, viz. three times stiffer then the best synthetic polymeric film produced by (biaxial) stretching. Probably the hydrogen bonding between the nano cellulose fibrils gives rise to these impressive properties. Cellulose gels produced by bacteria is a popular indigenous food in Asia but to process films is quite cumbersome. However, recently a novel process has been developed , the spinning of cellulose from nematic phosphoric acid solutions. Cellulose dissolves in (super)phosphoric acid. In instead of spinning fibers, the next step might well be casting of high-performance films.

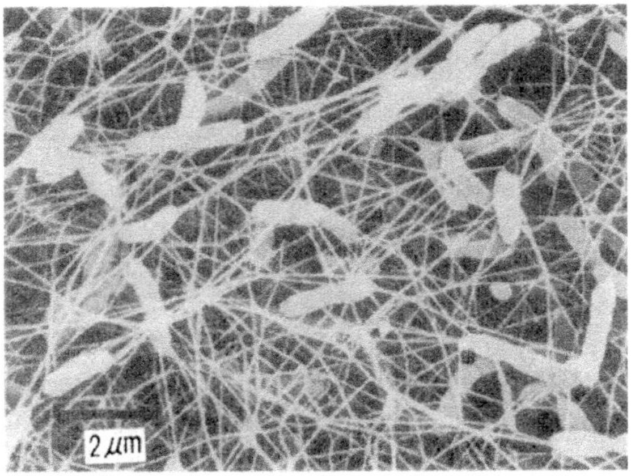

Figure 6. Scanning electron micrograph of freeze-dried surface of bacterial cellulose gel

Stiffness Matrix C_{ij} for PE crystal (GPa)

7.99	-3.28	1.13	0	0	0
3.28	9.92	2.14	0	0	0
1.13	2.14	316	0	0	0
0	0	0	3.19	0	0
0	0	0	0	1.62	0
0	0	0	0	0	3.62

Compliance matrix S_{ij} for PE crystal (10^{-2} GPa^{-1})

14.5	-4.8	-0.02	0	0	0
-4.8	11.7	-0.06	0	0	0
-0.02	-0.06	**0.032**	0	0	0
0	0	0	31.4	0	0
0	0	0	0	61.7	0
0	0	0	0	0	27.6

5. ULTIMATE PROPERTIES IN 3-D

For the production of 3-D(imensional) products many techniques are available such as injection-moulding, extrusion (pipes, tubes), blow-moulding (containers) etc. To meet increasing demands on performance, especially in the automotive industry, and in order to compete with classical construction materials, properties such as stiffness, toughness, temperature resistance (T_g) should be maximised. Properties such as stiffness, temperature resistance and to some extent toughness are material properties and hence other methods are needed to meet the criteria.

In order to increase the stiffness, additives are used in particulate (silicates, nano-clays) and fibrous form, notably glass fibers, both as short (injection-moulding) and continuous fibers (glass mats). In fact, most of the polymers used for engineering purposes are reinforced to meet the requirements for stiffness. For advanced applications, e.g.. aeronautics, high-modulus fibers such as carbon fibers are used. The woven carbon fiber fabric is impregnated with a resin, usually epoxy resins, and cured to obtain a thermoset composite structure. Cured epoxy resins are thermosets and, consequently, rather brittle. The use of high-T_g thermoplasts like PPO (polyphenylene ether) or Polyimides, is rather cumbersome because impregnating a woven carbon-fiber fabric is cumbersome due to the high melt-viscosity. Solvents are used to facilitate the impregnating process, but these solvents have to be removed completely afterwards in view of environmental legislation and product quality. A possible solution for this problem is the use of so-called reactive solvents which dissolve the thermoplastic polymer, hence lower the viscosity, and after impregnating the reactive solvents is polymerised, hence remains in the composite structure (figure 7). For example, PPO (polyphenyleen ether), a high T_g (215 ^{0}C) polymer, dissolves in epoxy resin at elevated temperatures. The solution can be impregnated and, upon curing (heating), phase separation and phase inversion occurs (the system becomes immiscible), and the epoxy resin is transferred into a dispersed thermoset phase. By tuning the volume fraction and polarity of the surface of the carbon fibers, the epoxy will migrate to the surface of the fibers, acting as a glue. The advantage is that no removal of the solvent used is needed, the (reactive) solvent is transferred into a part of the composite system and can even fulfil an important role (adhesion promoter, toughening agent etc.).[1]

In the case of automotive and other more down to earth applications, carbon fibers are too expensive and cheap glass fibers are used as reinforcing elements, notably in combination with PP. The problem is that glass fiber reinforced composites can not easily be recycled nor incinerated. In the latter case, glass poses problems in the furnaces. A recent novel approach[8] is to use fully recyclable composites, the so-called "all-PP" composites (figure 8). In this case, polypropylene tapes are co-extruded with a thin layer of PP-copolymer which possesses a lower melting point. After weaving a structure, and upon heating, the surface copolymer layer will melt and upon cooling the PP fibers are glued together. In this case a composite structure is obtained consisting of nearly 100% fibers (100% PP fiber – a small amount of copolymer) and recycling is no problem anymore since the composite is "all-PP".

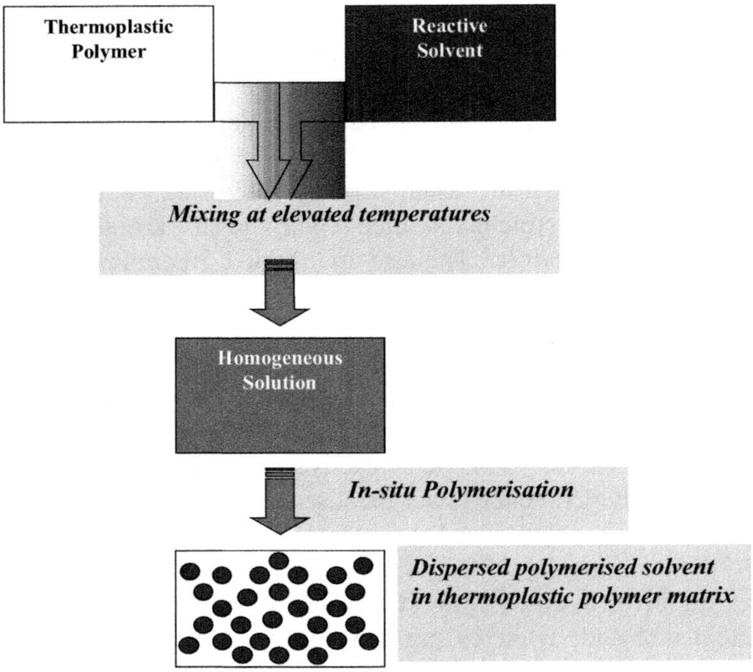

Figure 7. Schematic representation of using reactive solvents

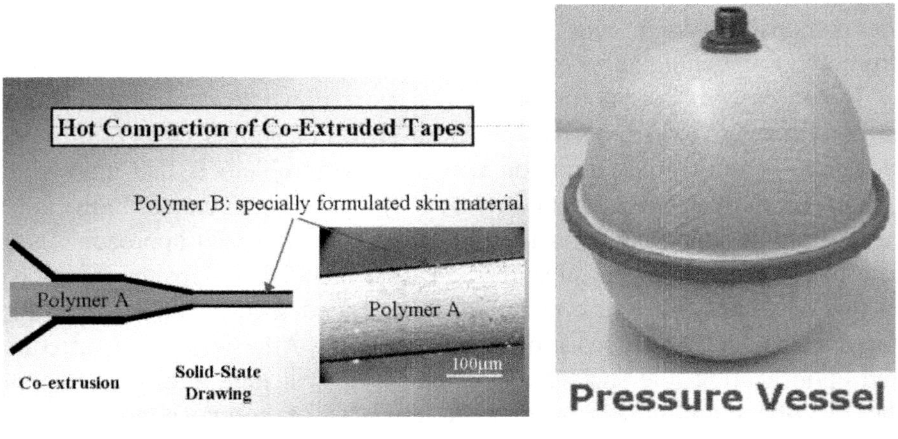

Figure 8. Examples of all-PP composites

6. CONCLUSION

A few examples have been shown to explore the ultimate properties of polymers taking into account more environmentally friendly processing routes and Nature as an example, if possible. The examples shown above might not be too convincing yet but hopefully inspire young scientists to do a better job along these lines of thinking.

REFERENCES

1. P.J. Lemstra,J. Kurja and H.E.H. Meijer in Materials Science & Technology vol. 18, Eds. R.W.Cahn, P.Haasen and E.J. Kremer, VCH, ISBN 3-527-26829-4, Chapt. 10 (1997)
2. L. Utracki, Polym. Eng. Sci. , 35, 2 (1995)
3. L.R.G. Treloar, Polymer 37, 95 (1960)
4. I.M. Ward in Development of Oriented Polymers, 2nd Ed., Elseviers N.Y., 1988
5. G. Capaccio, A.G. Gibson and I.M. Ward in Ultra-High Modulus Polymers, Eds. A. Ciferri and I.M. Ward, Elseviers Applied Science, London, 1979, Chapt.1
6. P.J. Lemstra in Structure Formation in Polymeric Fibers, Ed. D. Salem, Hanser Verlag, 2000, Chapt.5
7. M. Iguchi, Journal of Mater. Sci. 35, 261 (2000)
8. T. Schimanski, Ph.D. thesis TU-Eindhoven, 2002

DEVELOPMENT AND PROCESSING OF THERMOPLASTIC STARCH MATERIALS

ROBERT F.T. STEPTO
Polymer Science and Technology Group, Manchester Materials Science Centre, UMIST and University of Manchester, Grosvenor Street, Manchester, M1 7HS, UK

Abstract: The thermoplastics processing of native starch in the presence of water is a recent development with very wide possible applications. Eventually, oil-based polymer materials have to be replaced in many applications by sustainable, inexpensive, natural materials from renewable resources. As with conventional thermoplastics, starch-water melts may be processed by injection moulding and extrusion. The present contribution focuses on injection moulding. The bases of the processing and the thermal and molecular changes occurring are described. In addition, the rheological behaviour of the starch-water melts during processing is analysed quantitatively to give apparent melt viscosities. The dimensional, thermal and mechanical properties of moulded thermoplastic starch polymer (TSP) materials and the products presently being produced from them and from their blends with other thermoplastics are discussed.

Key words: Thermoplastic process, injection moulding, gelatinisation, biodegradability

1. INTRODUCTION

One of the emerging major themes in polymer science for the 21st Century is the preparation of sustainable polymeric chemicals and materials from renewable resources, rather than from petrochemicals. Eventually, oil-based polymer materials will be replaced in many applications by inexpensive, natural-based products. When designed properly, such products have a useful life and properties and are biodegradable with natural degradation products. Within this field, the thermoplastics processing of

F. Ciardelli and S. Penczek (eds.),
Modification and Blending of Synthetic and Natural Macromolecules, 219–240.

natural hydrophilic polymers, particularly starch, in the presence of water is a recent development with very wide possible applications[1-9].

It has been found that by heating hydrophilic polymers in closed volumes in the presence of given amounts of water for given times, homogeneous melts may be formed. If such melts are produced in injection-moulding machines and extruders then they may be processed like thermoplastics. The processing of various starches and of gelatin and other hydrophilic polymers and blends to form useful thermoplastics materials has been achieved in this way[1-5]. Essential features are that limited amounts of water are used and a confined volume is maintained throughout the process if a solid rather than a foamed product is to be formed.

The present contribution is the latest in a series of papers describing the injection moulding of starch[6-9]. First, comparisons with the conventional processing of celluloses and starch are made and the thermal and molecular changes on heating starch-water mixtures are discussed. Second, the rheological behaviour of starch-water melts during the refill part of the injection-moulding cycle is analysed quantitatively to give apparent melt viscosities. Finally, the dimensional, thermal and mechanical properties of moulded thermoplastic starch polymer (TSP) materials and emerging TSP-based products are reviewed.

2. THE BASES OF THERMOPLASTIC STARCH MELT FORMATION

2.1 Comparison with the Processing of Cellulose and Cellulose Derivatives and Conventional Starch

In polymeric terms, a main distinction between starch and cellulose is that the former contains highly branched molecules, whereas the latter contains only linear molecules. The branching means that crystalline sequences are shorter in starch and fibres do not form. Accordingly, native starch is more readily destructured than native cellulose in the presence of water. Indeed, native cellulose cannot be processed as a thermoplastic and it has to be converted to derivatives, e.g., esters and ethers, to reduce the strength of intermolecular forces so that molecular flow can occur under the action of heat and shear. The thermoplastics processing of cellulose derivatives is well-established and well-understood and does not involve water as an integral part of the processes.

The destructuring of starch under the action of heat, water and shear is, of course, the basis of much food preparation and the processing of starch for food and adhesives dates back several millennia into human history. Such conventional processing of starch is in the presence of heat and *excess* water. Initially, a process occurs that is termed gelatinisation[10,11], resulting in a breaking down of the structures in the starch granules to different extents, depending on the starch and the processing conditions. The structures within starch granules are complex and hierarchical, and are partly crystalline. Importantly, the crystallinity and the supramolecular structure are based on the amylopectin component, and not on the amylose. The native granules can be up to 100 μm in diameter. During processing in excess water, the amylopectin crystallinity is lost, some hydrolytic degradation occurs, granules swell and eventually disappear, and the linear amylose molecules diffuse into solution. On aging, the starch solution or suspension undergoes so-called retrogradation to a swollen network material with a structure now based principally on associations between sections of amylose molecules.

From a structural polymer materials point of view, the preceding, conventional processing of starch uses too much plasticiser (water) and eventually lays emphasis on the wrong component, namely, the lower molar mass amylose, of native $M_n < 10^6$ g mol^{-1}. Native amylopectins, on the other hand, can have M_n and M_w in excess of 10^6 g mol^{-1} and 10^8 g mol^{-1}, respectively. Amorphous materials of superior mechanical properties will be obtained if molecular, solid or network structures are formed at lower water contents and are based on the branched component of higher molar mass. In this respect, important break-throughs occurred in the 1980s[1,2,12], culminating in the thermoplastics processing of starch at approximately its natural water content (≈ 15 %) in a closed volume at temperatures above 100 °C. Using conventional injection moulding, glassy, amorphous, thermoplastic starch polymers (TSPs) were obtained. An important characteristic of TSP formation is the thermal and mechanical (shear) destructuring of the starch granules to form a homogeneous melt, the formation being unaccompanied by swelling. Many of the key references are to be found in the patent literature[12-17].

2.2 Compatibility with Water

Starch is a hydrophilic polymer, that is, for present purposes, a polymer whose uptake of water in equilibrium with pure water is unlimited. Hydrophilic polymers are characterised by water-vapour adsorption isotherms of sigmoidal shape, indicating the presence of bound and unbound water, tending to an infinite amount of water adsorbed in the presence of pure water[2]. Such behaviour ensures no phase separation will occur during processing. Both gelatin and starch show the required form of adsorption isotherm, see Figure 1, and both can be injection moulded successfully in the presence of water[1,2].

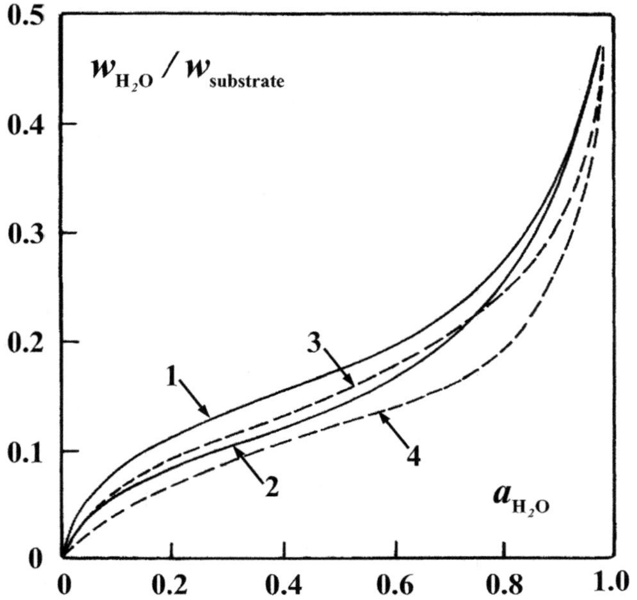

Figure 1. Adsorption isotherms of gelatin and starch in equilibrium with water vapour of activity a_{H_2O} .Curves: 1 – gelatin 20 °C; 2 – gelatin 60 °C; 3 – starch 20 oC; 4 – starch 67 °C.

2.3 Thermal Changes During Processing

The thermal changes occurring on heating starch-water mixtures can be followed using differential scanning calorimetry (DSC) employing completely filled pans with seals designed to withstand the pressure generated by the sample (up to 30 bar). Figure 2 shows examples of the endothermic changes occurring in a potato starch at two water contents. The endotherm for the higher water content occurs at less than 100 °C and is characteristic of the gelatinisation of conventional starch processing, in which the starch granules become swollen and destructured and lose amylose by diffusion. The endotherm for the lower water content is characteristic of melt formation, namely, a thermal and aqueous destructuring of the amylopectin crystallites and molecular order in the granule without the mass diffusion of water[1,3,19,20]. A similar variation in the temperature of the destructurisation endotherm with water content also occurs with gelatin-water mixtures[2].

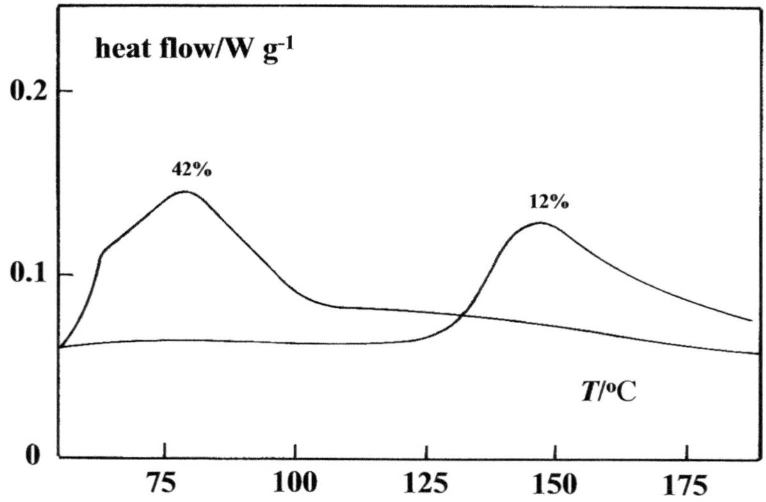

Figure 2. Examples of DSC endotherms for a potato starch at 42% and 12% water contents $(= 100W_{H_2O}/(W_{H_2O} + W_{starch}))$.

Figure 3 shows the variation of the sizes of the low-temperature and high-temperature endotherms with the water content of potato starch expressed as moles of water per mole of anhydroglucose unit (AGU). The results are taken from Donovan[20]. The low-temperature (gelatinisation) endotherms were centered at about 66 °C and the high-temperature endotherms occurred at between 90 °C and 120 °C, the temperature increasing with decreasing water content. The water contents that Donovan used did not go as low as 12 % (= 1.22 mol H_2O (mol AGU)$^{-1}$) because his DSC pans failed, due to excess pressure, below 170 °C.

Donovan's results show that only melt formation occurs at less than about 4 mol H_2O (mol AGU)$^{-1}$ (31 % water content) and that only gelatinisation occurs at greater than about 15 mol H_2O (mol AGU)$^{-1}$ (63 % water content). In addition, from about 10 – 40 mol H_2O (mol AGU)$^{-1}$, the sum of the high-temperaure and low-temperature endotherms remains approximately constant at about 700 cal (mol AGU)$^{-1}$.

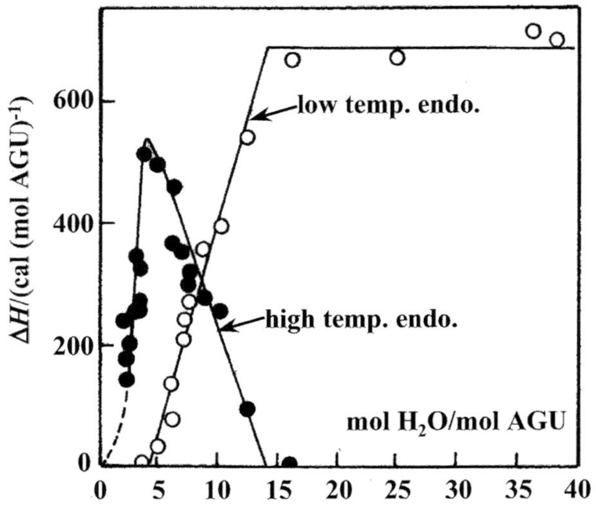

Figure 3. Sizes of low-temperature (gelatinisation) and high-temperature (melt-formation) endotherms of potato starch versus molar ratio of water to starch structural unit (AGU). Ref. [20].

The temperature range and size of the melt-formation endotherm for starch-water mixtures depend on the type of starch and also on the particular batch of starch. For example, different trace amounts of metallic ions in potato starches can affect the temperature range and, hence, the processing conditions.[14] In general, the whole temperature range of the highest endotherm has to be exceeded before destructuring is complete and a homogeneous melt can be achieved.[12]

2.4 Molecular Changes on Heating Starch-Water Mixtures

Figures 4 and 5 illustrate the molecular changes that occur on heating potato starch-water mixtures in closed glass ampoules for various lengths of time at 140-160 °C. Generally, due to hydrolysis, molar mass reduces as the length of time of heating increases. Figure 4 shows the logarithmic (Mark-Houwink) plot of intrinsic viscosity ($[\eta]$) versus mass-average molar mass (M_w), as determined by Rayleigh light scattering. To within experimental error, there was no detectable difference between the values of $[\eta]$ determined in the two solvents used. The low Mark-Houwink exponent of

0.39 is due to the hydrodynamic dominance of the highly branched amylopectin species. The intrinsic viscosity of the native starch was found to be about 280 $cm^3\ g^{-1}$ consistent with $M_w > 10^8\ g\ mol^{-1}$.

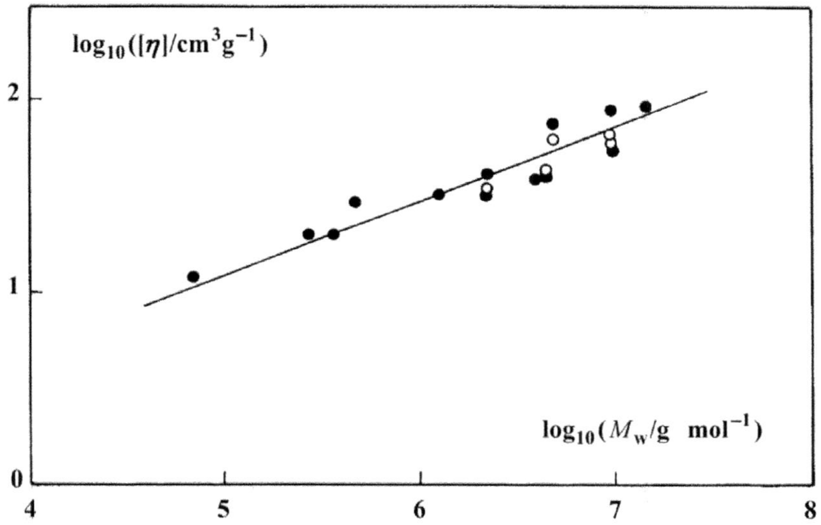

Figure 4. $\log_{10}[\eta]$ versus $\log_{10}M_w$ (Mark-Houwink plot) for hydrolysed potato starches. • $[\eta]$ measured in dimethyl sulfoxide at 40 °C; o $[\eta]$ measured in dimethyl sulfoxide / 6M urea at 40 °C; M_w measured in dimethyl sulfoxide. Least-squares line gives $[\eta]/cm^3\ g^{-1} = 0.23(M_w/g\ mol^{-1})^{0.39}$.

Figure 5 gives the logarithmic plot of number-average molar mass (M_n) versus M_w. The values of M_n were determined from assays of the numbers of reducing end-groups per unit mass of sample. There is only one reducing end-group per molecule. Hence, the number of reducing end-groups present in a sample is equal to the number of molecules present and $M_n/g\ mol^{-1}$ = (dry mass of sample/ moles of end-groups). It can be seen that $M_n \propto M_w^{0.28}$ and $M_w/M_n \propto M_w^{0.72}$, showing that M_w/M_n decreases rapidly as more hydrolysis occurs in sample. At the highest molar mass shown in Figure 5,

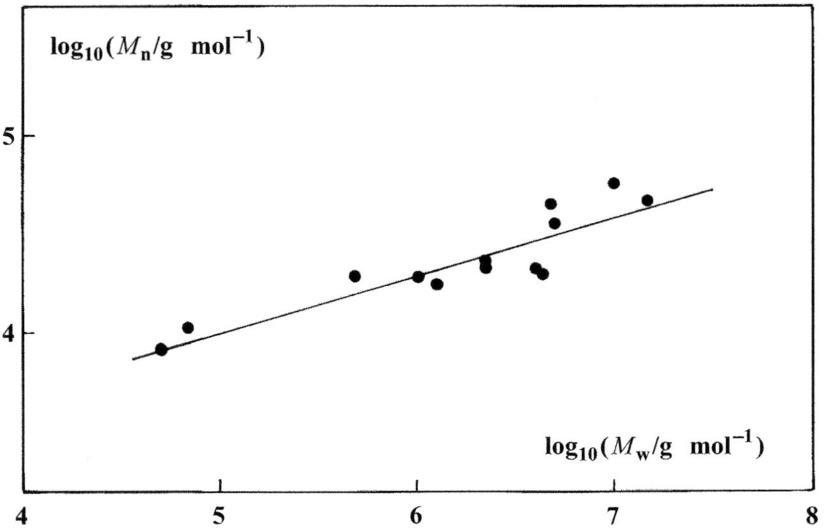

Figure 5. $\log_{10}M_n$ versus $\log_{10}M_w$ for for hydrolysed potato starches. Least-squares line gives $M_n/\text{g mol}^{-1} = 405(M_w/\text{g mol}^{-1})^{0.28}$, from which $M_w/M_n = (1/405)\times(M_w/\text{g mol}^{-1})^{0.72}$.

$M_w \approx 1.5\times10^7$ g mol^{-1} and $M_w/M_n \approx 300$, whilst for the lowest molar mass shown, $M_w \approx 5\times10^4$ g mol^{-1} and $M_w/M_n \approx 6.2$. This large decrease in dispersity, M_w/M_n, is consistent with random chain-scission through hydrolysis. The larger molecules are the more likely to be attacked, causing M_w to be reduced more quickly than M_n. To obtain injection-moulded TSPs with satisfactory mechanical properties it is found that M_w should lie in the approximate range 10^6-10^7 g mol^{-1}. Figure 5 indicates that M_w/M_n then lies in the approximate range 100-300.

2.5 Injection-Moulding Behaviour and Apparent Melt Viscosity

2.5.1 General Considerations

The thermoplastic processing (by heat and shear) of starch-water mixtures is essentially a reactive processing. The starch granules need to be destroyed to form a homogeneous polymer melt. In addition, the molar mass, particularly of the amylopectin component, needs to be reduced to give a polymer melt of manageable viscosity yet a final material with satisfactory mechanical properties. The processing conditions, therefore, need to be more carefully controlled, in terms of temperature profile, shear rate and residence time, than they do for conventional thermoplastics. It is also important to be able to use injection-moulding machines of conventional design and specifications.

Many of the processing experiments have used standard Arburg injection-moulding machines, employing conventional 22-25 mm, reciprocating three-zone (feed zone, compression zone and metering zone) screws with non-return valves. The metering zones were about 90 mm long, the flight depths 1.7-1.8 mm and the flight angles about 16.5°. The length-to-diameter (L/D) ratios were equal to about 20. Starch and water were thoroughly mixed, together with small amounts of a powder-flow additive and a melt lubricant, to give moulding powders of water contents of about 17%. The initial part of the feed zone of the injection-moulding machine was liquid-cooled to counteract the effects of frictional heating and the latter part was typically set at a temperature of 90 °C. The compression zone was set at about 165 °C and the metering zone was set at about 170 °C.

By taking photomicrographs of moulded starch materials it was found that a certain minimum total residence time had to be exceeded to obtain homogeneous melts and, hence, products of acceptable properties. Using the stated processing conditions, a residence time of about 500 s was needed. Further, by carrying out processing with different back-pressures applied to the screw (i.e. to the melt) and at different screw speeds it was possible to show that, under the correct processing conditions, starch-water melts behave like conventional thermoplastic melts. The bases of these processing experiments are now described. They were key to developing understanding and control of the thermoplastic processing of starch.

2.5.2 The Two-Plate Model of Injection-Moulding Behaviour

The concept of the two-plate model is illustrated in Figure 6. The two plates between which flow is supposed to occur are the unfolded inner surfaces of the screw and barrel. The model is used to analyse flow in injection-moulding machines and screw extruders[21,22].

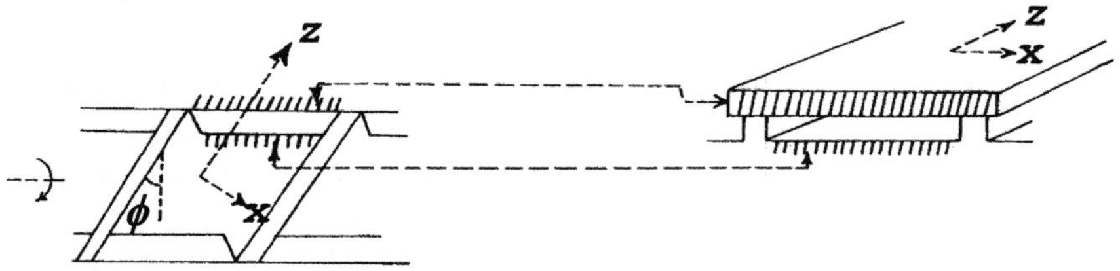

Figure 6. Illustrating the basis of the two-plate model used to analyse flow in injection-moulding machines and screw extruders. ϕ is the flight angle of the screw and flow occurs along the z direction. Ref. [21,22]

The model makes four basic assumptions:

1. The shear energy in the feed and compression zones of the screw is used solely for melt formation.
2. Only melt exists in the metering zone.
3. Material throughput is governed by the flow in the metering zone.
4. Flow can be analysed on the basis of a constant (Newtonian) viscosity.

The total rate of flow of polymer along the screw ($\dot{V}_{\text{total}}$) is equal to the forward drag-flow rate due to screw rotation ($\dot{V}_{\text{s}}$) less the backward viscous-flow rate ($\dot{V}_p$) resulting from the pressure difference (Δp) along the flow channel of the metering zone (the back-pressure), that is,

$$\dot{V}_{\text{total}} = \dot{V}_{\text{s}} - \dot{V}_{\text{p}} \; . \tag{1}$$

The detailed expressions for $\dot{V}_{\text{s}}$ and $\dot{V}_p$ are

$$\dot{V}_{\text{s}} = \frac{\pi^2 D^2 h_{\text{m}} n \cdot \sin\phi \cdot \cos\phi}{2} \tag{2}$$

and
$$\dot{V}_p = \frac{\pi^2 D h_{\text{m}}^3 \sin^2\phi}{12\eta} \cdot \frac{\Delta p}{L_{\text{m}}} \; , \tag{3}$$

D is the outer screw diameter, h_{m} the flight depth along the metering zone, n the screw rotational speed in revolutions per second, ϕ the flight angle (see Figure 6), η the apparent Newtonian viscosity of the melt in the metering zone and L_{m} the length of the metering zone.

For injection moulding machines, $\dot{V}_{\text{total}}$ can be measured by knowing the volume of the mould being used (the shot volume) (V_{sh}) and measuring the refill time (t_{ref}), during which the screw rotates against Δp to feed the next shot past the front of the screw ready for injection. Thus,

$$\dot{V}_{\text{total}} = \frac{V_{sh}}{t_{ref}} \tag{4}$$

Using equations (2), (3) and (4) to substitute for $\dot{V}_{\text{total}}$, $\dot{V}_{\text{s}}$ and $\dot{V}_{p}$ in equation (1) gives

$$\frac{V_{\text{sh}}}{t_{\text{ref}}} = \frac{\pi^2 D^2 h_{\text{m}} n \cdot \sin\phi \cdot \cos\phi}{2} - \frac{\pi^2 D h_{\text{m}}^3 \sin^2\phi}{12\eta L_{\text{m}}} \cdot \Delta p \tag{5}$$

and, dividing equation (5) by $V_{\text{sh}} \cdot n$,

$$\frac{1}{t_{\text{ref}} \cdot n} = \frac{\pi^2 D^2 h_{\text{m}} n \cdot \sin\phi \cdot \cos\phi}{2V_{\text{sh}}} - \frac{\pi D h_{\text{m}}^3 \sin^2\phi}{12\eta L_{\text{m}} V_{\text{sh}}} \cdot \frac{\Delta p}{n} \tag{6}$$

Equation (6) can also be written as

$$\frac{1}{t_{\text{ref}} n} = \frac{\dot{V}_{\text{total}}}{n V_{\text{sh}}} = \frac{\dot{V}_{\text{s}}}{n V_{\text{sh}}} - \frac{\dot{V}_{p}}{n V_{\text{sh}}} \tag{7}$$

$\frac{1}{t_{\text{ref}} n} = \frac{\dot{V}_{\text{total}}}{n V_{\text{sh}}}$ is equal to the (fractional) number of mould shots per screw revolution, $\frac{\dot{V}_{\text{s}}}{n V_{\text{sh}}}$ is the fractional shot volume per screw revolution due to screw rotation (drag flow) and describes the transport of a solid ($\eta = \infty$) and $\frac{\dot{V}_{p}}{n V_{\text{sh}}}$ is the reduction in fractional shot volume per screw revolution due to backward viscous flow.

In order to evaluate η, equation (6) is written as

$$\frac{1}{t_{\text{ref}} \cdot n} = a_0 - a_1 \cdot \frac{\Delta p}{n} \quad , \tag{8}$$

where
$$a_0 = \frac{\pi^2 D^2 h_{\text{m}} n \cdot \sin\phi \cdot \cos\phi}{2V_{\text{sh}}} \tag{9}$$

and
$$a_1 = \frac{\pi D h_{\text{m}}^3 \sin^2\phi}{12 L_{\text{m}} V_{\text{sh}}} \cdot \frac{1}{\eta} \quad . \tag{10}$$

In equation (8), a_0 depends only on screw and mould dimensions and a_1 is inversely proportional to the viscosity of the melt. Thus, by carrying out injection moulding with a given mould (V_{sh}) and a given screw and given temperature profile, it is possible to measure the variation of t_{ref} for different values of $\Delta p/n$. For this procedure, it should be remembered that the shear rate ($\dot{\gamma}$) is determined by n, with[21,22]

$$\dot{\gamma} = \frac{\pi D}{h_{\text{m}}} \cdot n \quad . \tag{11}$$

Hence, experiments are best carried out under different back-pressures and at constant n. As back-pressure is increased for a given screw speed, refill time increases because the backward (viscous) flow rate increases, detracting more from the forward, drag flow and giving a lower total flow rate.

2.5.3 Effective Viscosity of Starch-Water Melts

Figure 7 shows the results of the measurements of refill times under different applied back pressures and at constant screw speed. In accordance with equation (8), $1/(t_{ref} \cdot n)$ is plotted versus $(\Delta p/n)$. The linear behaviour predicted by equation (8) is observed with the intercepts at $\Delta p/n = 0$ agreeing

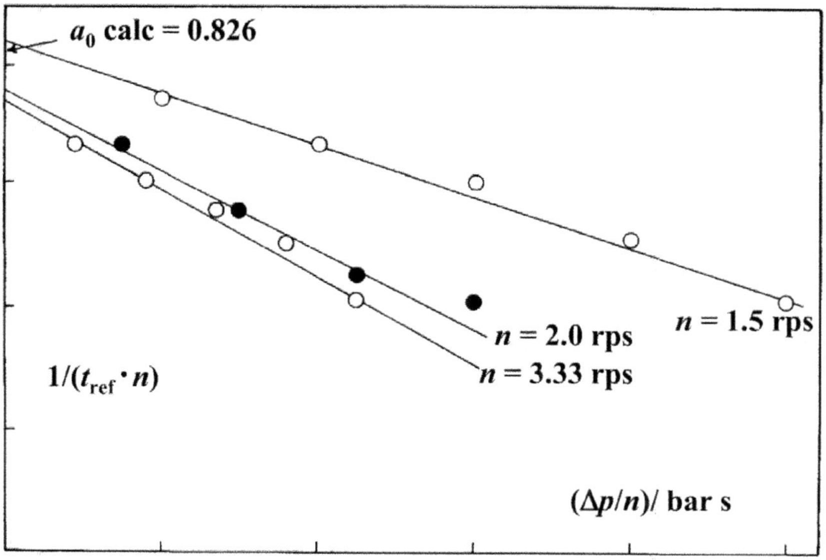

Figure 7. Injection moulding of potato starch at 17 % water and a metering zone temperature of 175 °C. Results of experiments measuring refill times (t_{ref}) at constant screw rotation speeds (n) under different back-pressures (Δp). Results analysed according to equation (8) with $1/(t_{ref} n)$ plotted versus ($\Delta p/n$).

well with the value of a_0 calculated from the screw and mould dimensions using equation (9). The slightly lower intercepts for the two higher screw speeds probably result from under-feeding of the initial moulding powder along the screw reducing the drag flow.

Using equation (10) to analyse the slopes of the lines in figure 7 gives an apparent melt viscosity, η, of about 1000 Pa s at the lowest screw rotation speed of 1.5 rps and about 700 Pa s at the highest speed of 3.33 rps. Such values are typical of those for conventional thermoplastic melts, as can be seen from Figure 8 where logarithmic plots of η versus shear rate, γ, are shown for the starch-water melts investigated and for polyethylene melts. The latter were taken from a manufacturer's literature. From the similarity of the values of η for starch-water mixtures and polyethylene obtained for a given shear rate, it can be seen that, once the processing parameters, such as water content, temperature profile and screw characteristics, are properly defined, starch processes like a conventional thermoplastic. The similarity of the viscosity-shear rate behaviour of starch-water melts to that of polyethene shown in Figure 8 is corroborated by the measurements of viscosities of pre-processed corn starch-water melts by Willet, Jasberg and Swanson[5].

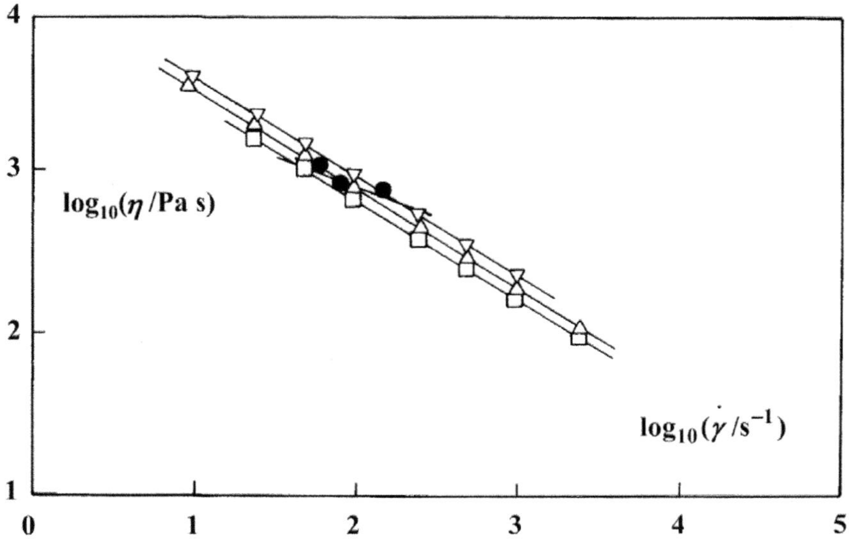

Figure 8. Comparison of melt viscosities of a medium density polyethylene and potato starch-water mixtures. Polyethene: □ 230 °C, Δ 210 °C, ∇ 190 °C. • Starch-water, 17 % water, 175 °C (metering zone).

2.6 Molar-Mass Changes on Injection Moulding and Extrusion

In addition to the reductions in molar mass of the native starch that occur on heating starch-water mixtures, such as those described in section 2.4, further reductions will occur during the refill part of the injection-moulding cycle when the melt undergoes backwards viscous flow under the influence of the applied back-pressure. Such flow aids homogenisation of the melt but the reductions in molar mass it produces must be controlled. However, it should be possible to use extrusion to aid the formation of homogeneous melts without excessive molar-mass reduction and to produce granules for subsequent injection moulding.

Examples of the changes in molar mass that occur after extrusion and after extrusion and subsequent injection moulding are given in Table 1. It is clear that injection moulding leads to much larger reductions in molar mass and that pre-extrusion can indeed be used to aid the production of homogeneous melts without excessive reductions in molar mass. The table also shows that larger reductions during injection moulding are apparent at higher metering-zone temperatures and longer residence times. Variations in the behaviours of native starches are also apparent from the different behaviours of potato starches (1) and (2). The beneficial effects of pre-extrusion on the mechanical properties of injection moulded starch are discussed later in section 3.3.

Table 1. Molar-mass changes on the injection moulding and extrusion of potato starch-water mixtures. extrusion conditions: 22 % H_2O / 150 °C / 0.833 rps. Injection moulding conditions: 17 % H_2O / 2.083 rps / Δp = 50 bar; metering-zone temperatures listed in table.

Starch	native	extruded	pre-extruded & injection moulded		
	10^{-6} M_v / g mol^{-1}	10^{-6} M_v / g mol^{-1}	T / °C	t_{res} / s	10^{-6} M_v / g mol^{-1}
potato starch (1)	670	540	170	300	34
				450	30
				600	20
				750	15
			175	300	16
				450	15
Potato starch (2)	650	270	155	300	11
				450	10
				600	5
				750	2
			165	300	4

3. PROPERTIES OF INJECTION-MOULDED TSPS

3.1 Dimensional Stability

Obviously, the dimensions of moulded objects from hydrophilic polymers depend on their water contents. If precise dimensions are required, processing needs to be carried out so that products are formed at approximately the equilibrium in-use water content. For potato starch, for example, this means water contents of around 14% for use under ambient conditions (50% relative humidity, 20-25 °C). If higher water contents are used in processing, distortion and shrinkage will occur as the equilibrium water content is achieved naturally after processing. In addition, higher water contents can induce more hydrolytic degradation of the starch chains during processing and they favour gelatinisation rather than melt formation. On the other hand, if lower water contents are used, thermal degradation can occur during processing, as well as swelling after processing.

3.2 Thermal Properties

Figure 9 shows the glass-transition temperatures of injection-moulded starch-water and gelatin-water mixtures. Similar values of T_g (60 °C to 80 °C) are observed for the two materials under normal ambient conditions, when values of w_{H_2O} are in the range 0.12 to 0.14. The materials are then in their glassy states.

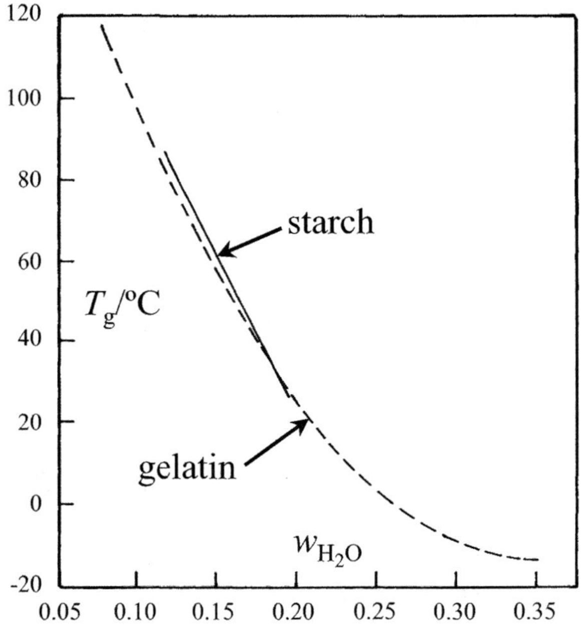

Figure 9. Glass-transition temperature (T_g) versus weight fraction of water (w_{H_2O}) for injection-moulded gelatin-water and starch-water mixtures.

3.3 Stress-Strain Behaviour

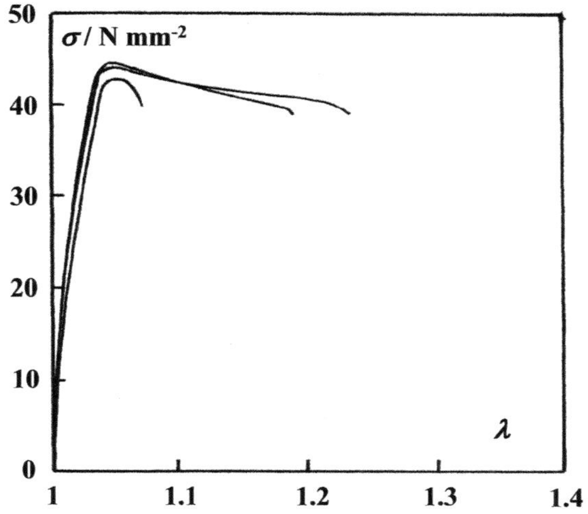

Figure 10. Tensile stress (σ) versus deformation ratio (λ) at ambient temperature of injection-moulded native potato starch conditioned to a water content of 13.5 %.

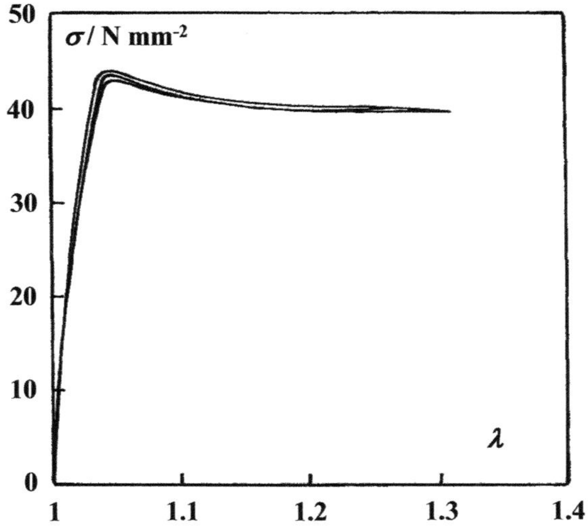

Figure 11. Tensile stress (σ) versus deformation ratio (λ) at ambient temperature of injection-moulded pre-extruded potato starch conditioned to a water content of 13.5 %.

Figure 10 illustrates the stress-strain behaviour at ambient temperature of tensile test-pieces moulded from native potato starch at 17% water and conditioned to 13.5 % water and Figure 11 demonstrates the reproducibility and improvements in properties that can be achieved by using pre-extrusion (see section 2.6.)

Figure 12 shows the effects of water content on stress-strain properties. According to Figure 9 the glass-transition temperatures at the water contents used vary from about 60°C to 100°C and the behaviour shown in Figure 12 is typical of that of glassy thermoplastics. The initial moduli are about 1.5 GPa, similar in value to those of glassy polyolefins (polypropylene and high-density polyethylene) and the materials show yield points at between 5 to 10% extension. The changes in properties with decrease in water content are consistent with the loss of free water, which has a plasticising action on the materials.

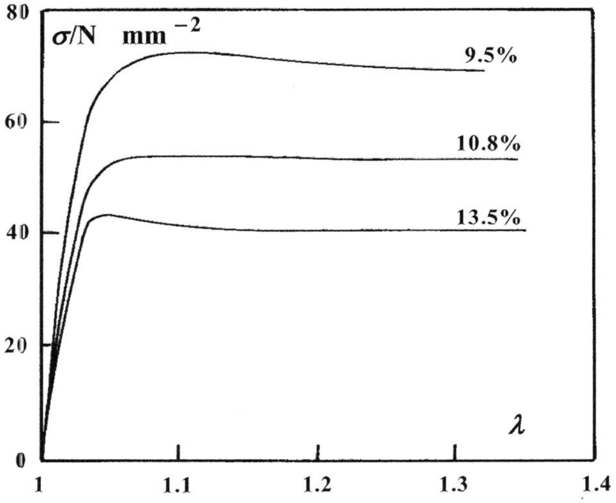

Figure 12. Tensile stress (σ) versus deformation ratio (λ) at ambient temperature of injection-moulded pre-extruded potato starch conditioned to the water contents shown with the curves.

It should be emphasised that the results presented in the present paper are preliminary. It is important that extensive systematic and fundamental studies on TSP processing and TSP materials are pursued in the future. In comparison with that of synthetic polymers, knowledge of the processing-structure-property relationships for TSP-based materials is still in its infancy. It is an exciting field with many possibilities for discoveries and developments.

4. TSP-BASED PRODUCTS

The first commercial product made of injection-moulded TSP was the drug-delivery capsule, Capill[1,6,19,23,24]. Figure 13 compares Capill with a conventional hard gelatin capsule (HGC) made by dip moulding. The more precisely controlled dimensions of the injection-moulded starch capsule are immediately apparent (linear dimensions are reproducible to better than 0.1 mm) and, because of the good adhesive properties of starch, the cap and body can easily be sealed to make the capsule tamper-proof. In addition, Figure 14 shows that Capill is bioequivalent to HGC in its release behaviour of Aspirin.

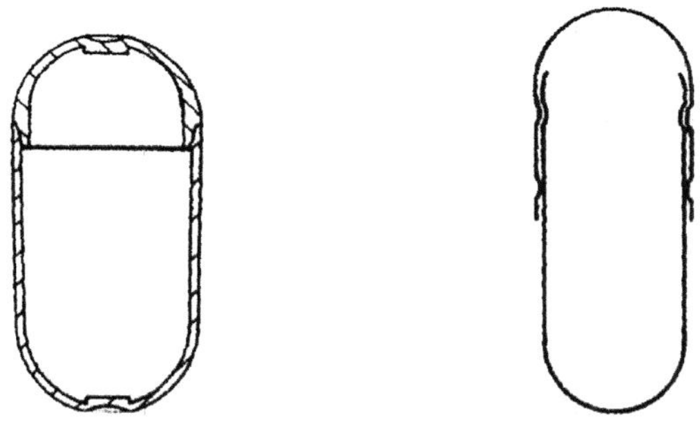

(a) **(b)**

Figure 13 (a) An injection-moulded starch capsule (Capill) and (b) a hard gelatin capsule (HGC).

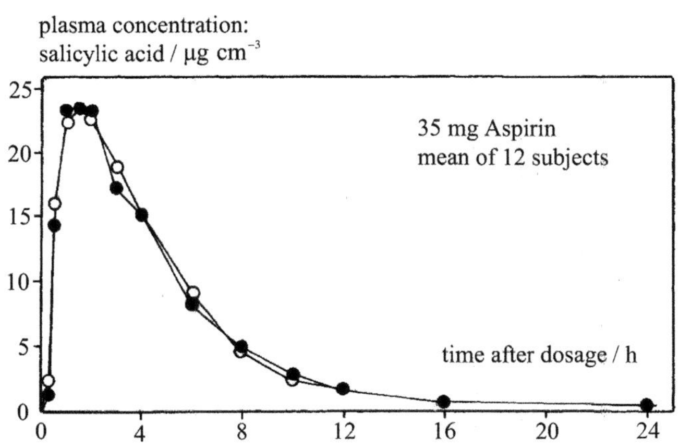

Figure 14. Release of Aspirin from Capill (•) and HGC (o).

More TSP products are gradually appearing, e.g., golf tees, cutlery, plates, food containers. In addition, extrusion has been applied to produce rigid foams, suitable for loose-fill packaging. Generally, the polymers are dimensionally and mechanically stable under ambient, indoor conditions and are completely biodegradable and compostable. They break-down in water. Hence, TSPs can be considered as a new class of inexpensive, green polymers that, after use, can be returned to the natural cycle with no pollution.

Further developments include the thermoplastic processing of blends of starch with hydrophilic synthetic polymers[4,16,25], e.g., poly(ε-caprolactone) and poly(ethylene-*co*-vinyl alcohol), to give the possibility of flexible films and materials with improved mechanical properties, lower water sensitivity, but also lower biodegradability.

Tables 2 and 3 summarise the TSP-based materials, blends and types of products currently available.

Table 2. Examples of TSP-based materials and products.

PRODUCT OR MATERIAL NAME	*PRODUCT TYPE*	*MANUFACTURER*
CAPILL	***pharmaceutical capsules***	***Capsugel/Warner Lambert; West Pharmaceutical Services***
ECO-FOAM	***foam packaging pellets***	***National Starch***
Mater-Bi ***(starch and starch blends with hydrophilic polymers)***	***moulding & extrusion powders for flexible films, rigid packaging, moulded articles, etc.***	***Novamont, Italy***

Table 3. Examples of TSP-based blends and their applications. The first two blends are marketed by Mater-Bi

BLEND	*PROCESS*	*USE*
starch/poly(ε-caprolactone) (Mater-Bi)	***film blowing blow moulding extrusion casting injection moulding***	***packaging film mooulde articles paper laminate bags mulch films, etc.***
starch/poly(ethene-co-vinyl alcohol) (Mater-Bi)	***injection moulding***	***moulded articles***
starch/poly(vinyl pyrrolidone)	***injection moulding film blowing (?)***	

ACKNOWLEDGEMENTS

My sincere thanks to the research group of Capsugel AG at Riehen, Switzerland for making this research possible and to UMIST for an extended leave of absence during the main period of the work. I should also like to thank Dr. C. Bastioli and her colleagues at Novamont S.p.A. for information and interesting discussions.

REFERENCES

1. Eith, L., Stepto, R.F.T., Tomka, I. and Wittwer, F. (1986) The injection-moulded capsule, *Drug Dev. & Ind. Pharm.* ***12,*** *2113-2126.*
2. Stepto, R.F.T. and Tomka, I. (1987) Injection moulding of natural hydrophilic polymers, *Chimia* ***41,*** *76-81.*
3. Tomka, I. (1991) in H. Levine, L. Slade (eds.), *Water Relationships in Food,* Plenum Press, New York, *p.627.*

4. Lay, G., Rehm, J., Stepto, R.F.T., Thoma, M., Sachetto, J.-P., Lentz, D.J., and Silbiger, J. (1992) Hydrophilic polymer processing, *US Patent 5,095,054*
5. Willet, J.L., Jasberg, B.K. and Swanson, C.L. (1994) Melt rheology of thermoplastic starch, in *ACS Symposium Series 575*, *Polymers from Agricultural Coproducts*, eds. M.L. Fishman, R.B. Friedman and S.J. Haag, Amer.Chem.Soc., Washington D.C., *Chapter 3, pp. 51-68.*
6. Stepto, R.F.T. (1997) Thermoplastic starch and drug delivery capsules, *Polymer International **43**, 155-158.*
7. Stepto, R.F.T. (2000) Thermoplastic starch, *Macromol. Symp.* ***152**, 73-82.*
8. Ross-Murphy, S.B. and Stepto, R.F.T. (2001) Green polymers for the 21st century: real prospects and virtual realities, in A.J. Ryan (ed.), *Emerging Themes in Polymer Science*, Special Publication No. 263, The Royal Society of Chemistry, Cambridge, 2001, *Chapter 13.*
9. Stepto, R.F.T. (2003) The processing of starch as a thermoplastic, *Macromol. Symp.* ***201**, 203-212.*
10. *Starch Chemistry and Technology*, R.L. Whistler, J.N. BeMiller and E.F. Paschall (eds.), Academic Press, New York, 1984.
11. Lapasin, R. and Pricl, S. (1995) *Rheology of Industrial Polysaccharides: Theory and Applications*, Blackie Academic and Professional, Glasgow.
12. Stepto, R.F.T. and Dobler, B. (1988) A process for producing destructurised starch, *UK Patent 88 01562.*
13. Stepto, R.F.T., Tomka, I. and Dobler, B. (1987) Destructurised starch and process for making same, *U.K. Patent 87 05442.*
14. Sachetto, J.-P., Stepto, R.F.T. and Zeller, H. (1987) Destructurised starch essentially containing no bridged phosphate groups and process for making same, *U.K. Patent 87 1594.*
15. R.F.T. Stepto, M. Thoma and I. Tomka (1987) Shaped articles made from pre-processed starch, *U.K. Patent 87 19485.*
16. Lay, G., Rehm, J., Stepto, R.F.T. and Thoma, M. (1988) Starch blends, *U.K. Patent 88 02313.*
17. Silbiger, J. and Stepto, R.F.T. (1988) New polymer composition, *U.K. Patent 88 12502.*
18. Sachetto, J.-P., Egli, M. and Stepto R.F.T. (1989) Washing process for starch, *U.K. Patent 89 07459.*
19. Eith, L., Stepto, R.F.T., Tomka, I., and Wittwer, F. (1986) Injection-moulded drug-delivery systems, *Proc. Interphex '86 Conference*, Cahners Exhibitions Ltd., Brighton, *pp.2-22 to 2-31.*
20. Donovan, J.W. (1979) Phase transitions of the starch-water system, *Biopolymers* ***18**, 263-275.*
21. *Extrusion and Other Plastics Operations*, ed. N.M. Bikales (ed.), Wiley-Interscience, New York, 1971.
22. Agassant, J.-F., Avenas, P., Sergent, J.-Ph. and Carreau, P.J. (1991) *Polymer Processing*, Hanser Publishers, Munich.
23. Augart, H., Borgmann, A and Stepto, R.F.T. (1987) Pharmaceutical properties and behaviour of capill: the new starch drug container made by injection moulding, *Proc. 6th Pharmaceutical Technology Conference, Canterbury, p.257-274.*
24. Vilivalam, V.D., Illum, L. and Iqbal, K. (2000) Starch capsules: an alternative sysytem for oral drug delivery, *Pharmaceutical Sci. & Tech. Today 3, 64-69.*
25. *Mater-Bi Brochure*, Novamont S.p.A., Novara, 1999

THEORETICAL CONSIDERATIONS ON THE REACTIONS IN POLYMER BLENDS

NICOLAY A. PLATE', ARKADY D. LITMANOVICH, and YAROSLAV V. KUDRYAVTSEV

A.V. Topchiev Institute of Petrochemical Synthesis, Russian Academy of Sciences Leninsky prosp. 29, Moscow B-71, 119991, Russia

Abstract: Advances in theory of reactions in polymer blends are reviewed. For polymer-analogous reaction, the evolution of the blend structure under concerted action of the reaction and interdiffusion is described for the first time. For end-coupling reaction, the reaction kinetics both in a homogeneous melt and at the interface is considered. An influence of the diblock copolymer formed both on the reaction kinetics and on the thermodynamic equilibrium is analyzed. Theoretical models account for interchain interactions between the block copolymer and homopolymers and analyze such phenomena as segregation of the copolymer to the interface leading to the significant change in the interface properties and the blend stability. For interchain exchange reaction proceeding in a homogeneous melt, an analytical description of the molecular weight and block length transient distributions of the reaction product has been developed. The competition between the reaction and spinodal decomposition has been described. Some unsolved important problems are formulated. The role of experimental studies of the reactions in relatively simple model systems, which stimulate formulating appropriate theoretical problems, is emphasized.

Key words: polymer blends, macromolecular reactions, theory, neighbor effect, unit distribution, interchain interactions, interdiffusion, diblock copolymer, compatibilization, end-coupling, particle growth, interchange reactions

F. Ciardelli and S. Penczek (eds.),
Modification and Blending of Synthetic and Natural Macromolecules, 241–281.

1. INTRODUCTION

In this lecture, we will discuss the reactions of polymer modification trying to elucidate the main features of this type of reactions. Chemical modification of natural polymers has been carried out in practice even earlier than the scientific basics for polymer synthesis via polyaddition and polycondensation were developed.

For example, nitration of cellulose was described by Schenbein even in 1846. Approximately at the same time Goodyear succeeded in vulcanization of natural rubber.

Polymer-analogous reactions constitute one of the principal methods of modification. Such a term was introduced by Staudinger in 1934 for the reactions that involve pendant groups keeping polymer backbone unchanged. This way allows one to apply the whole variety of methods of organic chemistry for obtaining new high-molecular compounds.

Moreover, modification is the only possibility for synthesizing certain polymers. For example, polyvinyl alcohol can be obtained exceptionally by hydrolysis of polyvinyl esters, mostly polyvinyl acetate (see Figure 1).

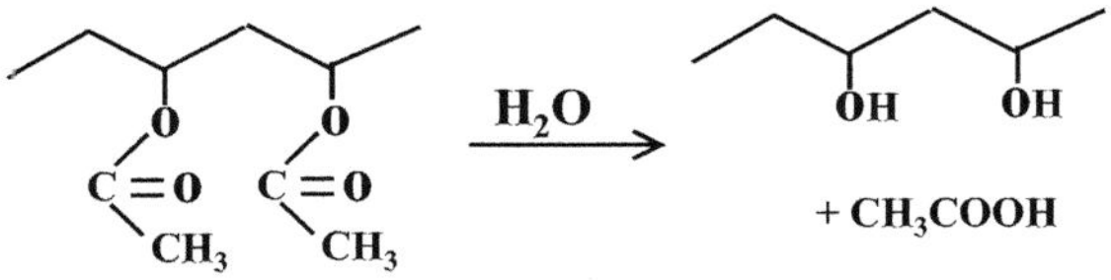

Figure 1. Polyvinyl acetate hydrolysis.

Copolymers attract special interest both of academic science and industry. They may be obtained by copolymerization from a mixture of different monomers and by polymeranalogous reactions terminated at the prescribed conversion as well. However, stereoregular copolymers cannot be obtained by copolymerization while in many cases it can be easily done by modification methods. For example, by hydrolysis of stereoregular polymethacrylates it is possible to obtain stereoregular copolymers of methacrylic acid (MAA) and the corresponding ester (see Fig. 2).

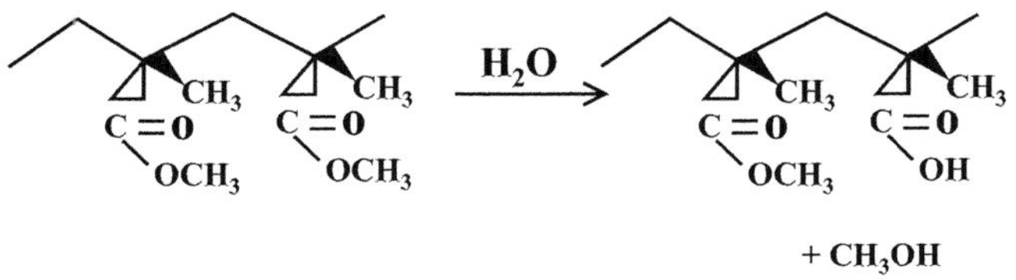

Figure 2. Hydrolysis of isotactic polymethyl methacrylate.

At given composition, properties of a copolymer depend on the distribution of units along the chain. Changing the proceeding conditions of a polymer-analogous reaction, it is possible to vary this distribution and obtain copolymers different in structure and therefore in properties from the products of copolymerization.

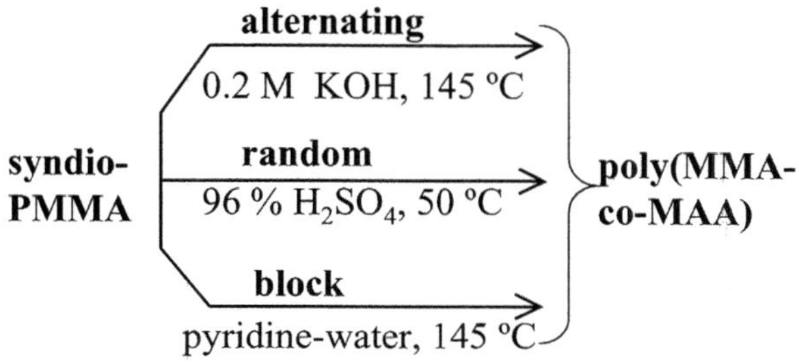

Figure 3. Hydrolysis of syndiotactic PMMA.

In particular, hydrolysis of polymethyl methacrylate (PMMA) (see Figure 3) may yield copolymers of MMA and MAA of different structure: random, with tendency to alternation or blockiness of units as well[1,2].

Reactions in polymer blends are widely used both to improve the blend properties, first of all compatibility, and to synthesize new high-molecular compounds. Hundreds examples of chemical transformations have been considered in the books and reviews devoted to the reactive blending. The publications indicate however that the impressive achievements related to the production of new polymer composites via reactive processing are based mainly on empirical approaches. Certainly, such approaches are necessary and fruitful for this rapidly developing field. At the same time it is clear that contemporary technologies should be grounded on more solid scientific base. This means that a thoroughgoing investigator should search for the regularities in the processes he studies.

However, it is difficult to reveal all the regularities by studying what is really going on in extruders. The situation was realized clearly by some experimentalists who began to study the peculiarities of the reactions in polymer blends under relatively simple model conditions.

The following three types of the model systems are mainly used[3]:

a) Homogeneous blends of the compatible polymers, mostly of similar chemical structure (say, polystyrene (PS) and deuterated PS (dPS), bearing different end or pendant reactive groups. Such blends keep homogeneity during the reaction between functional groups and are used to estimate the kinetic parameters: rate constants, activation energies.
b) Quasi-homogeneous blends prepared from mutual solutions of the components by casting or freeze drying. The initial blends being fine

dispersed or even highly interpenetrated are thermodynamically unstable and may be used to study a competition between the reaction and phase separation. Studying also an early stage of the reaction it is possible to estimate an influence of the initial blend structure on the reaction kinetics.

c) Heterogeneous blends with sharp, in particular planar, interface are used to estimate the contributions of the reactivity and diffusion in the process. A comparison of the data for homogeneous and heterogeneous systems enables one to reveal some peculiarities of the reaction within the interface layer.

Another advantage of studying model reactive systems is the possibility of theoretical analysis. Being of little use at the moment for reactive blending in mixers and extruders, theory can successfully describe processes in model systems at a quantitative level. It appears that reactions in polymer blends proceed very differently from the same reactions in the corresponding mixtures of low-molecular analogs.

In general, experimental studies of the reactions in model systems stimulate a formulation of the appropriate theoretical problems and vice versa. Let us emphasize that just studying model systems serves as a base to elucidate peculiarities of the true reactive blending in extruders and other mixers.

In this lecture, concrete examples of such studies will be considered.

First, a polymeranalogous reaction leading to a statistical copolymer. The main problem to be analyzed is an evolution of the reacting blend structure under mutual influence of the reaction and interdiffusion.

Second, an end-coupling reaction leading to a block copolymer. Various aspects of the block copolymer compatibilizing effect will be considered.

Third, an interchain exchange, mainly transesterification, yielding multi-block copolymers. Some problems concerning the reaction kinetics and competition between the reaction and phase separation will be discussed.

The analysis of interchain interactions is a key approach in the theory of reactions in polymer blends. These interactions determine miscibility of the blend components and affect strongly the dynamics of the reactive blending by changing both reactivity of macromolecules and their mobility.

Let us begin with polymeranalogous reactions. We will follow the evolution of the theory of these reactions from diluted solutions up to polymer blends.

2. POLYMER-ANALOGOUS REACTIONS

Polymer-analogous reactions, appertaining to macromolecular reactions, are widely used for chemical modification of polymers. In the course of a polymer-analogous reaction, A units of a polymer transform into B units, the main chain being unchanged, and an AB copolymer is formed as the reaction product.

Random copolymers formed in the course of reactive blending enhance the blend compatibility, thus influencing the physical properties of polymer composites.

For example, during a foaming the polyvinylchloride (PVC) and PMMA mixture at 200 °C, a partial dehydrochlorination of PVC occurs generating mobile allylic H atoms capable to form hydrogen bonds with PMMA carbonyl groups. As a result the impact-resistant foam is formed having a single glass temperature[4].

The theory of macromolecular reactions started to be developed from the early sixties. Such phenomena as the effect of neighboring units, electrostatic, conformational and configurational effects were described[1].

2.1 Neighbor effect

The term neighbor effect denotes a change in the reactivity of the given functional group or unit under the action of an already reacted group which is located adjacent to the given one. This effect is observed in a number of cases of hydrolysis, cyclization, quaternization, epoxydation and many other reactions. The reactivity of a group having the reacted neighboring group can increase by hundreds and even thousands times or decrease by ten times so both accelerating and retarding neighbor effect are possible.

Consider the following reaction model: A units are irreversibly transformed into B units, the reaction is of the first order but the reactivity of A units depends on the nature of their nearest neighbors (see Fig. 4).

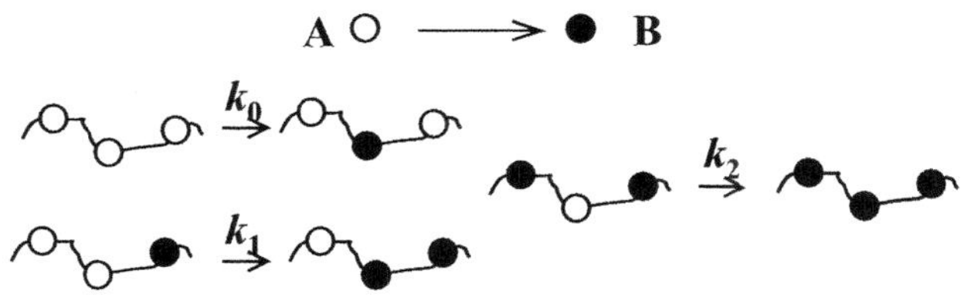

Figure 4. Scheme of the neighbor effect.

Let us denote the rate constants of the transformation of A units having 0, 1, and 2 reacted neighbors as k_0, k_1, and k_2, respectively. These constants do not depend on the concentration of reactants or on the degree of conversion.

The investigation of such reaction model includes derivation and solution of the kinetic equations, description of the unit distribution along the chain, and, finally, calculation of the compositional heterogeneity of the ensemble of reacting chains.

The most convenient form of the kinetic equations was proposed by Keller[5]. Let the initial chain be a homopolymer A and introduce N_0, N_1, and N_2 as the time-dependent average fractions of A-centered triads, that is AAA, AAB+BAA, and BAB, respectively. The triad fractions obey the following equations:

$$\begin{aligned} &dN_0/dt = -(k_0 + 2\bar{k})N_0 \\ &dN_1/dt = -(k_1 + \bar{k})N_1 + 2\bar{k}N_0 \qquad \bar{k} = \frac{k_0 N_0 + k_1 N_1}{N_0 + N_1} \\ &dN_2/dt = -k_2 N_2 + \bar{k}N_1 \end{aligned} \tag{1}$$

Solving these equations, one gets the dependence of the fraction of A units, $N = N_0 + N_1 + N_2$, on time that constitutes the solution of the kinetic problem.

It is worth noting that Eq. (1) are derived under the assumption that any two sequences of units separated by AA dyad are statistically independent. The mathematical formulation of this property is

$$P(XAAY) = \frac{P(XAA)P(AAY)}{P(AA)}, \tag{2}$$

where $P(\ldots)$ denotes the sequence probability, X, Y may be A or B. Later it was rigorously proved by the Russian mathematician Mityushin [6] that the considered reaction model actually obeys Eq. (2) and only since then the expressions for A-centered triads are considered to be exact formulas. The same statistical property enables one to find the time-dependent probability of A sequence of length j:

$$P(A_j) = e^{-jk_0 t} \exp\left(2(k_0 - k_1)\left(t - \frac{1 - e^{-k_0 t}}{k_0}\right)\right), \quad j \geq 2. \tag{3}$$

However, the complete description of the evolution of unit distribution should include the sequences containing reacted units B. The solution of this problem appears to be rather sophisticated task because sequences separated by units B are statistically dependent. The general algorithm of calculating the probabilities of any sequence in a reacting chain was suggested by Platé, Litmanovich, and coworkers (see[1] for details). Since using exact formulas leads to the time-consuming calculations, they developed also approximate methods. These methods are simple in use and lead to the results that are in a very good agreement with exact ones for all practically important cases.

Supposing that any two sequences of units separated by AB dyad are statistically independent one gets the so-called B-approximation[1,7]

$$P(XABY) = \frac{P(XAB)P(ABY)}{P(AB)} \tag{4}$$

That assumption enables derivation of the kinetic equation for the probability of any sequence of B units in the form suitable for numerical integration. Having this information and already evaluated probabilities of sequences of A-units, one can easily found the probability of any other sequence (see[1] for details).

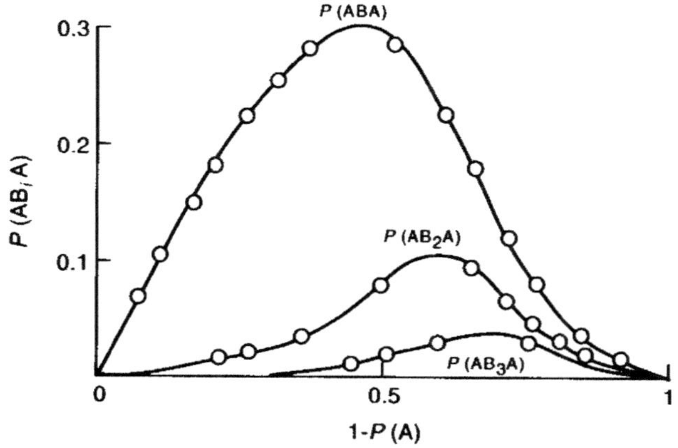

Figure 5. Probabilities of B-centered sequences: B-approximation (points) and the exact solution (curves) for the rate constant ratio $k_0{:}k_1{:}k_2 = 1{:}0.2{:}0.01$. Ref. [7].

It appears that B-approximation describes the distribution of units rather well in the wide range of the rate constant ratios. The example is shown in Fig. 5, where probabilities of various B blocks are plotted against conversion. Curves correspond to the exact solution of the problem, points to

B-approximation. Only for the strong accelerating effect, when $k_2 >> k_1$, the difference between the exact solution and the B-approximation is detectable.

Now let us consider the dispersion of composition of reacting chains

$$D_n \equiv \frac{n^2}{m} \sum_{i=1}^{m} (y_i - \bar{y})^2 . \tag{5}$$

The summation in Eq. (5) is performed over all chains in the system. It can be demonstrated that for long chains, block structure of which is far from homopolymer, the compositional distribution tends asymptotically to the normal, Gaussian form. In that case the dispersion of this distribution may be calculated analytically[8].

Good approximation for the exact formula, which is rather extended, may be obtained using the following method. If assume that the polymer chain structure may be described by the first-order Markovian chain, the dispersion of composition distribution is easily found from the formula

$$\lim_{n\to\infty} \frac{D_n}{n} = \frac{(1-P_{A/A})P_{A/B}(1-P_{A/B}+P_{A/A})}{(1-P_{A/A}+P_{A/B})^3} \tag{6}$$

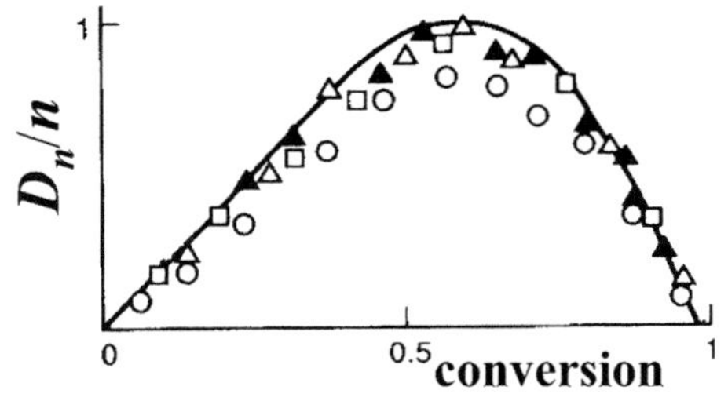

Figure 6. Dispersion of the compositional heterogeneity: modified Markovian approximation (curve) and Monte Carlo data for m = 100, n = 50 (circles); 100, 100 (rectangles); 200, 100 (empty triangles); 100, 200 (filled triangles). Ref.[9]

The key idea is to calculate transitional probabilities encountering Eq. (6), $P_{A/A} = P(AA)/P(A)$ and $P_{A/B} = (P(A)-P(AA)/(1-P(A))$, from the exact relations described above. In other words, one assumes that in the time moment under consideration, the chain is a first-order Markov one, but all its prehistory is described by the exact equations. Such a method called the modified Markovian approximation (see[1] for details) gives the results, which are in a good agreement with the exact solution, and asymptotically fits Monte Carlo data[9] for the system of m chains of length n each (see Fig. 6). Thus, it is very suitable for practical application.

So we may conclude that the utmost extension of the theory of polymer-analogous reactions of non-interacting chains has been reached. Thus, it is time now to speak about the benefit that one can get from the formal theory presented above.

The proposed neighbor effect model is proved to work well for a number of reactions. There exist a sufficiently clear-cut criterion of it: the experimental data on reaction kinetics, as well as the units distribution and the compositional heterogeneity of the products should be well fitted by the theoretical relations using one and the same set of individual rate constants k_0, k_1, k_2.

Let us consider an example.

The acidic hydrolysis of polyacrylonitrile proceeds with acceleration presumably due to the interaction of unreacted nitrile and reacted amide groups facilitating the first step of the hydrolysis, as shown at the scheme (see Figure 7). Rate constant set of 1:18:36 was found[10] as the best fit to NMR data on the average block length L_A, L_B and the probability of AB-boundary R, the same rate constants describe well the kinetic curve P(A).

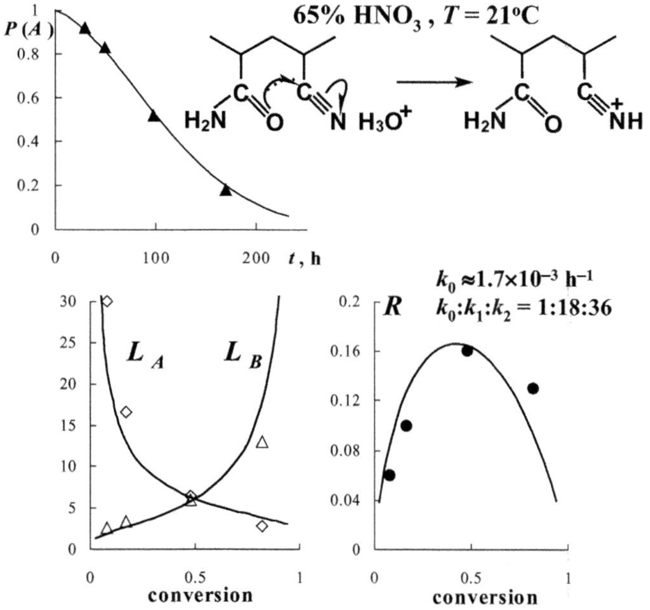

Figure 7. PAN acidic hydrolysis. See text for details. Ref. [10].

Sometimes NMR spectra are poorly resolved and a special computation procedure is necessary to extract the data on the triad distribution. In that case, it makes sense to verify via independent study the rate constants obtained by fitting NMR data. Just this was done for the hydrolysis of MMA-MAA copolymers (initially 89 % MMA) carried out at T = 145 °C in pyridine-water mixture[11]. It is seen from Fig. 8 that the rate constant ratio by NMR (k_0:k_1:k_2 = 1:2.5:3.4) is in a good agreement with one (k_0:k_1:k_2 = 1:3:5) obtained by the so-called method of polymeric models including kinetic studies on MMA-MAA copolymers of different composition (see[1] for details).

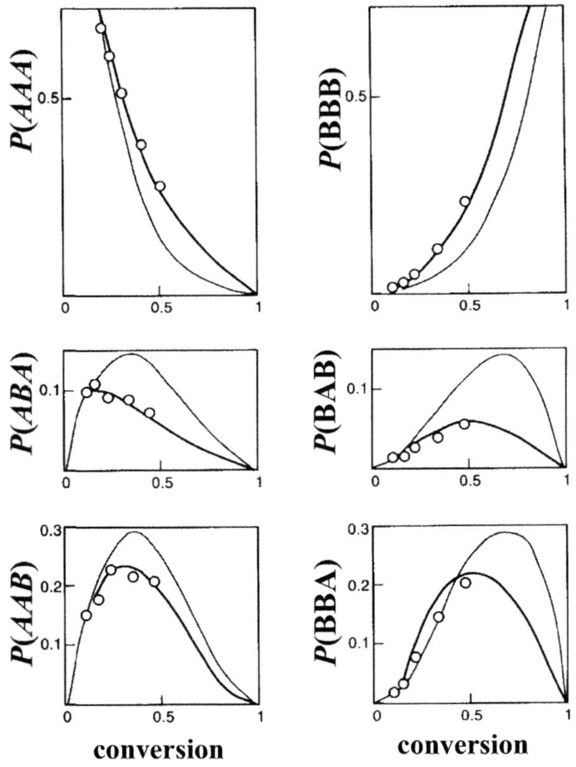

Figure 8. Hydrolysis of MMA-MAA copolymer. Distribution of triads measured by NMR points) and calculated using neighbor effect model (solid curves) and neglecting neighbor effect (thin curves). Ref.[11].

The individual rate constants provide valuable information on the process, since they make possible describing the reaction kinetics as well as the unit distribution and compositional heterogeneity. Moreover, these data provide an insight into essential features of the reaction, namely, the influence of the reaction medium, nature of neighboring units, stereochemical configuration and conformational characteristics of the chain on the reactivity.

2.2 Interchain effects

Now let us pass from dilute solutions to more concentrated systems. Since in melts, blends, glasses interchain interactions cannot be neglected, the theory of polymeranalogous reactions developed for quasi-isolated macromolecules should be accordingly modified.

Thermal decomposition of poly(tert-butyl acrylate) (PTBA) is especially convenient to study the interchain effect in a melt. During the reaction, isobutylene splits off and leaves the system, so that the reaction medium consists exclusively of macromolecules containing unreacted t-butyl acrylate and formed acrylic acid units. It was found that the decomposition proceeds with a sharp acceleration.

Basing on detailed kinetic studies, the interaction of one ester and two carboxylic groups, one or two of which are external neighbors, was proposed[12] as the most probable acceleration mechanism (see Figure 9).

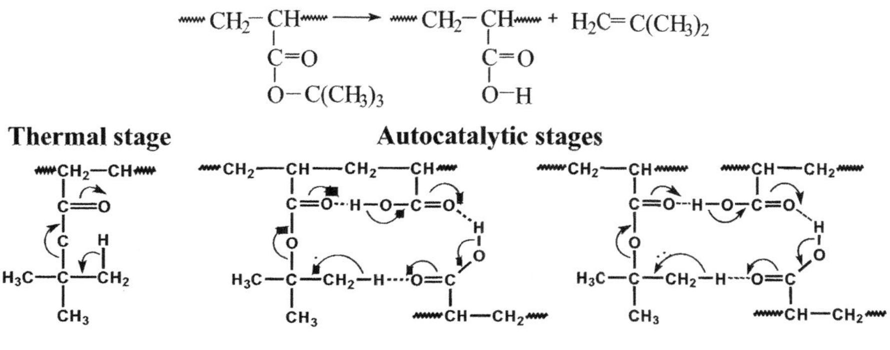

Figure 9. PTBA decomposition in melt.

Such mechanism may be described by the modified model of the neighbor effect. The results of calculations at various temperatures fit very well the kinetic curve plotted in Figure 10.

Since the modified kinetic equations contain rate constants that depend on the concentration of reacted units belonging to other chains, it seems instructive to verify the theory for the reaction in copolymers where reacting units A are diluted by inert units C.

Such verification can be carried out for the decomposition of ester groups in tert-butyl acrylate - styrene copolymers of different composition, styrene units being both intra- and interchain diluents. Again, good coincidence of the theory and experiment is found[13] (see Figure 11).

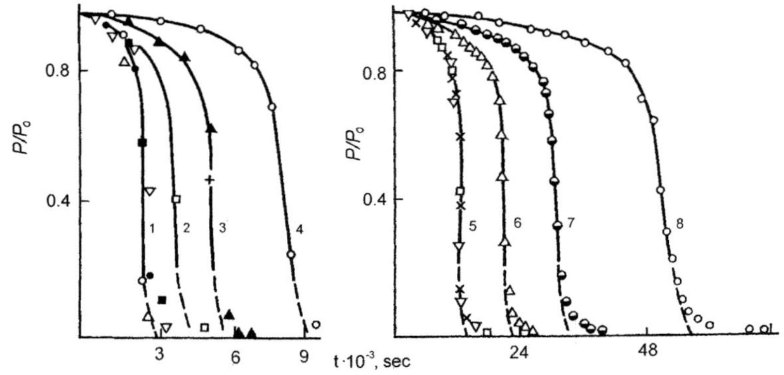

Figure 10. Kinetics of PTBA decomposition in a melt. Experimental data (points) and the modified neighbor effect model (theory). Curves correspond to different temperatures from 200 to 165 °C by 5 °C. Ref.[12].

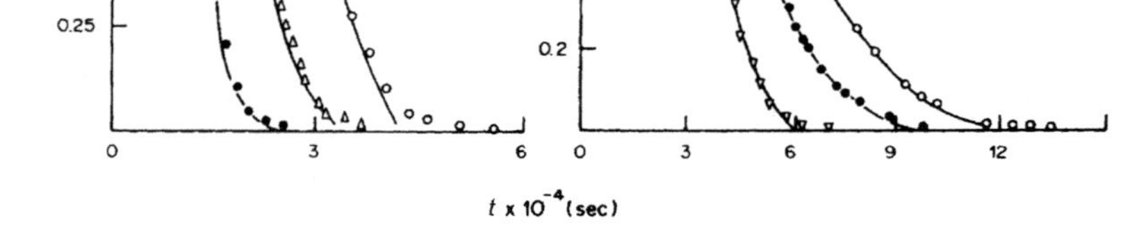

Figure 11. Kinetics of of P(TBA-*co*-S) decomposition in melt [13]. Experimental data (points) and the modified neighbor effect model (theory). Curves correspond to the different initial mole fraction of TBA: 0.89(1), 0.69(2), 0.53(3), 0.46(4), 0.38(5), 0.32(6). Ref.[13]

The interchain effect is very pronounced for PTBA decomposition in blends as well.

In particular, the initial reaction rate increases markedly in blends of PTBA with poly(acrylic acid) because carboxylic groups of PAA accelerate the reaction[14]. On the contrary, polyethyleneimine binds carboxylic groups formed during the reaction and reduces significantly the acceleration[15] (see Figure 12).

A new effect revealed itself when the reaction of PTBA decomposition was studied in blends with PAA. The blends were of the same composition, prepared by the same procedure but from different solvents. It was proved by the non-radiative energy transfer between the fluorescent labels incorporated in the polymers and also by IR measurements that the difference between kinetic curves of PTBA decomposition plotted in the figure (see Fig. 13) is caused by the different initial structure of blends[16,17].

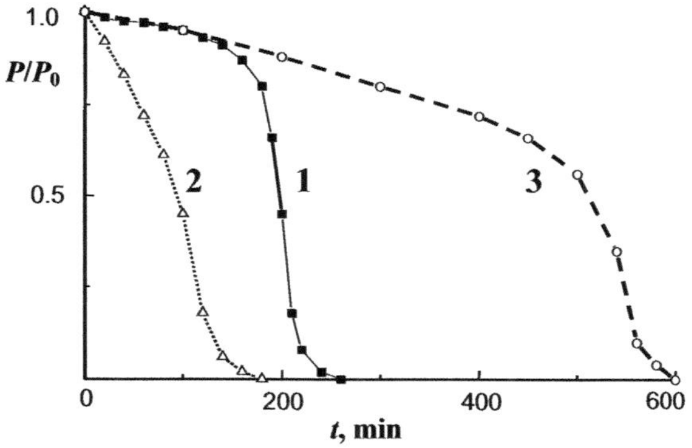

Figure 12. Decomposition of PTBA in pure state (curve 1) and in blends with PAA (2:1 unit/unit, curve 2) and PEI (3:1 unit/unit, 3). Ref. [14,15].

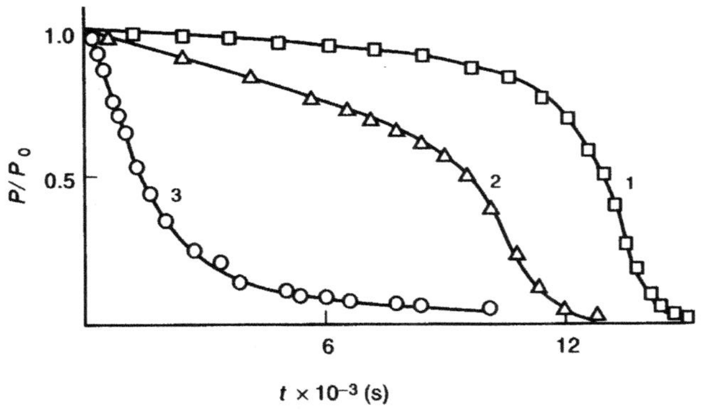

Figure 13. Decomposition of PTBA in pure state (curve 1) and in blends with PAA (1:10 unit/unit) prepared from dioxane-water mixture 50:1 (2) and *t*-butanol-water 7:1 (3). Ref.[16]

That effect stimulated the development of the theoretical models taking into account inhomogeneous structure of the reacting blend. Primarily the diffusion of polymer chains was not considered (see[1] for details), however, such models were unable to describe the reaction kinetics quantitatively.

The necessity to include the interdiffusion in the theory of macromolecular reactions in polymer blends became evident. Indeed, chains do not only react, they move one toward another and the interdiffusion process should be also taken into consideration.

2.3 Reaction and diffusion

A binary blend of compatible homopolymers A and B initially separated by a narrow interface is considered (see Fig. 14). An evolution of the blend structure is described by a time-dependent profile of the volume fraction of A chains, ϕ, shown at the left plot.

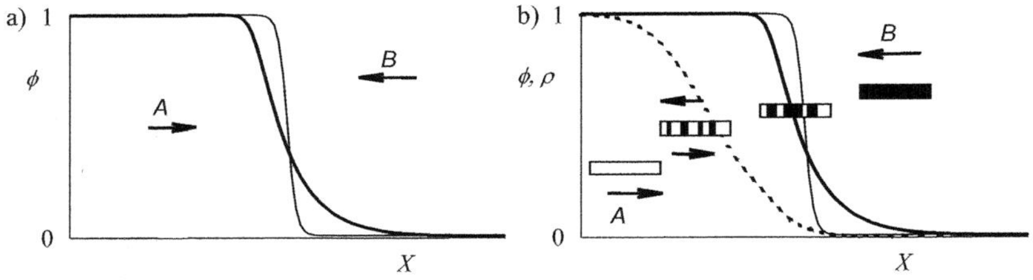

Figure 14. Interdiffusion in a binary blend without reaction (a) and in a reacting blend (b). Empty, black, and striped rectangles relate to A, B, and partly reacted macromolecules, respectively.

Let now a reaction with interchain effect proceed in the blend: A units transform into B ones, B units accelerating the reaction. It appears that in the reacting blend a number of new factors should be taken into account.

Firstly, in polymer systems these are chains that diffuse but these are units that react. Therefore, the concentration profile of reacting chains, ϕ, and that of transforming A units, ρ, do not coincide as demonstrated by the right plot of Fig. 14 thus affecting both the interdiffusion and reaction kinetics.

Secondly, in the course of the reaction AB copolymer is formed that is heterogeneous in composition, so an initial binary blend becomes essentially multi-component. It leads to the appearance of a new type of interdiffusion that is completely absent in binary blends. Indeed, a partially transformed chain (drawn as striped rectangle on the plot) may diffuse either to the left or to the right. Therefore, aside from intermixing of reacting and accelerating chains, a new type of diffusive movement arises, namely, intermixing of reacting chains of different composition.

Thus describing the evolution of the reacting blend structure appears to be a rather complicated task so it makes sense to start in theory with relatively simple models that however reflect peculiarities of the process.

We considered[18] a spatially inhomogeneous blend of compatible homopolymers A and B, where irreversible reaction A→B proceeds; B units accelerate the reaction, so that A-units transform via following independent pathways:

- spontaneously, that is independently of the microenvironment, with the rate constant α;
- with neighbor effect: one inner neighbor B increases the rate constant by β, the influence of two inner B neighbors being additive;
- with interchain effect: an external neighbor B provides the reaction with the rate constant γ.

The formulated model corresponds to the routes of PTBA decomposition mentioned above.

The theoretical task is to describe the evolution of the model blend structure under concerted action of the reaction and interdiffusion.

If the system is locally in equilibrium, the linear non-equilibrium thermodynamics may be used. The local equilibrium will be maintained in a system if two requirements are fulfilled:

– $\tau_{react} \geq R_0^2/D_s$, i.e., the reaction is rather slow such that the characteristic time of a unit transformation, τ_{react}, is not less than the time of chain diffusion at its mean-square radius R_0, D_s being the self-diffusion coefficient of the chain;
– $R_0|\nabla\phi|/\phi << 1$, i.e., blend inhomogeneities are weak so the variation of the volume fraction of reacting chains, ϕ, at the distance R_0 is small in comparison to its mean local value.

Using non-equilibrium thermodynamics includes several steps:

a) choice of relevant variables for the shortened description;
b) derivation of the free energy of mixing;
c) calculation of the Onsager kinetic coefficients;
d) solution of the reaction-diffusion equations under certain initial conditions.

Let us introduce three macroscopic variables ϕ, ρ_1, ρ_2 that are the volume fractions of reacting chains, of A units, and of AA dyads, respectively.

The mean-field expression for the free energy given by Eq. (7) includes several contributions.

$$\frac{F(\phi,\rho_1,\rho_2)}{kT}=\frac{\phi}{N_A}\ln\phi+\frac{1-\phi}{N_B}\ln(1-\phi)$$

$$+\rho_2\ln\rho_2+2(\rho_1-\rho_2)\ln(\rho_1-\rho_2)+(\phi-2\rho_1-\rho_2)\ln(\phi-2\rho_1-\rho_2) \quad (7)$$

$$-\rho_1\ln\rho_1-(\phi-\rho_1)\ln(\phi-\rho_1)+\chi\rho_1(1-\rho_1)$$

The first two are the entropy of mixing of reacting and accelerating chains, the last contribution gives the interaction energy of units in terms of the Flory-Huggins parameter χ. The remaining terms originating from the non-identity of reacting chains constitute the structural entropy contribution, which is absent in the homopolymer blend. Calculations are facilitated by the proportionality of the structural entropy and the information entropy, which is known from the theory of information as a measure of ordering in random sequences.

Onsager kinetic coefficients may be found using Kubo-Green relations. It is convenient to write down the kinetic coefficients in the matrix form:

$$\Lambda=\Lambda_{inter}\begin{pmatrix}1 & p_1 & p_2\\ p_1 & p_1^2 & p_1p_2\\ p_2 & p_1p_2 & p_2^2\end{pmatrix}+\Lambda_A\begin{pmatrix}0 & 0 & 0\\ 0 & \sigma_{11}^2 & \sigma_{12}^2\\ 0 & \sigma_{12}^2 & \sigma_{22}^2\end{pmatrix}, \quad (8)$$

where $P_1=\rho_1/\phi$ and $P_2=\rho_2/\phi$ are the probabilities to find A unit and AA dyad in the reacting chains, respectively, σ_{ij}^2 is the covariance of p_i and p_j, $\Lambda_{inter}=\phi(1-\phi)((1-\phi)N_AD_A+\phi N_BD_B)$, D_A and D_B are the self-diffusion coefficients of reacting and accelerating chains.

The first term of Eq. (8) accounts for interdiffusion of the accelerating chains B and the copolymer AB, whereas the second one describes the mutual diffusion of random copolymer chains of different composition.

The dispersion of the copolymer composition is characterized by the mean values of chosen variables and by the their covariances entering the second matrix. The covariances are calculated assuming the Markovian structure of the reacting chains (see[18] for details).

Now it is possible to write down the reaction-diffusion equations that describe the evolution of the macroscopic variables chosen above. The explicit form of these equations is found by substituting the kinetic coefficients and the second derivatives of the free energy of mixing[18]:

$$\begin{aligned}
\partial\phi/\partial t &= \vec{\nabla}\left(D_{coop}\vec{\nabla}\phi - 2\chi P_1\phi\Lambda_{inter}\vec{\nabla}P_1\right) \\
\partial(P_1\phi)/\partial t &= \vec{\nabla}\left(P_1(D_{coop} - 2\chi\sigma_{11}^2\Lambda_A)\vec{\nabla}\phi\right) \\
&\quad + \vec{\nabla}\left(\phi\left(D_A - 2\chi(P_1^2\Lambda_{inter} + \sigma_{11}^2\Lambda_A)\right)\vec{\nabla}P_1\right) + f_1 \\
\frac{\partial(P_2\phi)}{\partial t} &= \vec{\nabla}\left((P_2 D_{coop} - 2\chi P_1\sigma_{12}^2\Lambda_A)\vec{\nabla}\phi\right) \\
&\quad - 2\chi\vec{\nabla}\left((P_1P_2\Lambda_{inter} + \sigma_{12}^2\Lambda_A)\phi\vec{\nabla}P_1 + \phi D_A\vec{\nabla}P_2\right) + f_2
\end{aligned} \tag{9}$$

where

$$D_{coop} = \Lambda_{inter}\left(\frac{1}{N_A\phi} + \frac{1}{N_B(1-\phi)} - 2\chi P_1^2\right),$$

$$f_1 = -(\alpha + 2\beta(1-\rho_2/\rho_1) + \gamma(1-\rho_1))\rho_1,$$

$$f_2 = -2(\alpha + \beta(1-\rho_2/\rho_1) + \gamma(1-\rho_1))\rho_2.$$

It is seen that the equations are quite complex and may be solved only numerically. Performing the solution, one obtains the dependencies of the volume fractions of reacting chains, A units, and AA dyads on time. Thus, theory enables one to predict the evolution of reacting blend structure from the certain initial situation.

Consider two adjacent films of homopolymers A and B, where reaction described above starts at the initial moment of time[19]. Curve 5 of Figure 15 is the boundary between the films at that moment. Other curves represent a snapshot of this reacting system. One can clearly see the divergence of profiles corresponding to the fraction of reacting chains, ϕ, and the fraction of A units in reacting chains, P_1.

The theory enables calculating the detailed spatial structure of the reacting blend as well. In Figure 15, the dispersion of compositional heterogeneity of reacting chains, σ^2, and the probability of a boundary between the A and B sequences, $R = 2P(\mathrm{AB})$, are shown. It is worth to emphasize that such information is extremely difficult to obtain in experiments. Indeed, the spatial structure of the reacting blend determines the characteristics of final product – polymer composite, just as the distribution of units along the backbone determines the properties of a single macromolecule.

Thus theoretical considerations on simple model systems could help to understand the effect of diffusive intermixing on reacting polymer blends.

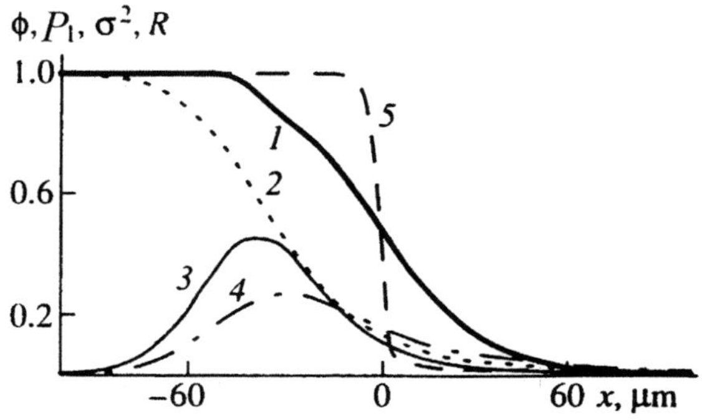

Figure 15. Detailed structure of the reacting blend: the volume fraction of the reacting chains ϕ (curve 1), the mean fraction of A units in the reacting chains P_1 (2), the dispersion of the reacting chains distribution in composition σ^2 (3), the probability of a boundary between the A and B sequences $R = 2P(\text{AB})$ (4), the initial interface profile (5). $N_A = 500$, $N_B = 2000$, $\chi = -0.01$, $\alpha = 0$, $\beta = \gamma = 2\text{x}10^{-5}\ \text{s}^{-1}$, time $1\text{x}10^5$ s. Ref.[19].

Summarizing this Section, we may conclude that the theory of polymeranalogous reactions, which has started with considering dilute systems, now successfully deals with rather complicated models related to the reactive processing of polymer composites.

3. STABILYZING EFFECT OF DIBLOCK COPOLYMERS

3.1 Experiment

Now let us discuss the problems of blend stabilization by diblock copolymers.

In fine dispersions usually prepared by mixing, the unique morphology with synergistic properties can be created[20]. For example, submicron size droplets of a rubbery phase can improve toughness 10-fold in a glassy or semicrystalline plastic[21]. The desirable blend morphology may be stabilized by adding suitable compatibilizers, either premade or formed in situ.

Diblock copolymer additives are mostly often used for this purpose. They can be melt-mixed with the main components of the blend, added to the artificially prepared quasi-homogeneous blend, or formed in situ via end-coupling although the latter case is most complex.

Static and dynamic stabilization are discriminated. In the former case a blend containing compatibilizer is annealed at elevated temperature.

Suppression of the blend coarsening means that the system has reached the thermodynamic equilibrium, therefore, various thermodynamic relations may be applied. On the contrary, the dynamic stabilization under permanent mixing in an extruder or batch mixer implies stationary state but not necessarily equilibrium. The interface in such systems is permanently renewed forcing copolymer chains perpetually move towards it. The dynamics of copolymer redistribution noticeably complicates the problem of developing the theory.

Macosko et al.[22] studied the static stabilization of the morphology of PS/PMMA blend by a premade block copolymer PS-PMMA. If the block lengths are quite different than the length of corresponding homopolymers, the copolymer cannot stabilize the blend, this is seen from Fig. 16.

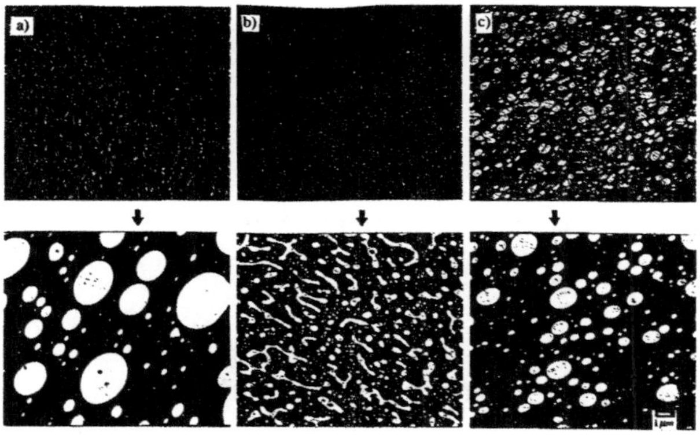

Figure 16. Effect of annealing on the morphology. Blend PS(95000)/PMMA(11000) 70/30 without compatibilizer (a), with 5% P(S-*b*-MMA) (85000, 55 wt% styrene) (b), and with 5% P(S-*b*-MMA) (160000, 50 wt% styrene) (c). All blends are unstable. Ref.[22]

The upper row represents the initial blends (after 20 min of mixing at 180 °C), the lower row corresponds to the same blends after 20 min of annealing at 195 °C. Stabilization was not achieved in any of the systems. Blend on the left does not contain copolymer at all, the rate of PMMA particle growth is maximum here. Two other blends contain 5% of copolymer, the copolymer chains on the right pictures are twice as long as in the center. Short copolymer chains are more mobile and form less micelles thus particles in the central pictures are smaller. If however a very short copolymer is taken, it would mix with homopolymers rather then adsorb at the interface.

The stable blend is most easily obtained if homopolymers and blocks are approximately of the same molecular mass. It is seen from Figure 17 that in that case the average particle size does not change in the course of annealing although without copolymer particles grow intensively.

Thus, any immiscible blend may be effectively stabilized during annealing by a diblock copolymer with properly chosen characteristics.

In blends with high concentration of the dispersed phase, from 10 to 50%, Brownian motion and molecular forces provide coalescence of particles and hence the blend coarsening. Under mixing, the rate of coalescence increases considerably due to the shear flow forces. At the same time, shear flow promotes particle deformation and breakup. As a result, the dynamical equilibrium characterized by a constant average particle size is set in the blend without any compatibilizer. However, diblock copolymer addition may essentially reduce the steady particle size under mixing conditions. What is the mechanism of the dynamic stabilization?

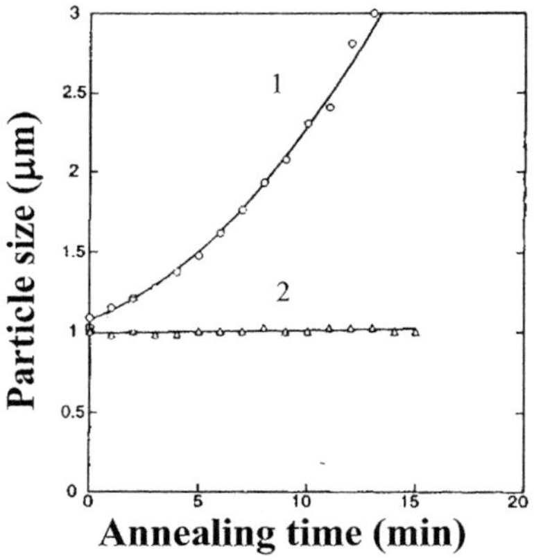

Figure 17. Effect of annealing at 180 °C on the average particle size in the 70/30 blend PS(43,000)/PMMA(43,000) without copolymer (curve 1) and with 5% of P(S-*b*-MMA) (85,000, 55 wt% styrene) (2). Ref.[22].

It appears that copolymer influences breakup and coalescence via different mechanisms. Lowering of the interfacial tension is the main effect causing easier droplet breakup. To explain why adding diblock copolymers suppresses coalescence two basic schemes are considered in the literature, they are presented in a sketch form in Figure 18. The key difference is related to the mobility of copolymer along the particle surface.

The left picture demonstrates the interaction of particles through so-called Marangoni force. When two droplets approach each other, the bulk phase between them is squeezed out. The viscous flow carries block copolymer chains along the interface to the back of particles thus forming a concentration gradient. According to estimations by Milner and Xi[23], the minimum coverage of block copolymer needed to completely suppress coalescence is proportional to the shear rate.

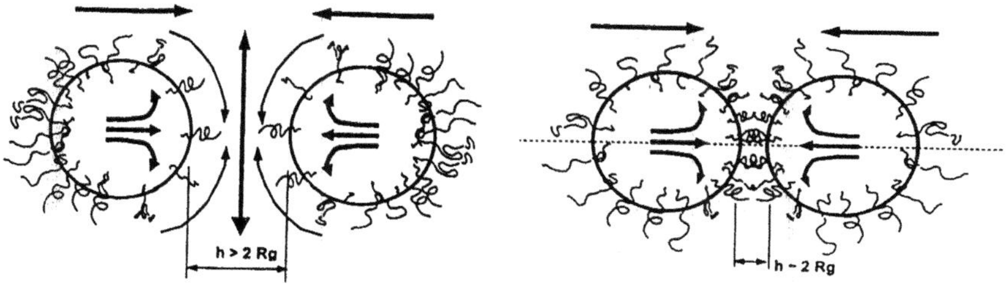

Figure 18. Mechanisms for suppressing coalescence by block copolymers. Ref.[25].

The right sketch of Figure 18 presents the mechanism of interaction via steric repulsion suggested by Macosko et al[24]. It is assumed that at the time scale of particle collision block copolymers are immobile so blocks that cover approaching particles will be compressed. In that case, the minimum coverage of block copolymer needed to suppress coalescence depends on the molecular mass of the copolymer rather than on the shear rate.

To check, which scheme is closer to reality, Lyu et al.[25] studied melt mixing of PS/high density polyethylene (PE) blend containing several percents of diblock copolymer P(S-*b*-E). The conditions were found when particle breakup does not occur, so coalescence was the only process that changes the particle size.

The dynamic stabilization was successfully achieved and it was found that the minimum surface copolymer density that completely prevents coalescence weakly decreases with shear rate - in contradiction to the mechanism of particle interaction via the Marangoni force. At the same time, measurements demonstrated that the effectiveness of diblock copolymer increases with its molecular mass, in agreement with the mechanism of steric repulsion. However, it should be noted that both mechanisms drastically underestimate the minimum copolymer coverage as compared to the experimental data, probably, due to neglecting the deformation of interacting droplets.

It seems hardly possible to understand peculiarities of reactions proceeding at interface layers if investigate them under mixing conditions. More promising are studies using specially prepared bilayer and trilayer films with well-defined interfaces.

Schulze et al.[26] studied the end-coupling reaction between amino-terminal deuterated PS and anhydride-terminal PMMA to find out whether this reaction is kinetically or diffusion controlled. Using bilayer geometry shown on the left picture of Figure 19 they monitored the concentration of dPS across the films by forward recoil spectrometry and found that the formation of copolymer occurs over several hours and approaches a steady state value after 24 hours of annealing. The limiting interfacial coverage in chains per square nanometer was determined, it decreased with the increase of molecular mass of reacting polymers, being in good agreement with the theoretical estimations.

Concentration profiles of dPS plotted in Figure 20 (left picture) remain flat outwards the reaction interfacial zone thus no depletion hole of reactants is found, which might indicate about diffusion-controlled regime.

Trilayer model system shown on the right picture of Figure 19 enables one to watch the diffusion and reaction independently. Indeed, in that case reactive dPS chains bearing NH_2 groups have to diffuse through the inert PS-layer before to react with PMMA chains at the interface between PS and PMMA.

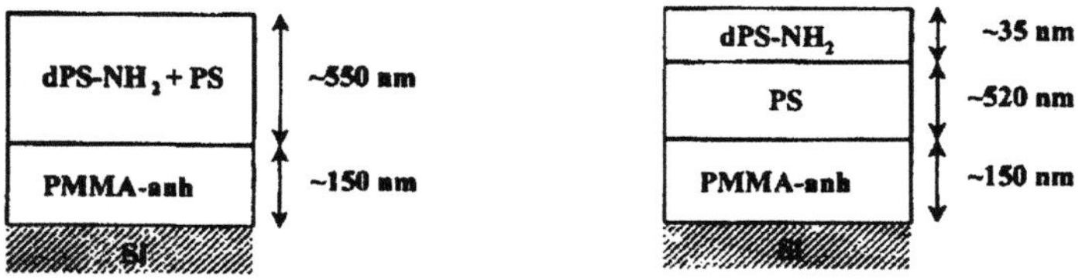

Figure 19. Geometry of bilayer and trilayer model systems. Ref.[26].

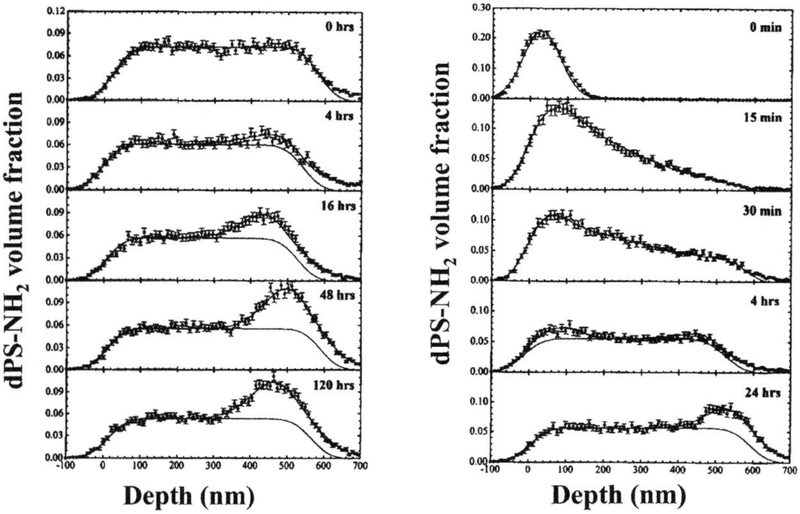

Figure 20. Concentration profiles of dPS in bilayer (left) and trilayer (right) systems. Ref.[26]

Concentration profiles of dPS in the trilayer model system are shown on the right side of Figure 20. It is seen that reactive dPS chains dispersed itself over PS-layer in about 1 h, the flattened profile indicating about diffusion equilibrium was established in 4 h, whereas the reaction became noticeable only after 24 h. Estimating the mobility of reactive dPS chains from their molecular mass of 92000, one finds that these chains cross the PS/PMMA interface many times before to react with PMMA. In other words, the reaction is kinetically controlled.

Artificial, model conditions are also realized when copolymers are added to the quasi-homogeneous blend of incompatible polymers prepared from their mutual solution. In that case one can investigate the influence of copolymer on the process of phase separation.

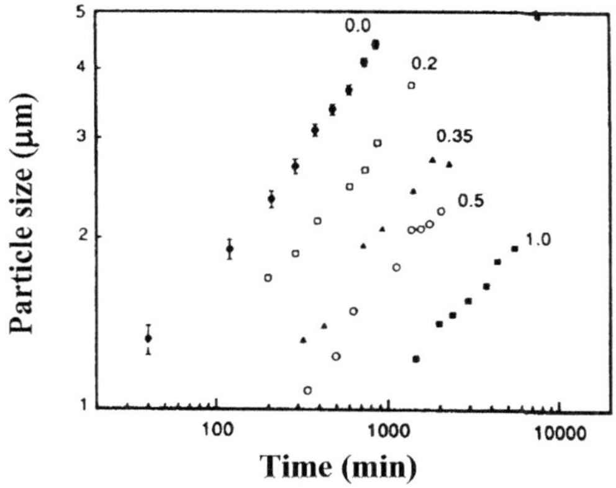

Figure 21. Dependence of the average particle size on time in the log-log coordinates. The 20/80 blend PS(1900)/PB(2350) is compatibilized by P(S-b-B) (25,000, 52.2 wt % styrene). Numbers on the plot denote the weight fraction of the copolymer. Ref.[27].

Roe and coworkers[27,28] studied the effect of adding diblock copolymer of styrene and butadiene (25,000, 52.2 wt % styrene) to the blend of corresponding homopolymers (PS(1,900)/PB(2350) 20:80 wt%) using light scattering and optical microscopy. Since the minority phase (rich in PS) occupied only 4% of the total volume, the particles grew mostly by the diffusion of chains dissolved in the continuous phase (so-called Ostwald ripening). That regime is characterized by the power dependence of the average particle radius on time, $R_{av} \sim t^{1/3}$, which gives the straight line in double logarithmic coordinates in the blend of two homopolymers (see curve labeled as 0.0 in Figure 21).

It is seen that the copolymer additives up to 1 wt% do not affect the dependence of the average particle radius on time but lead to the delay in establishing the regime of Ostwald ripening. It looks as if the block copolymer renders the early stage of phase separation, either nucleation-growth or spinodal decomposition, but ineffective at the stage of Ostwald ripening.

Summarizing, we note that the use of simple model systems gives one the possibility to reveal the important features of the block copolymer reactive formation and its influence on the equilibrium and dynamics in blends of incompatible homopolymers. These experimental results stimulate our and other people theoretical studies.

3.2 Theory

Stabilization of polymer blend structure by diblock copolymers, premade or formed in the course of end-coupling reaction in situ, was addressed in a number of theoretical studies. Thermodynamics of diblocks and their blends with homopolymers are now rather well understood, much less is still known about the dynamical properties of systems containing compatibilizers.

In annealed systems, the compatibilizing action of copolymer additives is attributed to the lowering of interfacial tension, which is the thermodynamic driving force for the phase separation.

The interfacial tension at the interface between two immiscible homopolymers A and B was first calculated by Helfand and Tagami[29] using the self-consistent mean-field approach. The free energy of that system consists of the conformational entropy that favors mixing and of the enthalpic term describing monomer units interaction that promotes separation. Minimizing the free energy it was possible to find the equilibrium concentration profile across the interface, $\rho_h(x)$, and to calculate the interfacial tension $\sigma_0 = (\chi/6)^{1/2}\rho_0 b k_B T$, where ρ_0 is the polymer density, χ is the Flory-Huggins parameter, b is the average length of a statistical segment.

Then, the concentration of chain ends at the interface, ρ_{he}, was predicted to be larger than in the bulk, being described by the relation $\rho_{he}(x)/\rho_{0e} = [\rho_h(x)/\rho_0]^{1/2}$, where ρ_{0e} is the total concentration of ends and $\rho_h(\boldsymbol{x})$ is the homopolymer A concentration.

Dependencies of the interfacial tension on the homopolymer lengths were calculated in a number of theoretical investigations as the entropic corrections to the Helfand-Tagami formula. The polydispersity also affects the interfacial properties lowering the interfacial tension due to the slight accumulation of small chains near the interface.

Leibler[30] was the first who studied the effect of diblock copolymer addition. He calculated analytically the decrease of the interfacial tension in two limiting cases, the weak and the strong segregation limits.

Using the results for a plane interface, Leibler considered the stabilizing effect of the diblock copolymer AB on a system of minor phase particles of homopolymer A immersed in the bulk phase formed by the homopolymer B[31]. Minimizing the interfacial free energy, he calculated the mean equilibrium radius of a particle depending on the copolymer concentration, block length and asymmetry, that is, the difference in the length of blocks.

It is the predictions of this theory that often successfully used by experimentalists for comparison with their data. However, the essential shortcoming of Leibler's consideration is that the equilibrium particle radius is predicted to be inversely proportional to the copolymer asymmetry. This leads to the impossibility of stabilizing the blend structure by symmetric diblocks, which is in contradiction to experimental observations.

This shortcoming was removed by Govorun et al.[32,33] who considered the Alexander-de Gennes ("wet-brush") model[34,35] taking into account the possibility of penetration of homopolymer into the interfacial copolymer layer. Swelling of this layer due to the osmotic pressure of homopolymers leads to the increase in the total interface area thus increasing the number of particles and decreasing their size. When this effect is balanced by the interfacial tension, the equilibrium is settled and the stable system of particles is formed.

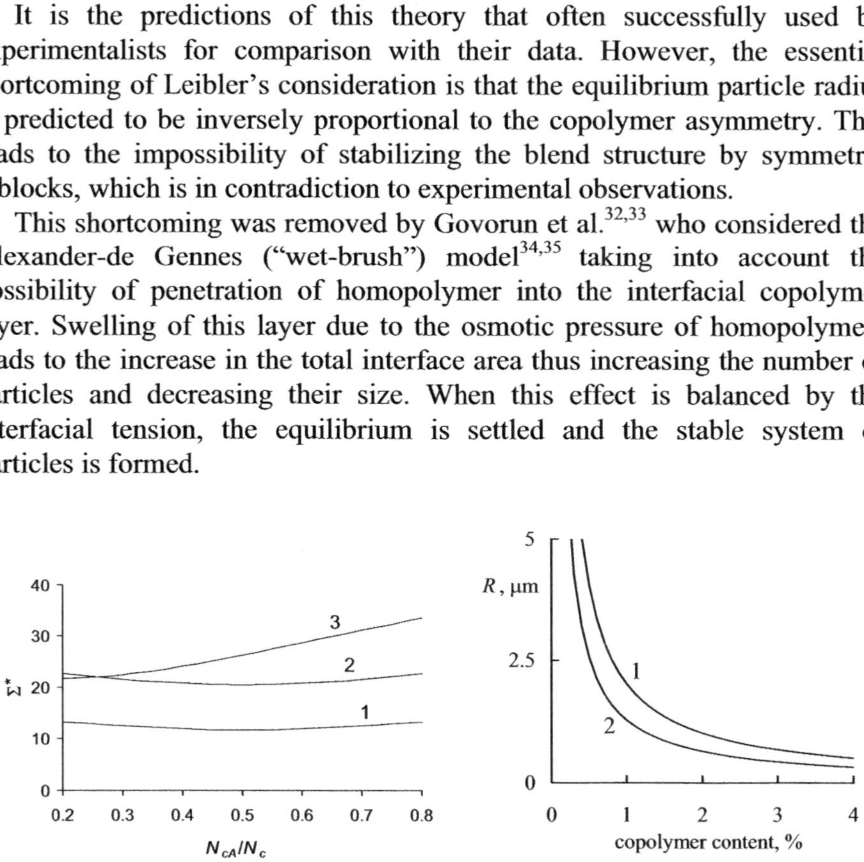

Figure 22. Equilibrium value of the interface area per copolymer molecule $\Sigma^* = \Sigma/a^2$ *vs* copolymer asymmetry N_{cA}/N_c. for the systems $N_A = N_B = N_c = 400$ (left plot, Fig.1), $N_A = N_B = 400$, $N_c = 1000$ (2); $N_A = 100$, $N_B = 700$, $N_c = 1000$ (3); $\sigma_0 = 0.1\ kT/a^2$. Equilibrium radius of disperse phase particles, R, vs diblock copolymer content (right plot) for $N_A = N_B = 400$ (curve 1), $N_A = N_B = 100$ (2). In both cases, $N_{cA} = N_{cB} = 400$ Ref. [35].

Symmetric diblock copolymer was demonstrated to stabilize particles of reasonable size. Moreover, it was shown[33] that the diblock copolymer asymmetry only slightly influences the equilibrium interfacial area per copolymer chain, $\Sigma^* = \Sigma/a^2$, as seen from the left plot of Figure 22.

The lowest value of Σ^* corresponds to the block length ratio approximately equal to the homopolymers length ratio. The corresponding numerical values of the copolymer coverage Σ^* are about several square nanometers that is in good agreement with experimental values. Such an agreement indicates that in the experiment the thermodynamic equilibrium is reached.

The average particle size R in equilibrium is inversely proportional to Σ:

$$R = 3N_c v/(\phi_c \Sigma^*) \tag{10}$$

Eq. (10) demonstrates that the particle size depends also on the copolymer length, N_c, and its concentration in the blend, ϕ_c. The latter dependence is shown at the right plot of Figure 22.

Diblock copolymer additives affect not only thermodynamic properties of incompatible polymer blends but also the rate of phase separation process.

If the volume fraction of a minor phase is relatively small and the diffusivity of dispersed component is high enough, then above mentioned Ostwald ripening is the most contributing mechanism of phase separation. There exists a time-dependent critical value of particle radius such that small particles dissolve while large particles grow due to the diffusion of macromolecules through the bulk. The critical radius r_{cr} is proportional to the interface tension σ_0 and inverse proportional to the supersaturation of bulk phase with minor component, Δ.

We studied the theory of Ostwald ripening in a blend of homopolymers A and B and diblock copolymer AB[36]. It was assumed that each growing A particle adsorbs copolymer chains uniformly distributed throughout the continuous phase so that the copolymer amount at particle surface is proportional to the particle volume. Properties of the interface layer were described using the wet-brush model.

The presence of diblock copolymer gives rise to two effects demonstrated in Figure 23. Firstly, it lowers the interfacial tension σ_c that is shown at the left plot, hence, the critical radius is decreased and the fraction of growing particles is increased. Secondly, the copolymer reduces the coefficient α describing the permeability of A particle surface for the diffusion of A-chains inwards or outwards; this is demonstrated at the right plot.

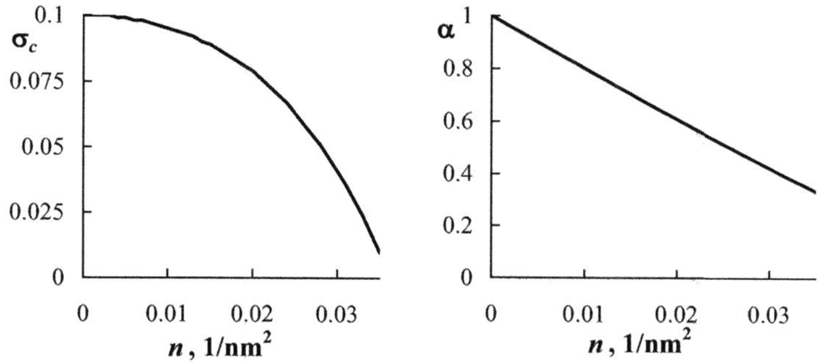

Figure 23. Effect of diblock copolymer on the Ostwald ripening. Dependence of the interfacial tension, σ_c, (right plot) and of the permeability coefficient, α, (left plot) on the copolymer coverage $n = 1/\Sigma$. Ref.[36].

Just these parameters α and σ_c were used to modify the Lifshitz-Slyozov equations describing the evolution of the particle size distribution:

$$\dot{r} = \frac{dr}{dt} = D\alpha(r)\frac{\Delta(t)r - 2\sigma_c(r)N_A v_0\phi_0/kT}{r^2}$$

$$\Delta(t) + \frac{4\pi}{3}\int r^3 f(t,r)dr = \Delta(t=0) \qquad (11)$$

$$\frac{\partial f(t,r)}{\partial t} + \frac{\partial}{\partial r}\left(f(t,r)\cdot\dot{r}\right) = 0$$

where $f(t,r)dr$ is the number of particles with radius r is the blend, $N_A v_0$ and D are the volume of A chain and its self-diffusion coefficient in bulk phase, respectively, ϕ_0 is the equilibrium volume fraction of homopolymer A in the bulk phase near a flat interface. The parameters necessary for performing the numerical solution of these equations can be taken from the above-cited experimental work by Park and Roe[28], where phase separation in the PS/PB/poly(S-*b*-B) blend was experimentally demonstrated to proceed via Ostwald ripening.

Results of the calculations are presented in Figure 24. Curves on the left graph are plotted for different copolymer content. It is seen that without a copolymer the average particle radius grows proportionally to $t^{1/3}$. 1 per cent of diblock copolymer within 200 minutes almost does not affect this dependence; only 3 per cent of the additive reduce essentially the rate of particles growth.

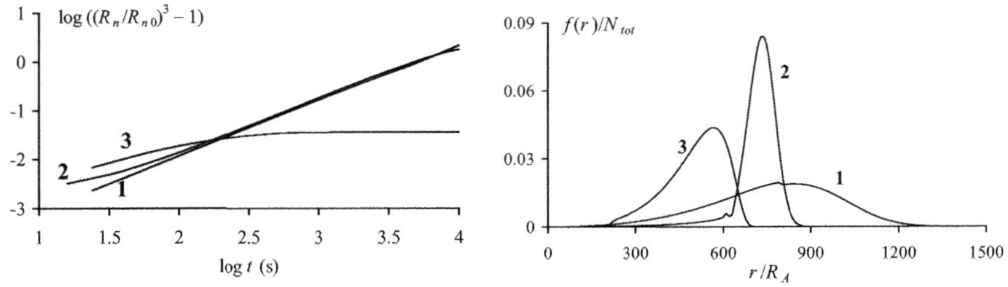

Figure 24. Time evolution of the average particle size in the course of Ostwald ripening (left plot). Particle size distribution at $t = 200$ min (right plot). Block copolymer content (%): 0 (curve 1), 1 (1), 3 (3), $N_A = N_B = 20$, $N_{cA} = N_{cB} = 40$, $R_A = 2$ nm, $D = 5x10^{-8}$ cm^2/s . Ref.[36].

Without a copolymer, the particle size distribution after 200 min is represented by curve 1. In a blend containing 1 per cent of copolymer, as soon as the growing particle reaches a certain size, the copolymer layer becomes dense enough to hinder both growth and dissolution, so resulting size distribution presented by curve 2 becomes considerably narrow. In a blend with 3 per cent of copolymer, only the smallest particles dissolve, so the distribution given by curve 3 almost coincides with the initial one (not shown in the figure).

Now let us look how the phase separation is affected by in situ formation of diblock copolymer via end-coupling. Consider an artificially prepared quasi-homogeneous blend of incompatible polymers A and B.

Let the spinodal decomposition proceed in a system after a proper temperature jump. It is convenient to describe the blend morphology in terms of structure factor S since this quantity is proportional to the experimentally measurable intensity of light or neutron scattering. The dependence of S on the wave vector k reflects non-homogeneity of the system. Permanent growth of the structure factor with time corresponds to the progressive phase separation (see upper plot of Figure 25).

Now suppose each A- and B-chain bears one reactive end-group, so that diblock copolymer AB is formed via end coupling. The reaction competes with phase separation since the diblock copolymer is a compatibilizing agent.

For a kinetically controlled reaction, the task of analyzing the initial stage of the spinodal decomposition was solved[19] using the approaches of linear non-equilibrium thermodynamics. Lower plot of Figure 25 shows that at early stage of the reaction, when volume fraction of the copolymer is small, the structure factor grows; then its maximum begins to decrease due to the compatibilizing effect of the block copolymer formed. The shift of the maximum to the left reflects the diffusive spreading of inhomogeneities.

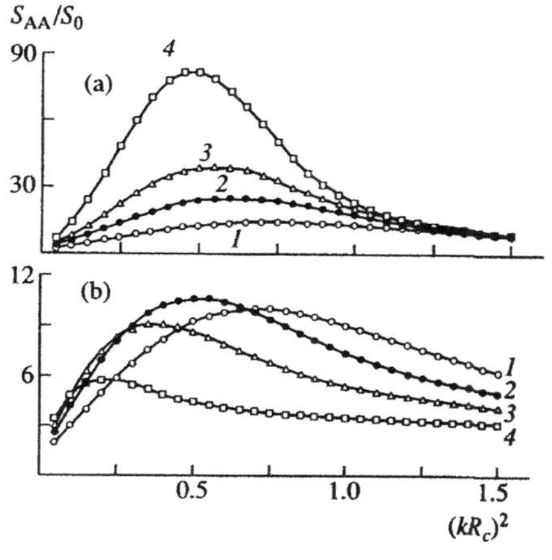

Figure 25. Change of reduced structural factor for an initially quasi-homogeneous blend of two incompatible homopolymers at time t = 20 (1), 35 (2), 50 (3), and 80 s (4). $S_0 = S_{AA}$ (t=0). (a) no reaction, (b) end-coupling proceeds with the rate $\alpha/v_0 = 0.01\ s^{-1}$. Ref.[19].

A large body of theoretical work has addressed the diffusion-controlled regime of reactions in polymers (see[3] for details). In that case, the reaction takes place at once when reacting groups meet each other so the ordinary equations of chemical kinetics based on the law-of-mass-action become inappropriate. Derivation of the reaction-diffusion equations that explicitly relate the apparent reaction rate to chain mobilities constitutes the main problem of this field. Despite the number of interesting theoretical results obtained for reactions in bulk, under flow, and at interfaces, they are still of purely academic interest since all the reactions employed so far for the diblock copolymer formation appeared to be insufficiently fast to reveal the diffusion-controlled behavior.

Summarizing this Section, we note that the maximal progress was achieved in the description of static stabilization by diblock copolymers. All fundamental regularities seem to be found. The dynamic effects due to the diblock copolymer addition are also intensively investigated both experimentally and theoretically but there it is still much work to do before the actual processes in mixers and extruders will be completely understood.

4. INTERCHANGE REACTIONS

Now we pass to interchange reactions that proceed in condensation polymers containing for example ester or amide bonds. Such reactions lead to the redistribution of unit sequences between different polymer chains and thus to the formation of multiblock copolymers. Depending on the position of reacting groups, all interchange reactions are divided into two classes.

One of them is direct interchange, in which two internal groups participate. Transesterification, transamidation and ester-amide interchange belong to this type. Another interchange mechanism is end-group interchange involving one internal and one terminal group. Depending on the type of the end-group it could be acidolysis, alcoholysis, or aminolysis.

Several aspects make interchange reactions important from the viewpoint of practical applications[37]. Firstly, such reactions give a chance to preparing novel copolymers with given composition and degree of randomness; secondly, they enhance the compatibility of polymer blends directly during processing; thirdly, interchange reactions provide for chemical healing of laminates of condensation polymers; and finally, they help to obtain more uniform polymer samples by minimizing molecular mass fluctuations in a melt stream during polycondensation and processing.

Dozens of polymeric systems, in which interchange reactions proceed, have been experimentally studied. It was found that any interchange reaction eventually transforms the initial blend into completely random copolymer having the most probable molecular mass distribution (MMD). Usually complete randomization takes several hours of treatment at elevated temperature.

The problem of describing interchange reactions consists in a characterization of the transient MMD and block mass distribution (BMD), the latter is known also as the distribution of block length.

First theoretical study on the subject was undertaken by Flory[38] who demonstrated that a polymer with the most probable MMD has the highest entropy of mixing provided the numbers of units and chains are fixed. Since interchange reactions just mix monomer units under constant numbers of units and chains, all of them will result in this distribution, which is called the geometrical or the Flory distribution in the case of linear chains. Flory has checked his conclusion by mixing two decamethylene adipate polyesters[39] of different molecular mass and demonstrating that the melt viscosity tends with time to the stationary value, which corresponds to the average chain polymerization degree predicted by the most probable MMD.

The next progress in theory was due to Kotliar[40] who developed a statistical approach to the description of structure of reacting chains. He supposed that the interchange reaction might be considered as a two-step process, namely, several random chain cleavages followed by the random redistribution of chain ends formed and by the same number of couplings between them. The number of cleavages per average chains is proportional to time and thus may be used as the measure of the reaction extent. This relatively simple approach enables one to calculate different averages over the transient distributions of molecular and block mass. However, this method accounts only for cleavages of the bonds present in the initial polymers and thus neglects the inverse reaction between the bonds formed just in the course of the interchange. Therefore it may be applied only at the initial stage of the reaction.

If consider the interchange as the second-order reaction between dyads, the dyad distribution and hence the average block length may be found as functions of time[41,42]. These parameters may be also obtained from experiment using various techniques[3,37]. Juxtaposing the theoretical and experimental data on the dyad distribution, it is possible to evaluate the reaction rate constant. From the temperature dependence of the rate constant, the energy of interchange activation may be found.

Unfortunately, experiment hardly can provide with information on the distribution of blocks longer than triads. Another problem is that with all mentioned experimental techniques only the total blend structure may be analyzed. Meanwhile it can essentially differ from the structure of the reaction product – the multiblock copolymer, at least at early stages of the reaction when the homopolymer content in the blend is yet considerable.

This difference was shown to be detectable experimentally by Montaudo et al[43] who investigated the interchange reaction in a blend of polyethylene terephtalate (PET) and polyethylene adipate (PEA) by a complicated procedure including the selective degradation of the product by aminolysis, mass spectrometry analysis, and synthesis of the model compounds. This technique made possible to detect the excess of adipic units in the copolymer at early stage of the reaction. Therefore it was concluded that the main mechanism of the reaction should be an attack of oligomer PEA ends on PET chains.

The aforementioned problems constitute the ground for the theoretical work: to describe the evolution of the MMD and BMD for the blend as a whole and for the reaction product – copolymer as well.

Let us formulate a model. The initial state is a homogeneous blend of two homopolymers A and B. Only groups belonging to different chain are able to react. The act of direct interchange includes the cleavage of two chains at the point of their contact immediately followed by the cross-recombination of newborn ends. For end-group interchange, a chain being attacked by an active end of another chain cleaves and one of the new ends joins at once to the attacking end while an end-group of the other part of the cleaved chain becomes active. In both cases the total numbers of units and chains are kept unchanged, so the number average chain length remains constant.

The kinetic equations for the MMD have been derived in the compact form for both direct interchange[44] (Eq. (12)) and end-group interchange[45] (Eq. (13)) by considering the detailed balance in the polydisperse system, where m_i is the number of chans of i units, k is the rate constant, $\overline{N}$ is the number average polymerization degree, n is the total number of units in the system.

$$\frac{\partial m_i}{\partial t} = k\left(\sum_{r=1}^{i}\sum_{j=r}^{\infty}\sum_{l=i-r+1}^{\infty}\frac{m_j m_l}{n} - i m_i\right) \tag{12}$$

$$\frac{\partial m_i}{\partial t} = k\left(\sum_{l=i}^{\infty}\frac{m_l}{\overline{N}} + \sum_{j=1}^{i}\sum_{l=i-j+1}^{\infty}\frac{m_l m_j}{n} - m_i - \frac{i m_i}{\overline{N}}\right) \tag{13}$$

Eq. (12), (13) may be explicitly solved in terms of generating function, which contains the complete information about all statistical moments of the MMD. Under specific initial conditions, it is possible to obtain analytical expressions for the functions of transient MMDs.

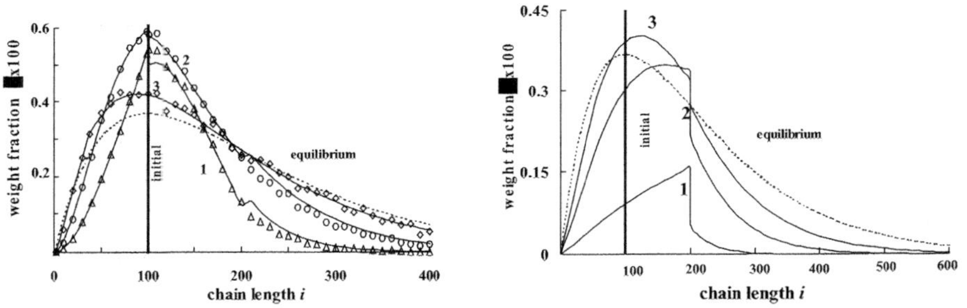

Figure 26. Transient MMDs for blend of polymers in the course of direct interchange (left plot, τ = 1 (curve 1), 3 (2), 10 (3)) and end-group interchange (right plot, τ = 0.2 (1), 1 (2), 2 (3)). Initially all chains consist of 100 units, the fraction of A-units is φ = 0.5. The final distribution is given by dashed line. Points at the left plot are the result of Monte Carlo modeling. Ref.[46].

In Figure 26 the mass functions of the transient MMD for the direct interchange are compared with those functions for the end-group interchange. The initial blend was taken to be monodisperse. Chain length is plotted along the horizontal axis, different curves correspond to the different conversion, which was measured in the number of interchanges per number average chain, τ, this quantity being proportional to real time. It is seen that although the initial and final distribution in the systems are the same, the transient distributions are essentially different.

In Figure 27 the case is presented, when both initial homopolymers have the Flory MMD so that the initial MMD of the blend is the sum of two such distributions. Even one interchange per average chain is enough to considerably change the initial distribution (curves 1). Several more interchanges per average chain bring the MMD of the system to the form close to the equilibrium one – the united Flory distribution. It is worth noting that this result does not depend on the absolute values of the chain length. Also the good agreement between theory and results of Monte Carlo modeling should be mentioned when looking at the left plot of Figure 27.

Kinetic equations for the BMD derived in ref.[47,48] have rather complicated form to discuss them in details. The only thing worthing to mention is that three equations were derived: for homopolymers, for end-blocks, and for internal copolymer blocks. This enables one to describe the evolution of copolymer composition separately.

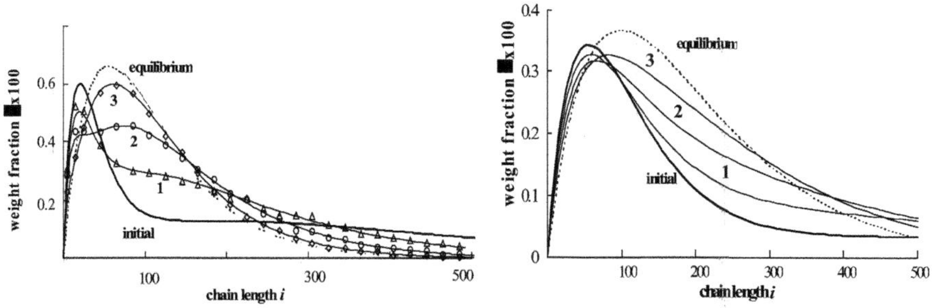

Figure 27. Transient MMDs for blend of polymers in the course of direct interchange (left plot, $\phi = 0.7$, $\tau = 1$ (curve 1), 3 (2), 10 (3)) and end-group interchange (right plot, $\phi = 0.5$, $\tau = 0.2$ (1), 1 (2), 2 (3)). The initial and final distributions are given by solid and dashed line, respectively. Points at the left plot are the result of Monte Carlo modeling. Ref.[46].

BMD for the copolymer and for the blend as a whole may be considerably different. Theoretical curves plotted in Figure 28 for the case of direct interchange correspond to different moments of time, while points represent the Monte Carlo results. Results of theory and modeling are in a good agreement. Then, corresponding curves in the left and right plots differ noticeably at small time although the BMDs eventually tend to the Flory form. At the very beginning of the reaction the copolymer composition is determined by the ratio of average homopolymer lengths while in the stationary state the copolymer composition should be equal to the average composition of the blend. The similar behavior is revealed for end-group interchange[47].

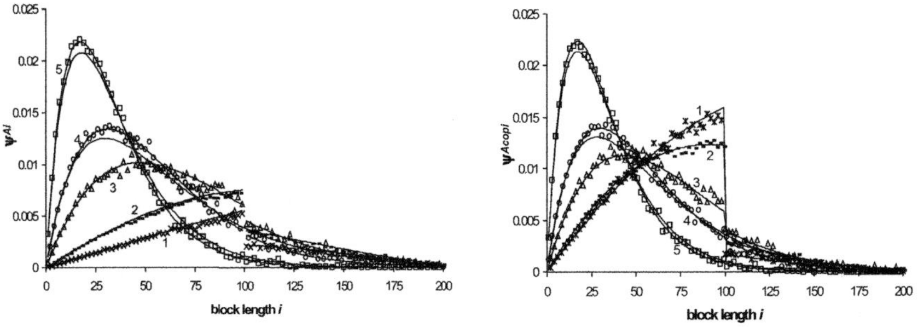

Figure 28. Time evolution of the BMD of A-units in the whole blend (ψ_{Ai}, left plot) and in the copolymer (ψ_{Acopi}, right plot). Ref.[47]. Solid lines correspond to the given number of interchanges per unit: τ = 0.005 (curve 1), 0.01 (2), 0.03 (3), 0.05 (4), 0.1 (5). Points denote the BWD for the corresponding τ generated by MC simulation. Ref.[46]. Lines near curves 4 and 5 are the Flory distributions for average N_A as for curves 4, 5. Initially all chains consist of 100 units, $\phi = 0.5$.

Calculations show that under arbitrary initial conditions, the BMD characteristic for the completely random copolymer is eventually formed. In several cases it can be explicitly found, for example, if initial homopolymers have MMD Flory. In that case the copolymer BMD has the Flory form throughout all the reaction, the average block length being decreased with time.

It is possible to derive from the BMD kinetic equations simple analytical expressions for the number of AB-dyads per unit, R_{AB}, and for the average length of A-sequence, $\overline{N}_A(\tau)$. For the initial conditions corresponding to the melt of two homopolymers A and B with average sequence lengths $\overline{N}_{A0}$, $\overline{N}_{B0}$ we get

$$\begin{aligned} R_{AB}(\tau) &= 2\phi(1-\phi)\left(\left(1-\frac{1}{\overline{N}}\right)\left(1-e^{-\tau}\right)+\left(\frac{1}{\overline{N}_{A0}}-\frac{1}{\overline{N}_{B0}}\right)(2\phi-1)\left(e^{-\tau/2}-e^{-\tau}\right)\right) \\ \overline{N}_A(\tau) &= \left(\left(1-\frac{1}{\overline{N}}\right)(1-\phi)\left(1-e^{-\tau}\right)+\frac{1}{\overline{N}}+\left(\frac{1}{\overline{N}_{A0}}-\frac{1}{\overline{N}}\right)\left(2\phi e^{-\tau/2}-(2\phi-1)e^{-\tau}\right)\right)^{-1} \end{aligned} \tag{14}$$

for direct interchange, where $\tau = kt$, k being the rate constant entering Eq. (12) and

$$R(\tau)=2\phi(1-\phi)\left(1-e^{-\tau/\bar{N}}\right)-2\phi(1-\phi)(1-2\phi)\left(\frac{1}{\bar{N}_{A0}}-\frac{1}{\bar{N}_{B0}}\right)\left(e^{-\tau/2}-e^{-\tau/\bar{N}}\right)$$
$$\bar{N}_A(\tau)=\left(\frac{1}{\bar{N}_{A0}}e^{-\tau/\bar{N}}+(1-\phi)\left(1-e^{-\tau/\bar{N}}\right)+2\phi(1-\phi)\left(\frac{1}{\bar{N}_{A0}}-\frac{1}{\bar{N}_{B0}}\right)\left(e^{-\tau/2}-e^{-\tau/\bar{N}}\right)\right)^{-1} \quad (15)$$

for end-group interchange, where $\tau = 2kt$, k being the rate constant in Eq. (13). The average length of B sequence, $\bar{N}_B(\tau)$, could be found from $\bar{N}_A(\tau)$ by substituting $1-\phi$ for ϕ and exchanging $\bar{N}_{A0}$ with $\bar{N}_{B0}$.

Summarizing our studying of interchange kinetics for both mechanisms and various initial conditions, it is worth emphasizing that the interchange process may be always divided into two stages: fast and slow. The fast stage takes the time enough for the average chain to undergo several interchanges; under the typical experimental conditions, it will be a few minutes. At this stage the MMD of the blend takes the form of the Flory distribution, homopolymer chains are almost exhausted, the copolymer composition is nearly equalized with the composition of the whole blend. In the course of the slow stage the gradual decreasing of the average copolymer block length takes place until it reaches the value characteristic of the completely random copolymer. The full randomization is attained at the time when each interchangeable group in the copolymer reacts one time on the average. In real experiments, it usually takes more than an hour.

From this behavior we may conclude that increasing average molecular weight of the system will accelerate the MMD evolution and have no effect on the BMD evolution if interchange proceeds via direct mechanism. In the case of end-group interchange, increasing average molecular weight leads also to the proportional decrease in the number of reactive end groups. As a result, no effect on the MMD evolution is predicted whereas the BMD is expected to change slower. That property may help one to determine the main interchange mechanism for a particular system.

It is worth noting that both theory for model systems and methods for interpreting experimental data reviewed above are related to one-phase systems. However most of polymer pairs are immiscible so at least for early stage of the reaction the competition between interchange reaction and phase separation must take place. So phase separation in blends undergoing interchange reactions appears to be an important factor.

Using phase contrast microscopy is the direct way to observe the competition between phase separation and interchange reaction. This method was applied by Tanaka et al.[49] to the initially homogeneous 50/50 blends of polycarbonate (PC) with PET and PC with polyarylate (PAr). In the course of annealing of PC/PET blend, a typical scenario of spinodal decomposition was realized, then particles appeared and grew, however, their boundaries broadened with time, as seen in Figure 29. On the contrary, in the PC/PAr blend the reaction dominated, as a result coarse particles were not formed from the fine dispersed blend structure and, finally, the blend became homogeneous. The difference in the blends behavior was explained by the greater mobility of PET chains in comparison with that of PAr chains.

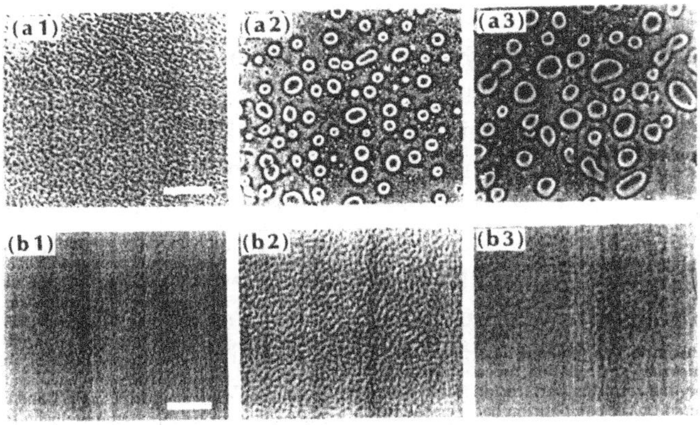

Figure 29. Pattern evolution PC/PET (5/5), T_{an}= 251°C, (a1) 10 s, (a2) 185 s, (a3) 920 s. The bar corresponds to 100 μm. (b) PC/PAr (5/5), T_{an} = 240°C, (b1) 10 s, (b2) 2290 s, (b3) 3020 s. The bar corresponds to 20 μm. Ref.[49]

Scattering techniques may be as well applied to monitor phase separation in reacting blends.

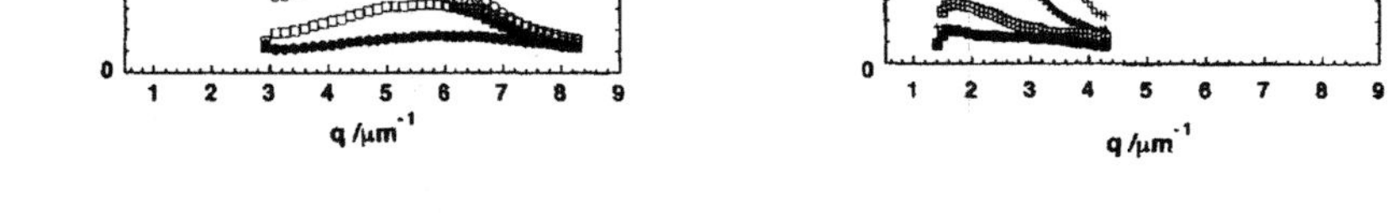

Figure 30. Spinodal decomposition in PEN/PET blend monitored by light scattering.Ref.[50].

For example, Okamoto and Kotaka[50] measured the intensity of light scattering in polyethylene naphtalate (PEN)/PET blend (see Figure 30). The initial growth of inhomogeneities (left plot) is followed by their decay (right plot), which was ascribed to the transesterification.

Recently Govorun and Kudryavtsev[51] considered in theory the linear stage of the spinodal decomposition in the blend, where an interchange reaction proceeds.

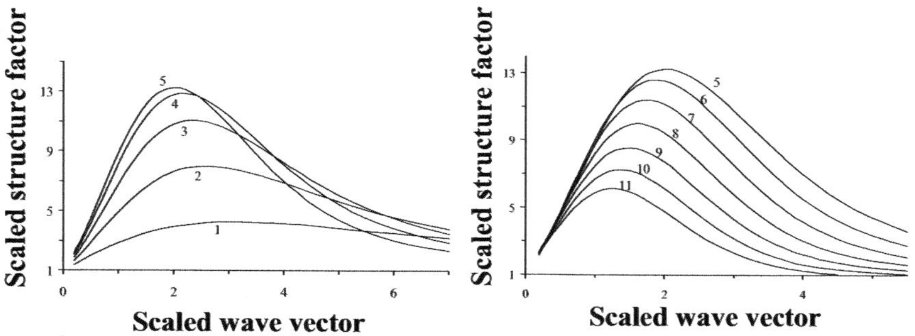

Figure 31. Evolution of the dynamic structure factor in the course of spinodal decomposition in the reacting blend. Curves 1-11 are plotted for the moments of dimensionless time from τ = 0.3 to 3.3 by equal intervals $\Delta\tau = 0.3$. $\overline{N}_A = \overline{N}_B = 20$, ϕ= 1/2, $\chi = 0.4$. Ref.[51].

As follows from Figure 31, the structure factor that is proportional to the scattering intensity first grows and then slowly decreases, in qualitative agreement with the experimental curves. Maximum value of the structure factor is less than the corresponding value of the structure factor in the non-reacting blend by the order of magnitude. Thus interchange reaction essentially slows down the phase separation from the very beginning of the process.

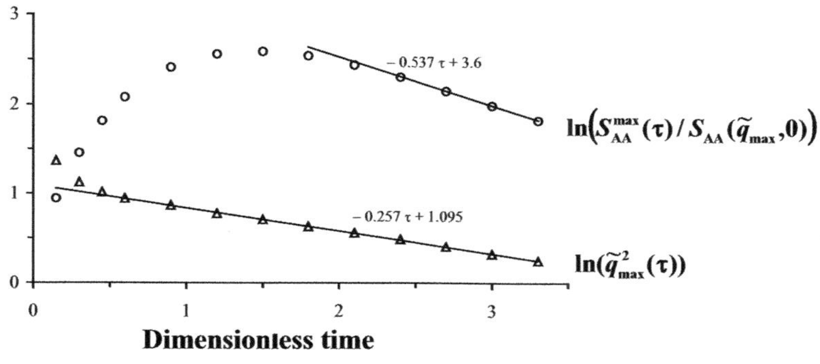

Figure 32. Temporal evolution of the maximum value of the scaled structure factor (circles) and of the corresponding wave number (triangles) in a reacting blend. Ref.[51].

The position of the structure factor maximum shifts with time to smaller wave numbers, so the typical inhomogeneity size in the blend grows. The corresponding dependencies are shown in Figure 32. The upper points represent the maximum value of the structure factor and the lower ones show its position. The theory predicts that the mean size of compositional fluctuations grows exponentially with time. It is interesting that after certain period of time both dependencies can be approximated by straight lines.

If the interaction energy between polymer units A and B is unfavorable and rather high, then sharp interfaces are formed in the blend before the interchange proceeds up to the measurable extent. There is still no theory for this case so let us consider the Monte Carlo results by Jo et al[52]. They simulated an interchange reaction on a cubic lattice of 60x60x60 filled with monodisperse homopolymers A and B at a blend ratio 25/75, where 60 % of the lattice sites having been occupied by segments. The reduced concentration of both A and B units was plotted *vs* spatial coordinate in Fig. 33. It is seen that initially the system consists of two A and two B domains in a highly segregated state. The reaction gradually reduces the composition difference between two phases without changing the initial domain sizes. Indeed, the arrow length in the figures corresponding to the total width of A- and B-rich domains is kept constant while the composition of domains levels off.

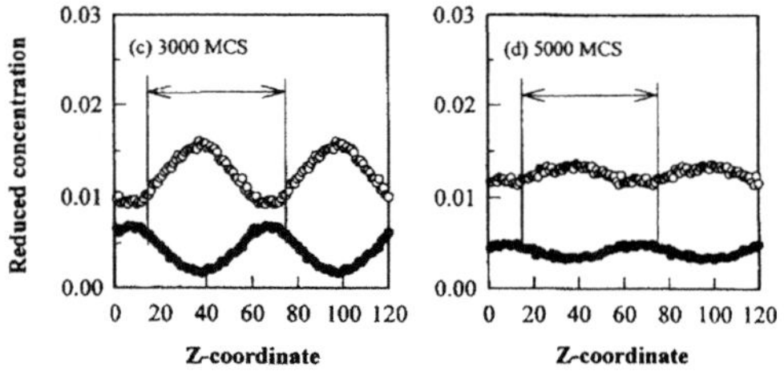

Figure 33. Interchange in the initially segregated system by Monte Carlo modeling. Ref.[52]. Concentration profiles of polyester A (filled circles) and polyester B (empty circles) are presented. Time is measured in Monte Carlo steps (MCS).

Summarizing, we note that quantitative studies of the phase behavior of polymer blends undergoing interchange reactions are still scarce. There are many problems to solve related to the thermodynamics and kinetics of phase separation and compatibilization in such blends.

At the moment there is no theory of interchange reactions in the strongly segregated systems. The interchange kinetics at interfaces may appear essentially different from that in the bulk, as it occurs for other reactions, for example, end-coupling.

5. CONCLUSIONS

The following achievements related to different kinds of the reactions are worth mentioning.

For polymer-analogous reaction, the evolution of the blend structure under concerted action of the reaction and interdiffusion has been first described. Due to the reaction, the initial binary blend becomes multicomponent one consisting of macromolecules of time-dependent composition and mobility. The theory accounts for the interchain interactions influence both on the reactivity of polymer functional groups and on the mobility of macromolecules, in other words, both chemical and physical interchain effects.

For end-coupling reaction, the reaction kinetics both in a homogeneous melt and at the interface is described. An influence of the diblock copolymer formed both on the reaction kinetics and on the thermodynamic equilibrium is analyzed. For an initial quasi-homogeneous blend, the competition between the phase separation and the compatibilizing action of the diblock copolymer is described; in particular, the diblock copolymer influence on the particle growth via Ostwald ripening is considered.

For interchain exchange reaction proceeding in a homogeneous melt, an analytical description of the molecular mass and block mass transient distributions of the reaction product has been worked out. An initial stage of phase separation is considered in theory as well. The reaction in a heterogeneous blend can be described using Monte Carlo simulations.

At the same time, such important problems as peculiarities of the reaction kinetics at interfaces, a widening of the interface layer in the course of the reaction, and a contribution of interdiffusion to the processes in heterogeneous blends are to be studied.

ACKNOWLEDGEMENT

Financial support from the grant of the RF President for young scientists and leading scientific schools (project no. НШ-1598.2003.3) is gratefully acknowledged. Y.K. is also thankful to the Russian Foundation for Science Support.

REFERENCES

1. Platé N.A., Litmanovich A.D., Noah O.V. Macromolecular reactions. New York: Wiley, 1995.
2. Klesper E., Gronski W., Barth V. Macromol Chem 1971, *150*, 223.
3. Litmanovich A.D., Platé N.A., Kudryavtsev Y.V. Prog Polym Sci 2002, *27*, 915.
4. Jayabalan M. J Appl Polym Sci 1982, *27*, 43.
5. Keller J. J Chem Phys 1962, *37*, 2584.
6. Mityushin L.G. Problemy Peredachi Informatsii (in Russian) 1973, *9*, 81.
7. Platé N.A., Litmanovich A.D., Noah O.V. Makromolekulyarnye Reaktsii (in Russian). Moscow, Khimiya, 1977, p.97.
8. Kuchanov S.I., Brun Y.B. Dokl AN SSSR 1976, *227*, 662.
9. Noah O.V., Litmanovich A.D., Platé N.A. J Polym Sci Polym Phys 1974, *12*, 1711.
10. Krentsel L.B., Kudryavtsev Y.V., Rebrov A.I., Litmanovich A.D., Platé N.A. Macromolecules 2001, *34*, 5607.
11. Platé N.A., Stroganov L.B., Seifert T., Noah O.V. Dokl Akad Nauk SSSR 1975, *223*, 396.
12. Litmanovich A.D., Cherkezyan V.O. Eur Polym J 1984, *20*, 1041.
13. Litmanovich A.D., Cherkezyan V.O. Eur Polym J 1985, *21*, 623.
14. Litmanovich A.D., Cherkezyan V.O., Khromova T.N. Vysokomol Soedin B 1981, *23*, 645.
15. Litmanovich A.D., Cherkezyan V.O., Khromova T.N. Vysokomol Soedin A 1985, *27*, 1865.
16. Ermakov I.V., Litmanovich A.D. Vysokomol Soedin 1988, *30*, 2595.
17. Ermakov I.V., Lebedeva T.L., Litmanovich A.D., Platé N.A. Polym Sci 1992, *34*, 513.
18. Yashin V., Kudryavtsev Y., Govorun E., Litmanovich A. Macromol Theory Simul 1997, *6*, 247.
19. Platé N.A., Litmanovich A.D., Yashin V.V., Ermakov I.V., Kudryavtsev Y.V., Govorun E.N. Polym Sci Ser A 1997, *39*, 3.
20. Utracki L.A. Polymer alloys and blends: thermodynamics and rheology. Munich: Hanser, 1989
21. Wu S. Polymer 1985, *26*, 1855.
22. Macosko C.W., Guegan P., Khandpur A.K., Nakayama A., Marechal P., Inoue T. Macromolecules 1996, *29*, 5590.
23. Milner S.T., Xi H.W. J Rheol 1996, *40*, 663.
24. Sundararaj U., Macosko C.W. Macromolecules 1995, *28*, 2647.
25. Lyu S.P., Jones T.D., Bates F.S., Macosko C.W. Macromolecules 2002, *35*, 7845.
26. Schulze J.S., Cernohous J.J., Hirao A., Lodge T.P., Macosko C.W. Macromolecules 2000, *33*, 1191.
27. Roe R.J., Kuo C.M. Macromolecules 1990; *23*, 4635.
28. Park D.W., Roe R.-J. Macromolecules 1991, *24*, 5324.
29. Helfand E., Tagami Y. J Polym Sci B 1971, *9*, 741.

30. Leibler L. Macromolecules 1982, *15*, 1283.
31. Leibler L. Macromol Chem Macromol Symp 1988, *16*, 1.
32. Erukhimovich I., Govorun E.N., Litmanovich A.D. Macromol Theor Simul 1998, *7*, 233.
33. Govorun E.N., Litmanovich A.D. Polym Sci Ser A, 1999, *41*, 1111.
34. Alexander S. J Phys 1977, *38*, 983.
35. de Gennes P.G. Macromolecules 1980, *13*, 1069.
36. Platé N.A., Litmanovich A.D., Kudryavtsev Y.V., Govorun E.N. Macromol Symp 2003, *191*, 11.
37. Transreactions in Condensation Polymers. *S. Fakirov*, ed. Weinheim: Wiley, 1999.
38. Flory P.J. JACS 1940, *62*, 1057.
39. Kotliar A.M. J Polym Sci Polym Chem Ed 1975, *13*, 973.
40. Devaux J., Godard P., Mercier J.P. Polym Sci: Polym Phys Ed 1982, *20*, 1875.
41. Benoit H.C., Fisher E.W., Sackmann H.G. Polymer 1989, *30*, 379.
42. Montaudo G., Montaudo M.S., Scamporrino E., Vitalini D. Macromolecules 1992, *25*, 5099.
43. Kudryavtsev Y.V. Macromol Theory Simul 2000, *9*, 675.
44. Kudryavtsev Y.V. Macromol Theory Simul 2001, *10*, 355.
45. Kononenko O.A., Kudryavtsev Y.V., Litmanovich A.D. Polym Sci Ser A 2002, *8*, 911.
46. Kudryavtsev Y.V., Govorun E.N. e-Polymers 2002, no. 033.
47. Kudryavtsev Y.V., Govorun E.N. e-Polymers 2003, to be published.
48. Tanaka H., Suzuki T., Hayashi T., Nishi T. Macromolecules 1992, *25*, 4453.
49. Okamoto M., Kotaka T. Polymer 1997, *38*, 1357.
50. Govorun E.N., Kudryavtsev Y.V. Polym Sci 2004, in press.
51. Jo W.H., Kim J.G., Jang S.S., Youk J.H., Lee S.C. Macromolecules 1999, *32*, 1679.

FUNCTIONAL COPOLYMER MACROMOLECULES: DESIGN, CHARACTERIZATION AND PROPERTIES

ALEXEI R. KHOKHLOV
Physics Department, Moscow State University, 119992 Moscow, Russia

Abstract: The paper reports about a detailed investigation of the methods of design of sequences of monomer units in copolymers with special emphasis on biomimetic approaches. The main objective of the described approach is to formulate new methods of synthesis of copolymers with sophisticated functional properties. In particular the design of protein-like copolymers is performed by computer simulation. The properties of these copolymers are compared with those of random and block copolymers having the same overall composition by the Montecarlo method using the bond fluctuation model. Experimental approaches aimed to obtain these protein-like copolymers carried out in various laboratories are finally reported.

Key words: Protein-like copolymers, MonteCarlo calculations, coil-globule transition, .monomer sequences, globule formation, computer simulation

1. INTRODUCTION

For a long time chemical industry was interested in polymers mainly from the viewpoint of obtaining unique construction materials (plastics, rubbers, fibers etc.). Couple of decades ago the main focus of interest shifted to functional polymers (superabsorbents, membranes, adhesives etc.). In the nineties scientific and industrial polymer community started to discuss "smart" of "intellectual" polymer systems (e.g. soft manipulators, polymer systems for controlled drug release, field-responsive polymers etc.); the meaning behind this term is that the functions performed by polymers

F. Ciardelli and S. Penczek (eds.),
Modification and Blending of Synthetic and Natural Macromolecules, 283–291.

become more sophisticated and diverse. This line of research concentrating on polymer systems with more and more complex functions will be certainly in the mainstream of polymer science in the 21st century.

One of the ways to obtain new polymers for sophisticated functions is connected with the synthesis of novel monomer units where the required function is linked to the chemical structure of these units. However, the potential of this approach is rather limited, because complicated and diverse functions of polymer material would then require a very complex structure of monomer units which normally means that the organic synthesis is more expensive and less robust.

The alternative approach is to use known monomer units and to try to design a copolymer chain with given sequence of these units. There are practically infinite possibilities to vary sequences in copolymers: from the variation of some simple characteristics like composition of monomer units, average length of blocks (for the chains with blocky structure), availability of branching, etc. to more sophisticated features like long-range correlations or gradient structure. Therefore, in this approach a wide variety of new functional copolymers can be tailored.

It is important to emphasize that the nature has chosen this way in the evolution of main biological macromolecules: DNA, RNA and proteins. These polymers in living systems are responsible for functions which are incomparably more complex and diverse than the functions which we are normally discussing for synthetic copolymers. The molecular basis for this ability to perform sophisticated functions is associated with unique primary sequences of units in biopolymers which emerged in the course of biological evolution.

Thus, one of the promising approaches in the sequence design of functional copolymers is biomimetic in its nature: it is tempting to look at the main features of sequences of monomer units in biopolymers, understand how these sequences define functional properties, and then try to implement similar ideas for synthetic copolymers.

Some time ago, we started a detailed investigation of the methods of design of sequences of monomer units in copolymers with special emphasis on biomimetic approaches. In this way we hope to formulate new methods of synthesis of copolymers with sophisticated functional properties. Also, one of the motivations for this investigation is the attempt to understand, at least partially, the principles of evolution of sequences of biological macromolecules at the early stages of this evolution.

2. PROTEIN-LIKE COPOLYMERS

The first ideas connected with biomimetic sequence design of functional copolymers were formulated by us in 1998[3-6]. They were based on the simple and well-known fact that the functioning of all globular proteins depends on two main factors: (i) they are globular; (ii) they are soluble in aqueous medium. It should be mentioned that the combination of these two factors is non-trivial, e.g. for homopolymers and random copolymers the transition to globular conformation is usually accompanied by the precipitation of globules from the solution[1-2]. Protein globules are soluble in water because of the special primary sequence: in the native conformation most of hydrophobic monomer units are in the core of the globule while hydrophilic and charged monomer units form the envelope of this core (Fig. 1). Of course, the division of 20 types of monomer units available in globular proteins in only two classes (hydrophobic and hydrophilic) is rather rough, but still Figure1 gives a correct general picture of the structure of a protein. Now, having in mind the biomimetic approach described above, we can formulate the following problem: whether it is possible to design such sequence of synthetic AB-copolymer (copolymer consisting of monomer units of two types, A and B) that in the most dense globular conformation all the hydrophobic B-units are in the core of this globule while hydrophilic A-units form the envelope of this core? This question was first addressed in ref.[3] (see also[4-6]) and the corresponding polymers were called protein-like AB-copolymers.

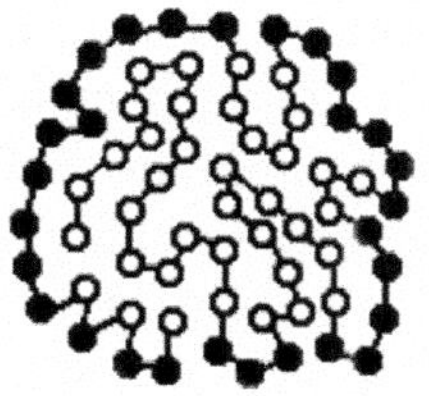

Figure 1 Schematical representation of globular protein. Hydrophobic units are shown by open circles, while hydrophilic units are represented as filled circles.

The protein-like AB-sequences were first obtained in computer experiments[3-6] which can be described as follows. We start with arbitrary homopolymer globule conformation formed due to the strong attraction of monomer units (Figure2a) and perform for it a "coloring" procedure (Figure 2b): monomer units in the center of the globule are called B-type (hydrophobic) units, while monomer units belonging to globular surface are assigned to be A-type (hydrophilic) units. Then this primary structure is fixed, attraction of monomer units is removed and protein-like copolymer is ready for the further investigation (Figure2c).

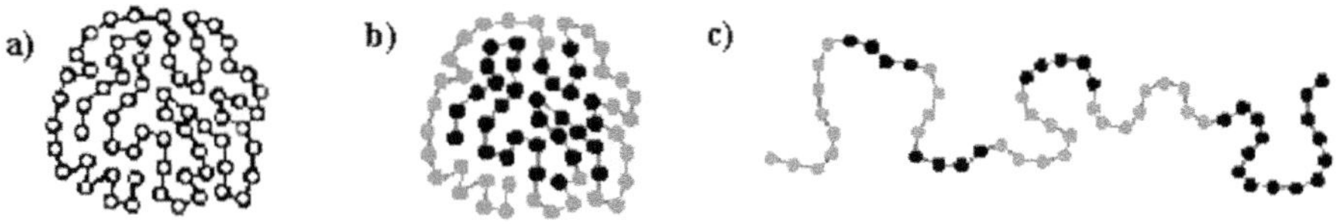

Figure 2 Main steps of the sequence design scheme for protein-like copolymers: (a) homopolymer globule; (b) the same globule after "coloring" procedure; (c) protein-like copolymer in the coil state.

In particular, in refs.[3-6] the coil-globule transition for thus generated AB-copolymers was studied. This transition was induced by the attraction of only B-units (the interactions A-A and A-B were chosen to be repulsive). The properties of this transition were compared with those for random AB-copolymers with the same composition and random-block AB-copolymers with the same composition and the same "degree of blockiness" as for designed protein-like AB-copolymers. The calculations were performed by MonteCarlo method using the bond fluctuation model[14].

It was shown[3-6] that the coil-globule transition in protein-like copolymers occurs at higher temperatures, is more abrupt, leads to the formation of denser globule and has faster kinetics than for random and random-block counterparts. The reason for this is illustrated in Figure 3 where the typical snapshots of globules formed by protein-like and random AB-copolymers with the same AB-composition are shown. One can see that the core of protein-like globule is much more compact and better formed, it is surrounded by the loops of hydrophilic units which stabilize the core. Apparently, this is due to some memory effect: the core which existed in the "parent" conformation (this is the term introduced in ref.[3] to describe the conformation of Figure 2b where the coloring is performed) was simply reproduced upon refolding caused by the attraction of B units. One may say that the features of "parent" conformation are "inherited" by the protein-like AB-copolymer. Looking at the conformations of Figure 3 it is natural to argue that protein-like copolymer globule should be soluble in water (see also[20]) and thus open to further modification in the course of biological evolution, while random copolymer globules will most probably precipitate and thus drop out of the evolution.

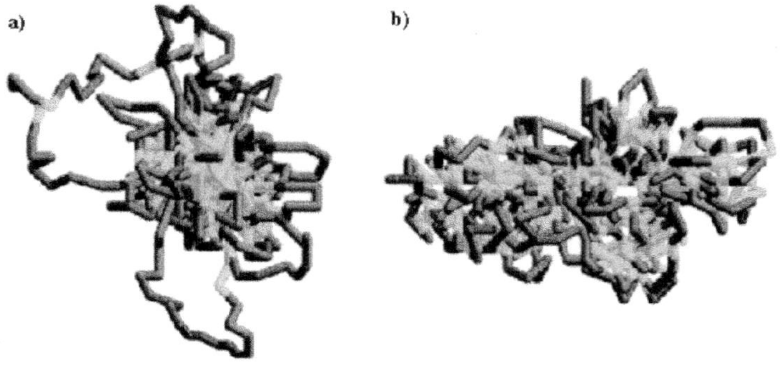

Figure 3. Typical snapshots of globular conformation for protein-like (a) and random (b) copolymers.

3. EXPERIMENTAL REALIZATIONS

After the idea of sequence design of protein-like copolymers was presented and realized in computer simulations, several teams started experimental research aimed to obtain such copolymers in synthetic chemical laboratory.

In particular, Professor H.Tenhu at the University of Helsinki studied grafting of short poly(ethylene oxide) (PEO) chains to the copolymer of thermosensitive N-isopropylacrylamide (NIPA) and glicydil methacrylate[9]. At room temperatures such copolymer is in the coil state and grafting takes place in a random manner. At elevated temperatures the transition to globule occurs, and grafting proceeds mainly in the globular surface, thus leading to its hydrophilization and to the creation of protein-like copolymer in the sense described above. Indeed, it was shown that protein-like copolymer prepared in this way exhibits solution turbidity at higher temperatures than the random one, and gives smaller aggregates in the turbid solution.

In the group of Professor V. Lozinsky (Institute of Organoelement Compounds of Russian Academy of Sciences) the redox-initiated free-radical copolymerisation of thermosensitive N-vinylcaprolactam (NVCa) with hydrophilic N-vinylimidazole (NVIz) was studied at different temperatures[10]. At room temperatures such polymerization gives a random copolymer. On the other hand, when polymerization takes place at elevated temperatures (ca. 65 °C) growing chains form globules, and the concentration of monomers around the active radical is influenced by this fact. The conditions were found when protein-like copolymers are emerging as a result of such synthesis. These copolymers were not precipitating at all when the solution is heated up to 80°C, on the other hand dense globules were formed already around 30°C.

In the group of Professor B.Mattiasson (University of Lund, Sweden) similar type of protein-like copolymers were obtained for the pair of monomers NIPA/NVIAz synthesized in aqueous solution[14]. It was demonstrated that the copolymer with virtually random distribution of NVIAz units along the chains did not interact with the metal chelate adsorbent, Cu^{2+}-iminodiacetate-Sepharose, whereas the copolymer, possessing the protein-like sequence, was absorbed specifically by the resin, since the hydrophilic pendant imidazole groups were accumulated in the outer hydrophilic shell of macromolecular coil.

Recently, we performed computer experiments specially designed to describe the process of copolymerization with simultaneous globule formation in order to give more careful theoretical foundation for this synthetic method proposed in[10]. We have shown that such copolymerization process does indeed lead to the formation of a globule with protein-like sequences[18].

4. STATISTICAL CORRELATIONS IN THE SEQUENCES

Returning to the computer-generated protein-like sequences (Figure 2) it is clear that they should exhibit long-range correlations along the chain, since the type of monomer unit (A or B) depends on the conformation of globule as a whole, not on the properties of some small part of the chain. In refs.[7,11,12] it was shown, both by exact analytical theory and by computer simulation, that this is indeed the case and that the long-range correlations in the protein-like sequences can be described by the so-called Levy-flight statistics[12].

5. SOME GENERALIZATIONS

The approach to sequence design of functional copolymers presented above can be generalized in many different ways.

Figure 2 represents only one of possible ways of realization of coloring procedure. If our intention is to mimic not globular proteins – enzymes, but rather membrane proteins, we can apply another coloring procedure for parent homopolymer globule, namely, we can assign the type B to monomer units lying inside the central cylindrical slice and the type A to monomer units lying in two hemispheres to both sides of B-layer. It was shown[7,8,11] that such coloring procedure leads to formation of AB-copolymer chain which exhibit a number of unusual properties. For example, such a chain shows the effect of stability of parent micro-segregated structure: after refolding the segregation of both hemispheres of A-units is reestablished.

Also, it is possible to use several colors in coloring of parent conformation, thus generating sequences with several (more than two) types of monomer units. For example, in refs.[7,8,11] this idea was implemented for modeling of proteins with active enzymatic center using the triple coloring procedure. This "active" center was modeled by small spherical core of C-units lying inside a protein-like globule consisting of larger core of B-units and a shell of A-units. Computer simulation has shown that such ABC-copolymers usually restore their original structure with B- and C-cores.

Furthermore, to design sequences with special properties it is not necessary to perform coloring of a dense globule. In fact, any special macromolecular conformation can play the role of a parent one. For example, in ref[13] the conformation of a polymer chain adsorbed on a plane surface was considered. The monomer units closest to the surface in some instant snapshot conformation were assigned to be A-units, others became B-units. The AB-chain thus obtained was called an adsorption-tuned copolymer. It was shown that this copolymer adsorbs on another plane surface (to which only A-units are attracting) more efficiently than random and random-block copolymers with the same AB-composition and the same degree of blockiness.

Another development is a computer realization of the idea of molecular dispenser capable to absorb colloidal particles of a given size from their mixture[19].

6. RELATION TO EVOLUTION PROBLEMS

In all cases described above some functional features of the "parent" conformation were "memorized" by the copolymers generated according to our sequence design scheme. These features are then manifested in other conditions. Such an interrelation can be regarded as one of the possible mechanisms of molecular evolution: polymer acquires some special primary sequence in the "parent" conditions and then (in other conditions) uses the fact that primary structure is "tuned to perform certain functions".

Recently, we have introduced explicitly the concept of evolution of sequences into the scheme of generation of protein-like copolymers[15]. Namely, after the formation of initial HP sequence we allowed the macromolecule to undergo a coil-globule transition to a new globule, mainly induced by the strong attraction between H units, and then we made "recoloring" in the newly formed globular conformation. The rules of this recoloring are the same as in original model: the units which have maximum contacts with solvent molecules are declared to be of P-type (polar), while the units which mainly contact with other monomer units are converted to the H-type. In this way we obtain a macromolecule with a new HP sequence. For this macromolecule, we can again perform globular folding induced by the attraction between H units in the new sequence, again perform recoloring, obtain new HP sequence, etc. Following this procedure, the protein-like HP sequences will undergo some evolution. The question is: whether this evolution leads to the increase of complexity or we will end up with some trivial sequence? We have shown[15] that the answer to this question depends on the interrelation between H-H, H-P and P-P interaction constants. For some parameters, the sequences become more complicated and long-range correlations more pronounced (model of the ascending branch of the evolution), while for the other parameters the degree of complexity decreases and we come to rather trivial HP block copolymer (model of descending branch of the evolution).

REFERENCES

1. Lifshitz I.M., Grosberg A.Yu., Khokhlov A.R. Some Problems of the Statistical Physics of Polymer Chains with Volume Interactions. Rev Mod Phys 1978; 50:683-713
2. Grosberg, A.Yu., Khokhlov, A.R., *Statistical Physics of Macromolecules*. New York: American Institute of Physics, 1994.
3. Khokhlov A.R., Khalatur P.G. Protein-like copolymers: Computer simulation. Physica *A* **249** (1998) 253-261.
4. Khalatur P.G., Ivanov V.I., Shusharina N.P., Khokhlov A.R., Protein-like Copolymers: Computer Simulation. Russ Chem Bull 1998; 47:855-860
5. Khokhlov, A.R., Ivanov, V.A., Shusharina, N.P., Khalatur, P.G. "Engineering of Synthetic Copolymers: Protein-Like Copolymers." In *The Physics of Complex Liquids,* Yonezawa F., Tsuji K., Kaij K., Doi M., Fujiwara T., Eds., Singapore: World Scientific, 1998.
6. Khokhlov A.R., Khalatur P.G. Conformation-Dependent Sequence Design (Engineering) of A B Copolymers. Phys Rev Lett 1999; 82:3456-3459
7. Ivanov V.A., Chertovich A.V., Lazutin A.A., Shusharina N.P., Khalatur P.G., Khokhlov A.R. Computer simulation of globules with microstructure. Macromol Symposia 1999; 146:259-270
8. Chertovich A.V., Ivanov V.A., Lazutin A.A., Khokhlov A.R. Sequence design of biomimetic copolymers: modeling of membrane proteins and globular proteins with active enzymatic center. Macromol Symposia 2000; 160:41-48
9. Virtanen J., Baron C., Tenhu H. Grafting of Poly(N-isopropyl acrylamide) with Poly(ethylene oxide) under Various Reaction Conditions. Macromolecules 2000; 33:336-341; Virtanen J., Tenhu H. Thermal properties of Poly(N-isopropyl acrylamide)-g-poly(ethylene oxide) in aqueous solutions: influence of the number and distribution of grafts. Macromolecules 2000; 33:5970-5975; Virtanen J., Lemmetyinen H., Tenhu H.

Fluorescence and EPR studies on the collapse of poly(N -isopropyl acrylamide)-g - poly(ethylene oxide) in water. Polymer 2001; 42:9487-9493
10. Lozinsky V.I., Simenel I.A., Kurskaya E.A., Kulakova V.K., Grinberg V.Ya., Dubovik A.S., Galaev I.Yu., Mattiasson B., Khokhlov A.R. Synthesis and Properties of "Protein-Like" Copolymer. Doklady Chemistry 2000; 375:273-276
11. Khokhlov A.R., Ivanov V.A., Chertovich A.V., Lazutin A.A., Khalatur P.G. "Conformation-Dependent Sequence Design of Copolymers: Example of Bio-Evolution Mimetics Approach." In *Structure and Dynamics of Confined Polymers*, Kasianowicz J.J., Kellermayer M.S.Z., Deamer D.W., eds. Kluwer Academic Publishers, 2002.
12. Govorun E.N., Ivanov V.A., Khokhlov A.R., Khalatur P.G., Borovinsky A.L., Grosberg A.Yu. Primary Sequences of Proteinlike Copolymers: Levy-Flight-Type Long-Range Correlations. Phys Rev E 2001; 64:040903-1-4
13. Zheligovskaya E.A., Khalatur P.G., Khokhlov A.R. Properties of AB Copolymers with a Special Adsorption-Tuned Primary Structure. Phys Rev E 1999; 59:3071-3078
14. Wahlund P.-O., Galaev I.Yu., Kazakov S.A., Lozinsky V.I., Mattiasson B. "Protein-Like" Copolymers: Effect of Polymer Architecture on Performance in Bioseparation Process. Macromol Biosci 2002; 2:33-42
15. Khalatur P.G., Novikov V.V., Khokhlov A.R. Conformation-Dependent Evolution of Copolymer Sequences. Phys Rev E 2003; 67:051901-1-10
16. Chertovich A.V., Ivanov V.A., Zavin B.G., Khokhlov A.R. Conformational-Dependent Sequence Design of HP Copolymers: An Algorithm Based on Sequential Modificatios of Monomer Units. Macromol Theory Simul 2002; 11:751-756
17. Van den Oever J.M.P., Leermakers F.A.M., Fleer G.J., Ivanov V.A., Shusharina N.P., Khokhlov A.R., Khalatur P.G. Coil-globule transition for regular, random, and specially designed copolymers: Monte Carlo simulation and self-consistent field theory. Phys Rev E 2002; 65:041708-1-13
18. Berezkin A.V., Khalatur P.G., Khokhlov A.R. Computer Modeling Of Synthesis Of Proteinlike Copolymer Via Copolymerization With Simultaneous Globule Formation. J Chem Phys 2003; 118:8049-8060
19. Velichko Y.S., Khalatur P.G., Khokhlov A.R. Molecular Dispenser: Conformation–Dependent Design Approach. Macromolecules 2003; 36:5047-5050
20. Khalatur P.G., Khokhlov A.R., Mologin D.A., Reineker P. Aggregation and Counterion Condensation in Solution of Charged Protein-like Copolymers: A Molecular Dynamics Study. J Chem Phys 2003; 119:1232-1247

HIGH PERFORMANCE ENGINEERING THERMOPLASTICS VIA REACTIVE COMPATIBILIZATION

DONALD R. PAUL

Department of Chemical Engineering and Texas Materials Institute, University of Texas at Austin, Austin, Texas 78712 USA

Abstract: The performance of engineering thermoplastic polymers often must be modified by additives to meet demanding requirements on stiffness, strength, toughness, heat resistance, etc. The dispersion of these additives, like rubbers, rigid polymers, glass fibers, etc., often needs to be facilitated by *in situ* chemical reactions during processing. Reactive compatibilization is a powerful tool for increasing the strength of the interface and for controlling and stabilizing blend morphology which is critical for outstanding performance.

Key words: Engineering thermoplastic, reactive processing, compatibilization, high mechanical performance, melt compunding.

1. INTRODUCTION

Engineering thermoplastics (ETP) are high performance polymers that can be fabricated into useful shapes by melt processing techniques like extrusion, injection molding, etc. Engineering polymers combine high strength and other mechanical performance with light-weight characteristics that allow these high-value, specialty polymers to replace metals in certain applications. In the automotive industry, replacement of metal parts with ETPs is an effective strategy for reduced fuel consumption. They are also widely used in electronic/electrical systems, consumer appliances, medical products, etc. because of their resistance to heat, corrosion and chemicals;

F. Ciardelli and S. Penczek (eds.),
Modification and Blending of Synthetic and Natural Macromolecules, 293–315.

their electrical properties; and their design flexibility and ease of fabrication. Some of the most common ETPs include various polyamides, polycarbonate (PC), polyesters, and poly(phenylene ether) (PPE). The world consumption of these particular materials was 2.8 billion kilograms in 2000.

Despite the inherent attributes of ETPs, it is often necessary to incorporate various kinds of additives (other plastics, rubbers, fillers, pigments, etc.) to improve performance or appearance[1-3]. It is frequently necessary to employ some type of "compatibilizer" in order to successfully compound additives into polymer matrices. The objective here is to review the use of *in situ* chemical reactions during melt compounding (or processing) to effect this compatibilization, i.e., "reactive compatibilization," for ETPs[4].

2. BASIC CONCEPTS

The additives used to modify engineering plastics can be divided into two types. The first are solids with fixed shapes, determined by their fabrication process, e.g., glass fibers, pigment particles, carbon nanotubes, etc. The challenge for this class of modifiers is to obtain adequate dispersion of these particles of fixed shape within the ETP matrix and adhesion between the filler and matrix. The second is polymers that become fluid during processing such that the size and shape of the dispersed phase are determined by the melt compounding step. The challenge in this case is to achieve the desired morphology, to stabilize this morphology against subsequent change, and to obtain some adequate level of interfacial adhesion. In both cases, interfacial interventions may be needed to achieve the desired results of compounding.

In the extreme case, melt compounding of two immiscible molten polymers has some similarities to mixing oil and water. The size of the dispersed phase represents a balance between the competing processes of drop break-up and coalescence as illustrated in Figure 1.

Break-up of large droplets into smaller ones can occur when the stress on a drop imposed by the shear field (stress = $\eta_m \gamma$) is great enough to overcome the interfacial resistance (stress = σ/R) to deformation of the drop. Break-up occurs by an instability mechanism which was elegantly analyzed by Taylor[5] and applied by others[6].

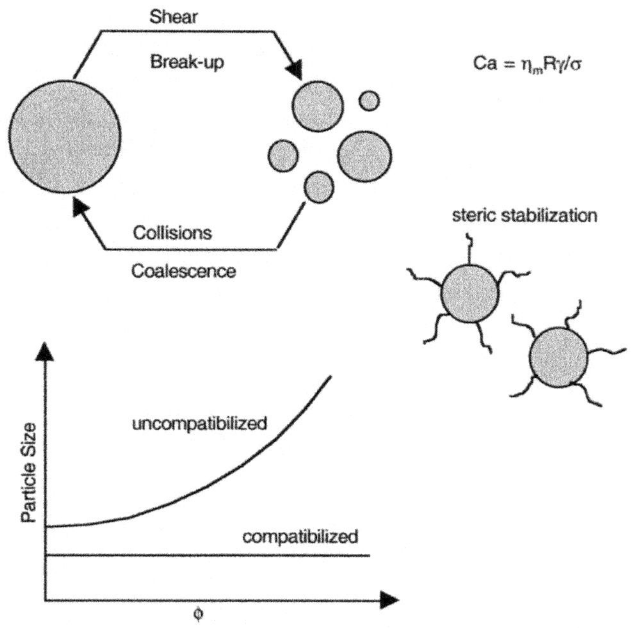

Figure 1. Mechanisms of morphology generation during compounding of polymer blends.

A drop of radius R is unstable when the capillary number (Ca) exceeds a given value that is a function of the ratio of the viscosity of dispersed phase (η_d) to that of the matrix phase (η_m), i.e.,

$$Ca = \frac{\eta_m R \gamma}{\sigma} > F(\eta_d / \eta_m) \quad (1)$$

where γ = shear rate and σ = interfacial tension. The compounding devices, like twin screw extruders, needed to mix high viscosity melts involve much more complex flow fields than envisioned by equation (1); furthermore, the rheology of the components involves both non-Newtonian and viscoelastic character. Thus, theories like that of Taylor only provide a qualitative understanding of the issues of break-up in melt compounding. Nevertheless, we see that high shear rates, low interfacial tension, high matrix viscosity, and having η_m and η_d well matched facilitate break-up. Coalescence of two particles may occur as they approach one another in a flow field and involves a complex set of hydrodynamic issues which have received a great deal of attention[7-9]. Coalescence is more frequent the higher the probability that one particle will encounter another, e.g., as concentration of the dispersed phase increases.

Some of what classic colloid science[10] teaches about mixtures of oil and water has application to dispersing one polymer (or a filler) in another. Addition of surfactant molecules which migrate to the oil-water interface lowers the interfacial tension and improves the stability of the dispersion, or emulsion. Block and graft copolymers, which self-assemble into microdomains because of the immiscibility of the different segments, are the polymeric equivalents of surfactants. If properly chosen, they can migrate to the interface in a blend of two immiscible polymers as suggested in Figure 2.

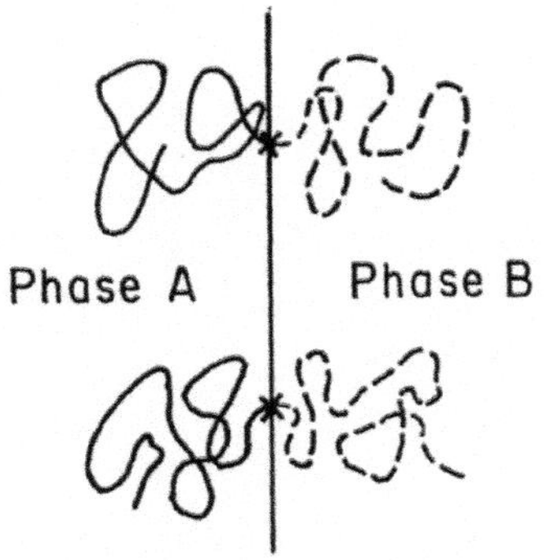

Figure 2. Idealization of how a block copolymer molecule locates at the interface between polymers A and B

The copolymer segments must be chemically identical with or miscible with the chains comprising the respective phases on either side of the interface[11,12]. Just as surfactants do, such copolymers can reduce the interfacial tension which makes break-up easier. More importantly, attaching polymer chains to the drop surface like hairs (see Figure 1) strongly repels two particles as they approach each other and greatly diminishes the frequency of coalescence[4,10]; this is known in colloid science as "steric stabilization." As a result, addition of copolymer compatibilizers can greatly reduce the size of the dispersed phase particles by slightly reducing the resistance to break-up but sharply decreasing the frequency of coalescence. For an uncompatibilized blend, the size of the dispersed phase increases significantly with the volume fraction (ϕ) of this phase[13] as suggested earlier. Compatibilization can reduce particle size for all ϕ but especially at higher values because of the increased stability against coalescence; these effects are illustrated schematically in the lower part of Figure 1.

Stability is an important issue since the manufacturer desires a robust material whose morphology is not very dependent on processing history. Morphological stability can be conveniently tested in a batch or static mixer[14,15] as illustrated in Figure 3. The experiment shown is one where the speed of the mixing rotor goes from a high level to a low level and back to high.

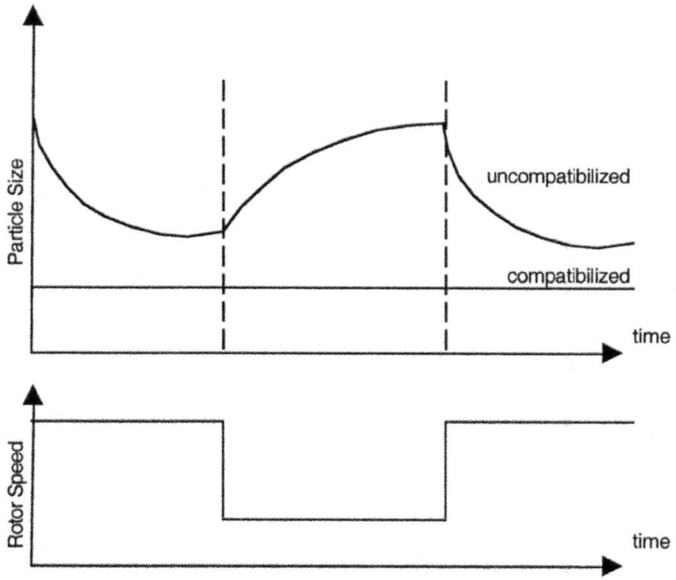

*Figure 3.*Use of a batch mixer to test the morphology stability of a two-phase polymer blend in the melt state.

The speed of the rotor (shear) affects the rate of drop break-up, and the size of the droplets in an uncompatibilized blend responds accordingly. At high speeds, the droplets become small, but at low speeds they become larger in time owing to coalescence. This illustrates the instability of the morphology. However, it has been shown for a well-compatibilized blend that not only are the dispersed particles smaller at high mixing speeds, but they also do not increase in size at low speeds. That is, the morphology is stable against coalescence.

Unlike surfactants at oil-water interfaces, block and graft copolymers can provide tremendous increases in the interfacial strength in the solid state owing to the chemical bonds traversing the interface. The extent of improvement in interfacial strength or toughness is a strong function of the number of chains traversing a unit area of the interface[16].

There are two ways to introduce block or graft copolymers into a blend. One is to preform the copolymer by an appropriate synthesis methodology and then mix it with the blend during compounding. This approach is used in some instances; however, it has proven to be of rather limited utility since chemical routes to block (or graft) copolymers generally will not yield the types of structures needed. Anionic polymerization is an excellent way to make block copolymers from styrene and dienes, but not the types of copolymers required for ETP systems[12]. A more practical route is to form these copolymers *in situ* by chemical reactions that occur during melt processing, see Figure 4.

3. CHEMISTRY OF REACTIVE COMPATIBILIZATION

Figure 4 suggests the basic chemical scheme of reactive compatibilization.

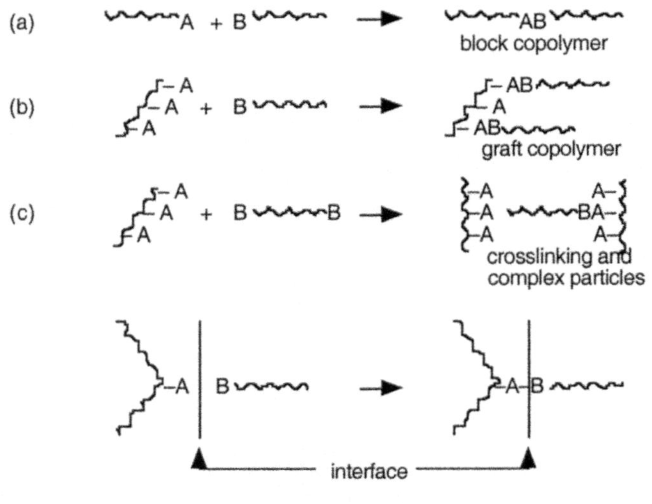

Figure 4. Schematic illustrations of various reaction topologies used in reactive compatibilization

One phase contains chains having functional groups A while the other phase contains chains having functional groups B. The functional groups A and B meet at the interface and react to form the desired copolymers. Of course, there are many important details that will determine whether this really works effectively or not. In what follows, some of the most important issues will be addressed in turn. Of basic importance is what are the functional groups A and B, how they are incorporated, and how readily they react.

Some engineering thermoplastics have functional groups at the end of the chains as a result of the chemistry by which they are made, e.g., polyamides have ~NH_2 and ~COOH ends while polyesters have ~OH and ~COOH ends but PC and PPE generally do not have reactive chain ends. Reactive compatibilization of polyamides generally takes advantage of its amine chain ends and the reaction that is possible with anhydrides[17], i.e.,

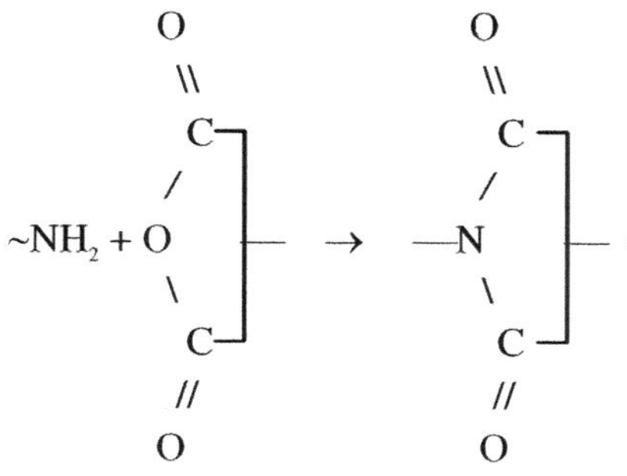

(2)

Anhydrides may be incorporated into chains of the other phase by copolymerization or grafting[17]. Carboxyl groups can react with the amine groups, but not nearly as rapidly as with the anhydride[18]. Other reactions are also possible and have been studied in some detail[19]. For polyesters, use is often made of the carboxyl group and its reaction with epoxide rings, i.e.,

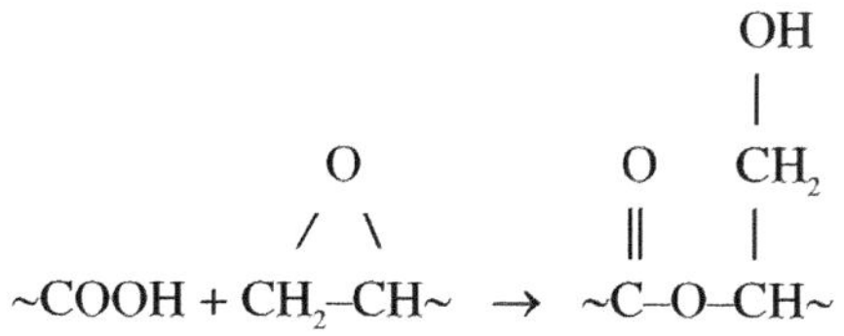

(3)

However, many other reactions are also possible[14,19,22]. Epoxide functionality can be introduced into chains of the other phase by copolymerization (or, less frequently, grafting) of monomers like glycidyl methacrylate[23]. Other functional groups are possible, but these are enough to explore some important issues.

First, how rapid does the reaction of A + B need to be? In typical compounding extruders, the residence time is 1 to 3 minutes of which a significant fraction involves heating to temperature, melting, and finally mixing to achieve particles with a high surface to volume ratio. Generally, the concentrations of the reactants are low (e.g., end groups or reactive comonomers that comprise no more than 1% of the phase volume), and finally these functional groups can only meet at an interface! The spatial and time opportunities for the A + B reaction are quite limited; thus, the intrinsic reaction rate must be very fast. It should be mentioned that the interface between two immiscible polymer phases is not a sharp boundary but rather a diffuse region the width of which also influences the opportunity for the A + B reaction. The initial thickness of the interface between these two phases depends on the thermodynamic interaction between the two polymer phases, and this greatly affects the extent of reaction achieved[24,25]. Of course, the high melt processing temperatures of the order of 300°C ± favor high rates of reaction. Clearly, experience teaches that the reaction shown in equation (2) is this fast. Similarly the reaction shown via equation (3) is also fast. Other reactions like that of an anhydride with either hydroxyl or carboxyl end groups do not seem to be fast enough. Similarly, in our hands, the potential reactions between oxazaline units and carboxyl groups seem ineffective.

Another chemical issue is what other reactions may be possible. For example, the anhydride unit may react with the amide linkages of polyamides in addition to the amine ends and the water produced from equation (2) may cause amide hydrolysis resulting in molecular weight degradation and regeneration of amines[19]. The epoxy group may undergo many more reactions than the one with carboxyl groups shown in equation (3). There is the potential for reaction with hydroxyl end groups; however, this does not seem to be as important as the one with carboxyl ends[14]. The presence of even trace quantities of strong nucleophiles can induce other reactions, including crosslinking, in addition to the desired grafting[21]. A straightforward reaction mechanism that leads only to the desired product rather than unpredictable side reactions works best. A major limitation currently for reactive compatibilization is the availability of practical means of incorporation of reactive functionalities that meet all the criteria mentioned.

Another important issue is the "topology" of the reaction possibilities. The reaction in part (a) of Figure 4 leads to a block copolymer which is an efficient compatibilizer; however, it is often difficult to achieve functional groups on only one end of polymer chains that meet all the requirements mentioned above. A more common situation is shown in part (b) of Figure 4. Here, there is only one reactive group B per chain; however, there may be multiple A groups on the other functional polymer. This leads to a graft copolymer which can be satisfactory if the number of A groups per chain is not too large; otherwise, some steric or asymmetry issues can arise[26].

Finally, we consider the case shown in part (c) of Figure 4. In addition to multiple A units on the chains in the phase on the left, some of the chains in the phase on the right contain two or more B groups. The latter arises quite naturally in some cases. PA6 and similar polyamides, as typically made, contain only one amine per chain; however, approximately one third of PA66 chains (and others of this type) contain two amines per chain. This leads to complex particle formation via crosslinking type reactions that occur[4,27,29]. In practical terms, this means that more intensive mixing is needed to achieve particles in a PA66 matrix like those that may be achieved more simply in a PA6 matrix where only end grafting occurs[30]. Most polyesters are of the type illustrated in part (c).

4. EXAMPLES OF MORPHOLOGY CONTROL

This section gives selected examples of how dispersed phase particle size can be significantly reduced via reactive compatibilization using the anhydride-amine reaction illustrated in equation (1). In these examples the amine units are the end groups of the chains forming the continuous phase, polyamide 6 (or nylon 6). In every case, a key variable is the amount of anhydride incorporated in the dispersed phase.

The first example is shown in Figure 5 which involves blends of various rubbers (20% by weight) with two different nylon 6 materials (80% by weight)[31].

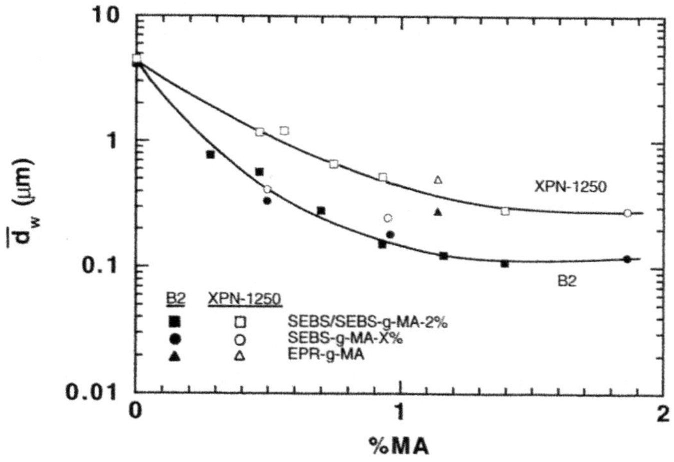

Figure 5. Weight average rubber particle size in nylon 6 blends versus amount of grafted maleic anhydride in the rubber phase; see text for description of materials. Ref.[31]. Reproduced with permission of John Wiley and Sons, Inc.

Each blend was prepared in a single screw extruder outfitted with an intensive mixing head; the extrudate was pelletized and then injection molded into test bars. Ultrathin specimens were cryomicrotomed from the test bars, stained with phosphotungstic acid and viewed by transmission electron microscopy (TEM). The weight average particle diameter was computed using image analysis techniques for each composition with the results shown in Figure 5. The rubbery materials include an ethylene/propylene copolymer to which 1.14% by weight of maleic anhydride was grafted, i.e., EPR-g-MA. The other elastomer was a styrene-butadiene-styrene triblock copolymer whose midblock was hydrogenated to resemble ethylene-butene sequences, SEBS, that were subsequently maleated to different extents, SEBS-g-MA. The maleation in the rubber phase was further varied by diluting an SEBS-g-MA containing 1.84% maleic anhydride with SEBS; the two form a single phase of rubber. The nylon 6 designated as B2 is a commercial product with $\overline{M}_n$ = 19,400 where each chain should have one amine and one carboxyl end. The nylon 6 designated as XPN-1250 is an experimental material with $\overline{M}_n$ = 18,300 but has more than four times as many amine chain ends as carboxyl ends resulting from adding hexamethylene diamine during polymerization of caprolactam. Thus, B2 represents case (b) in Figure 4 while XPN-1250 represents case (c). For both materials, there is a dramatic reduction in rubber particles size as the maleic anhydride content of the rubber phase increases; it does not seem to matter whether the %MA is achieved by maleating every chain, i.e. SEBS-g-MA-X%, or by mixing SEBS with SEBS-g-MA-2%. However, single ended grafting, case (b), clearly leads to smaller rubber particles than case (c) where many of the nylon 6 chains can graft onto the rubber from both ends; the latter leads to particles of complex shapes as described more fully elsewhere[4,27,29]. For materials like XPN-1250 or PA66, etc., production of smaller particle sizes requires use of more aggressive mixing via a properly designed twin screw extruder[30]. The scheme illustrated here is the basis for generating "super tough" polyamides.

The second example is shown in Figure 6 where the continuous phase contains 75% nylon 6 [case (b) with $\overline{M}_n$ = 22,000] and 25% of a dispersed phase consisting of a miscible mixture of a styrene/acrylonitrile copolymer (with 25% acrylonitrile, designated as SAN 25) and an imidized acrylic polymer, designated as IA-250-C, having about 1% by weight of glutaric anhydride in the backbone[32].

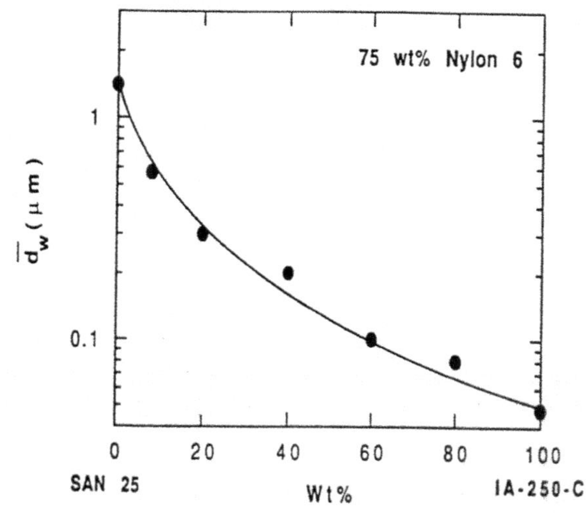

Figure 6. Weight average dispersed phase particle size in nylon 6 blends versus the amount of anhydride containing polymer in the miscible phase with SAN. Ref.[32]. Reproduced with permission of Elsevier Science Ltd.

Again, there is a dramatic reduction in the dispersed phase particle size (determined like that in the previous example) as the content of anhydride in this phase is increased. If SAN 25 and IA-250-C were not miscible, one might expect two populations of dispersed phase particles, small ones comprised of the imidized acrylic and large ones of SAN 25. This scheme is the basis for reactively compatibilizing commercially important blends of ABS materials with polyamides[33,43]; ABS is simply an SAN containing dispersed rubber particles.

5. RUBBER TOUGHENING

Addition of a rubber phase to an ETP like nylon 6 can lead to exceptional toughness, if proper compatibilization is used, as illustrated here[44,49]. Figure 7 shows that addition of EPR-g-MA (same as in Figure 5) to nylon 6 (same as Figure 6) leads to very large improvements in the room temperature Izod impact strength at high enough rubber content; however, as might be expected, there is a corresponding decrease in stiffness[50].

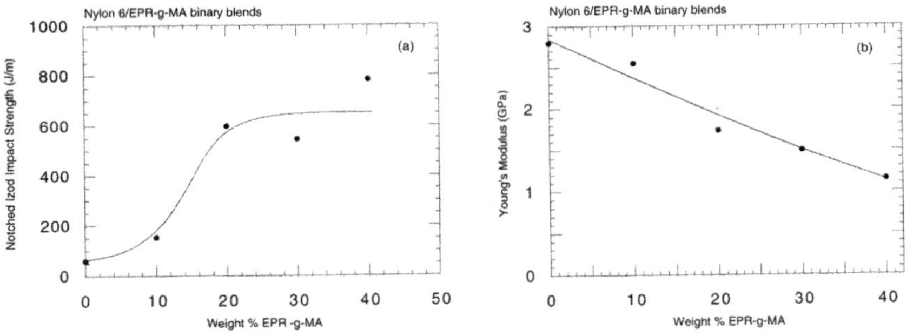

Figure 7. An example of how addition of a maleated elastomer to nylon 6 increases impact strength (part a) but decreases stiffness (part b). Ref.[50]. Reproduced with permission of Elsevier Science Ltd.

The extent of toughening achieved depends on the size of the rubber particles in addition to the amount of rubber as illustrated in Figures 8 and 9[31]. Figure 8 shows how the room temperature Izod impact strength depends on the size of the rubber particles; these are the same materials shown in Figure 5. These results demonstrate that superior toughness requires optimization of rubber particle size; the particles of rubber can be either too large or too small[45,47]. The slight differences between the B2 and XPN-1250 in Figure 8 may simply reflect differences in rubber particle shapes for these two matrices.

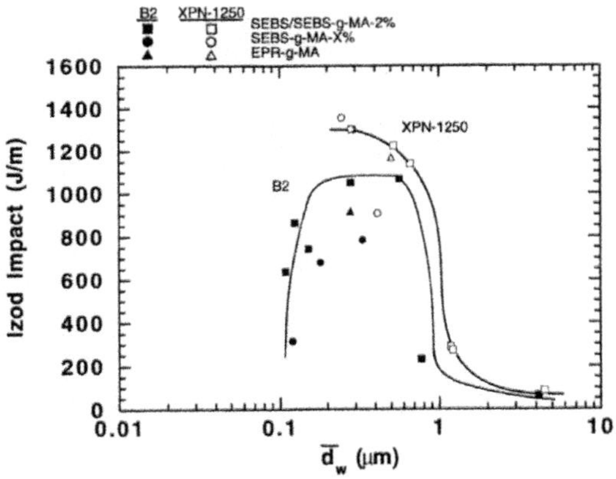

Figure 8. Effect of rubber particle size on room temperature impact strength for nylon 6; materials are the same as in Figure 5 Ref.[31]. Reproduced with permission of John Wiley and Sons, Inc.

Toughness, of course, depends on temperature, and Figure 9 shows how rubber particle size affects the so-called ductile-brittle transition temperature[49] for materials like those in Figure 7; these blends have excellent toughness at temperatures above this transition but are brittle below. Clearly smaller particles favor good low temperature toughness. The upper limit on particle size for room temperature toughness shown in Figure 8 simply reflects the fact that the ductile-brittle transition temperature is above room temperature. At very small particle sizes, the ductile-brittle transition temperature may reverse direction and once again exceed room temperature[47]. Ultimately the ductile-brittle transition temperature is bounded by the glass transition of the rubber since toughening by this mechanism requires the dispersed phase to have a low modulus[47].

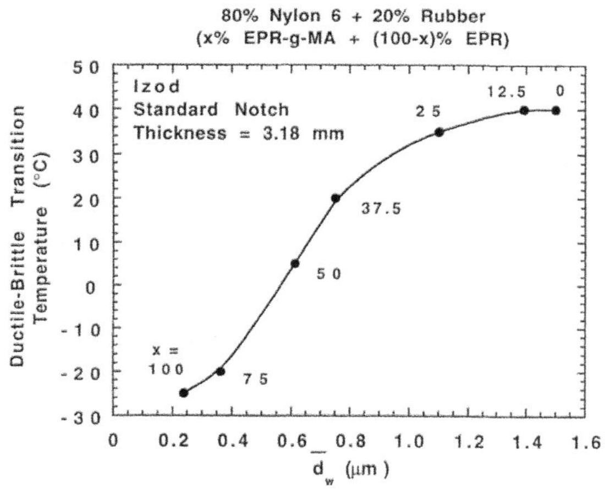

Figure 9 Effect of rubber particle size on ductile-brittle transition temperature for toughened nylon 6 . Ref.[49]. Reproduced with permission of Elsevier Science Ltd.

A great deal of literature has been devoted to understanding the cause for these upper and lower limits on particle size. It has been argued that the upper limit reflects a critical interparticle distance beyond which super toughness will not occur; however, this explanation has been somewhat controversial[2,3]. It is necessary to have a better understanding of the role of the rubber particles in the toughening mechanism to appreciate the trends with particle size[3]. The role of these particles can be to concentrate or to relieve stresses in the matrix, and for these pseudo-ductile ETP matrices stress relief is an especially important part of the toughening mechanism. Notches or cracks in the specimen create states of triaxial stress in the matrix that inhibit ductile yielding of the matrix during fracture tests[51]. At some point ahead of the propagating crack, the triaxial stress in the matrix may cause the rubber particles to cavitate which relieves the triaxial stress state by allowing the matrix to contract in the neutral direction and, thus, permits the matrix to undergo ductile deformation by shear yielding which absorbs a great deal of energy. The ability of the rubber to cavitate depends on its size in addition to its inherent mechanical properties; a semi-quantitative understanding of these size effects has been recently elucidated by theories developed by Bucknall et al[51].

Figure 10 illustrates synergistic toughening by blending nylon 6 with a high rubber

ABS material when appropriately compatibilized using the strategy suggested in Figure 6[33]. Without compatibilization, all of these blends are brittle at room temperature. For certain ratios of ABS and nylon 6 these blends are tougher than either material is by itself. Quite naturally, the ductile-brittle transition temperature decreases as the rubber (i.e., ABS

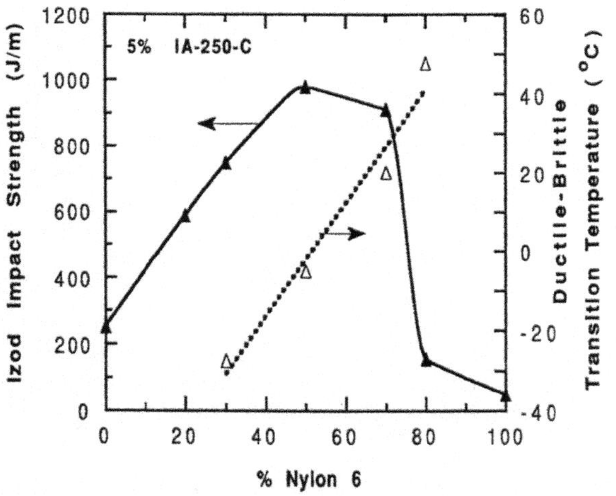

Figure 10.]Room temperature impact strength and ductile brittle transition for compatibilized nylon 6/ABS blends. Ref.[33]. Reproduced with permission of Elsevier Science Ltd.

component) content of the blend increases. The morphology of these blends is complex and cannot be described by a simple parameter like $\bar{d}_w$; however, it is clear that obtaining the proper morphology is key to toughness.

To this point, no mention has been made of the role that strengthening the rubber-matrix interface, via reactive compatibilization, may have on the toughening process. For the nylon 6/ABS blends, interfacial failure was noted in the post-mortem TEM analysis of failed specimens that were not compatibilized[43]. It has been suggested that adhesion is not essential to toughening and that even holes would be effective. However, as a practical matter, it has proved difficult, it not impossible, to make particles small enough for toughening without having good adhesion so no clear experimental assessment of this issue is available. The theories by Bucknall et al assume that the adhesion is good enough that cavities form in the interior of the rubber and not at the interface[51].

6. BLENDS WITH OTHER RIGID POLYMERS

Sometimes ETP materials are blended with very rigid glassy polymers, and often rubber is present as well; a good illustration of this are the commercial blends of polyamide/rubber/PPE[52]. The rubber provides toughness while the rigid polymer provides additional stiffness particularly at temperatures above the glass transition of the ETP. The latter is a well-known way of raising the "heat distortion temperature" or HDT as shown by Scobbo[52]. The HDT is essentially the temperature where the deflection of a specified part exceeds a critical value under load and corresponds to a certain value of modulus. For automotive applications, it is sometimes necessary to boost the HDT to avoid excessive deformation in the paint ovens. This could be accomplished by glass fiber fillers; however, this would deteriorate the surface finish. Adding a rigid polymer like PPE solves the HDT problem without sacrificing surface finish.

The hard polymer phase in ternary blends also provides an avenue of balancing toughness versus stiffness; indeed, rubber toughened nylon 6 can become both stiffer and tougher by adding a hard phase[50]. In simpler binary blends it has been shown that ordinarily brittle materials, like SAN 25, may deform in a ductile manner provided the particles are small enough; this is illustrated by data in Figure 11[40]. Again, it is clear that control of morphology is critical to performance. Of course, making the particles smaller in such cases requires more grafted chains across the interface, i.e., a stronger interface.

7. REINFORCEMENT WITH GLASS FIBERS

Glass fibers are often added to ETPs to increase stiffness and strength. In many cases, a form of reactive compatibilization is used to facilitate good dispersion of the glass fibers in the matrix polymer and to increase the interfacial strength.

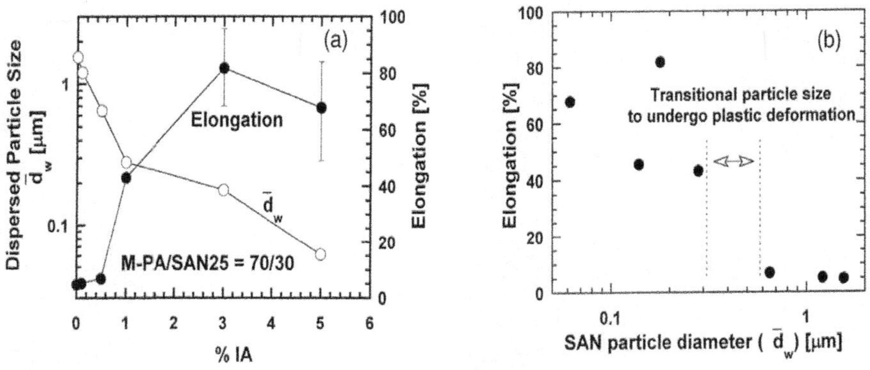

Figure 11. Demonstration that particles of a brittle polymer in a nylon 6 matrix undergo a brittle-to-ductile transition as the particle becomes smaller. Ref. [40]. Reproduced with permission of Elsevier Science Ltd.

Silane coupling agents can be effectively bonded to the surface of glass by well-known pretreatment techniques involving reactions of the silane with the hydroxyls on the glass surface[53]. The silane molecule may have pendant groups that can react chemically with the matrix polymer (amine, anhydride, epoxy, etc.) or not. For example, a silane with anhydride functionality, thus, provides an analogous pathway for dispersing and bonding glass fibers with a polyamide as discussed above for anhydride containing polymers.

There is growing interest in ETPs which are both reinforced, e.g., by glass fibers, and rubber toughened in order to obtain a balance of stiffness, strength, and toughness[53,56]. Until recently there was very little literature on how to optimize the dispersed rubber phase to achieve the best mechanical performance of glass fiber reinforced, rubber-toughened composites. Recent papers have focussed on how the reactive compatibilization chemistries for the two dispersed phases may interact and whether the optimum range of rubber particle size is the same with glass fibers as without[56].

The mechanical properties of nylon 6 materials containing maleated rubber and glass fibers with different silane coupling agents were recently explored[53]. The anhydride, epoxy, and amine functional silanes used are all capable of chemical reaction with the nylon 6 matrix. In the absence of rubber, there were no significant differences in the mechanical properties of the materials containing these three reactive coupling agents; however, significant differences were observed in the case of silane with only a nonreactive octyl group. The composite using the octyl containing agent has the lowest yield strength, elongation at break, and notched Izod impact strength, while the modulus is only slightly lower than that for the reactive coupling agents. Modulus is defined in the limit of zero strain and is not expected to be very sensitive to interfacial strength unless there is significant slippage at the fiber-matrix interface. In fact, models of composite stiffness do not generally include consideration of the nature of the interface.

Strength, on the other hand, is highly dependent on interfacial strength; therefore, the octyl silane produces the lowest strength since the fiber-matrix interface is weak. When there is good chemical bonding between the glass fibers and the polymer matrix, the effective interfacial strength becomes the shear strength of the matrix. Indeed, for each reactive coupling agent, failure occurs in the polymer matrix as opposed to the fiber-matrix interface, and mechanical properties are independent of the type of reactive silane coupling used. Since the interface is stronger than the matrix, failure occurs in the matrix in an identical manner for the reactive silane coupling agents. Thus, the mechanical properties of composites based on different reactive coupling agents are essentially similar in the absence of EPR-g-MA. For unreactive coupling agents, failure occurs at the interface and mechanical properties are dependent on the nature of the interface.

Materials reinforced with glass treated with reactive silanes provide large increases in strength relative to the pure nylon 6 or the rubber-toughened blend and very large decreases in ductility. Figures 12 and 13 compare the properties of reinforced, toughened nylon 6 composites using octyl versus anhydride silanes. The modulus is virtually the same for both silanes; however, the anhydride gives a much larger increase in strength (Figure 12). The glass treated with unreactive octyl silane provides a similar increase in stiffness but a smaller Izod impact strength[53].

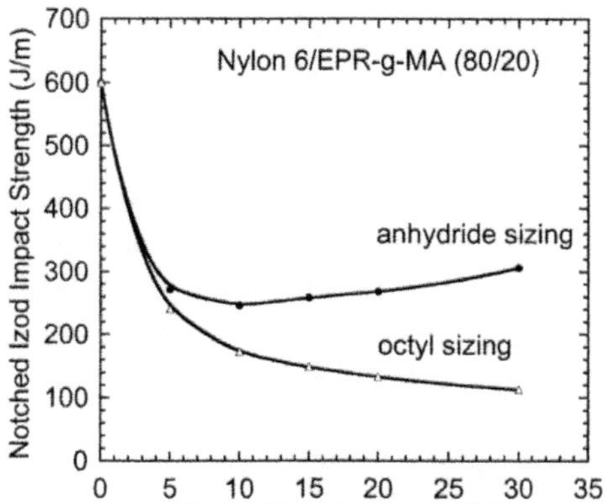

Figure 12. Effect of glass fiber sizing agent on strength of composites with a toughened nylon 6 matrix. Ref. [53]. Reproduced with permission of Elsevier Science Ltd.

Previous work has shown that loss in ductility is the main cause of decreased toughness in glass fiber reinforced, rubber-toughened polyamides compared to their unreinforced counterparts. The octyl silane was chosen to act as a release agent, so that the ductile matrix material would not be as tightly bound to the brittle fiber reinforcement with the hope of improving composite ductility at the expense of strength.

However, no improvement in ductility was observed so these composites show no improvement in toughness. The Izod impact strength of the rubber-toughened material decreases as glass fibers are added in all cases; however, the decline is more severe for the octyl sizing. For the anhydride sizing, the fracture energy goes through a minimum and then increases again as the fibers raise the strength as seen in Figure 13.

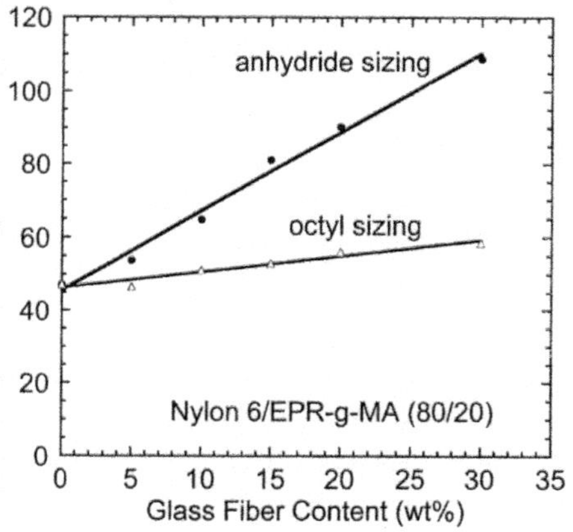

Figure 13. Effect of glass fiber sizing agent on impact strength of composites with a toughened nylon 6 matrix. Ref.[53]. Reproduced with permission of Elsevier Science Ltd.

Rubber particle size can be varied over a wide range by varying the ratio of maleated to unmaleated rubber for either the EPR or SEBS systems as shown earlier. The rubber particles are slightly smaller when glass fibers are compounded in the blend than without owing to rheological effects[56]. Figure 14a shows the room temperature notched Izod impact strength of the nylon 6/EPR/EPR-g-MA system versus the weight average rubber particle. When no glass fibers are present, there is a maximum in fracture energy at a weight average particle size between 0.3 μm and 0.4 μm. For materials containing 15 wt% glass fibers with an anhydride surface treatment, there is no maximum; instead the fracture energy steadily increases as the rubber particle size becomes smaller. The glass fiber reinforced materials do not follow the same qualitative trend as the unreinforced materials, and there seems to be no correlation between the toughness of the unreinforced materials and those containing 15 wt% glass fibers. The relationship between notched Izod impact strength for the nylon 6/SEBS/SEBS-g-MA system and the weight average particle diameter is shown in Figure 14b. When no glass fibers are present, there is a strong maximum in toughness when the rubber phase contains SEBS/SEBS-g-MA in a (25/75) ratio which corresponds to a weight average particle diameter of 0.33 μm; this is in the same range of the maximum toughness for materials made with EPR/EPR-g-MA mixtures. The maximum notched Izod impact value of 950 J/m is much higher than any material made with EPR/EPR-g-MA mixtures; however, when 15 wt% glass fiber is added, the SEBS/SEBS-g-MA toughened materials have a lower fracture energy than the EPR/EPR-g-MA based materials of the same particle size. Unlike the glass fiber reinforced EPR/EPR-g-MA toughened materials, where fracture energy continually increases as the rubber particles became smaller, the glass fiber reinforced SEBS/SEBS-g-MA toughened materials have a relatively constant value of notched Izod impact strength for

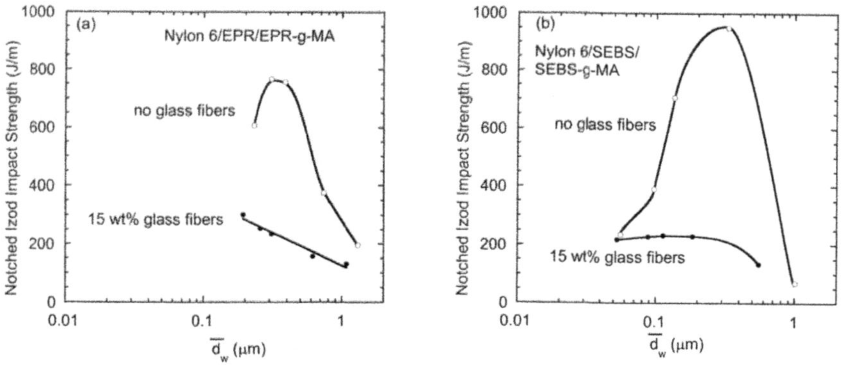

Figure 14. Comparison of the effect of rubber particle sizes effect on room temperature impact strength of nylon 6 blends with and without glass fibers for the EPR (part a) and SEBS (part b) systems. Ref.[56]. Reproduced with the permission of Elsevier Sciences Ltd.

compositions that contain maleated rubber. The unmaleated rubber has a coarse morphology and poor adhesion with the polyamide matrix and is expected to have a lower impact energy. In these reinforced materials, the type of rubber and the particle size clearly influence the toughness achieved and in different ways than seen in the absence of glass fibers[56].

8. SUMMARY AND FUTURE PROSPECTS

The purpose here has been to illustrate how interfacial reactions or compatibilization can be a useful technique for optimizing the effectiveness of a wide range of additives used to beneficially modify the properties of engineering thermoplastics. The principles have been illustrated for polyamide matrices since there is a larger body of information for this case; this is no doubt a reflection of the better developed chemistry for these materials. One of the challenges for the future is to develop equally effective approaches for a wider range of ETPs. Also, it is anticipated that other additives like carbon nanotubes, organoclays, and the like will be of increasing importance in the future and would benefit from well-designed interfacial reaction strategies for effectively compounding them in various ETP matrices.

REFERENCES

1. Miles, I.S. and Rostami, S., eds (1992) *Multicomponent Polymer Systems*, Longman Scientific and Technical, Essex.
2. Collyer, A.A., ed. (1994) *Rubber Toughened Engineering Plastics*, Chapman and Hall, London.
3. Paul, D.R. and Bucknall, C.B., eds. (2000) *Polymer Blends: Formulation and Performance*, Volumes 1 and 2, John Wiley, New York.
4. Majumdar, B. and Paul, D.R. (2000) Reactive compatibilization, Chapter 17 in D.R. Paul and C.B. Bucknall (eds), *Polymer Blends,* Volume 1: *Formulation*, John Wiley, New York.
5. Taylor, G.I. (1934) The formation of emulsions in definable fields of flow, *Proc. R. Soc. London Ser. A*. **146**, 501-523.
6. Han, C.D. (1981) *Multiphase Flow in Polymer Processing*, Academic, New York.
7. Chesters, A.K. (1991) The modelling of coalescence processes in fluid-liquid dispersions: a review of current understanding, *Trans. I. Chem. E*. **69**, Part A, 259-270.
8. Fortelny, I. and Zivny, A. (1998) Film drainage between droplets during their coalescence in quiescent polymer blends, *Polymer* **39**, 2669-2675.
9. Lyu, S.P., Bates, F.S., and Macosko, C.W. (2002) Modeling of coalescence in polymer blends, *AIChE J*. **48**, 7-14.
10. Hunter, R.J. (1989) *Foundation of Colloid Science*, Volumes I and II, Oxford Science Publishers, Oxford.
11. Paul, D.R. (1978) Interfacial agents ("compatibilizers") for polymer blends, Chapter 12 in D.R. Paul and S. Newman (eds.), *Polymer Blends*, Vol. 2, Academic Press, New York.
12. Paul, D.R. (1996) Polymer blends containing styrene/hydrogenated butadiene block copolymers: solubilization and compatibilization, Chapter 15C in G. Holden, N.R. Legge,

R. Quirk, and H. Schroeder (eds.), *Thermoplastic Elastomers: A Comprehensive Review*, 2nd Edition, Hanser Publishers, Munich.

13. Favis, B.D. (2000) Factors affecting the morphology of immiscible blends in melt processing, Chapter 16 in D.R. Paul and C.B. Bucknall (eds.), *Polymer Blends: Formulation and Performance*, John Wiley, New York.
14. Hale, W., Keskkula, H, and Paul, D.R. (1999) Compatibilization of PBT/ABS blends using glycidyl methacrylate/methyl methacrylate terpolymers, *Polymer* **40**, 365-377.
15. Wildes, G., Keskkula, H. and Paul, D.R. (1999) Coalescence in PC/SAN blends: effect of reactive compatibilization and matrix phase viscosity, *Polymer* **40**, 5609-5621.
16. Brown, H.R. (2000) Strengthening polymer-polymer interfaces, Chapter 23 in D.R. Paul and C.B. Bucknall (eds.), *Polymer Blends: Formulation and Performance*, John Wiley, New York.
17. Keskkula, H. and Paul, D.R. (1994) Toughening agents for engineering polymers, Chapter 5 in A.A. Collyer (ed.), *Rubber Toughened Engineering Plastics*, Chapman and Hall, London.
18. Lu, M., Keskkula, H. and Paul, D.R. (1994) Acrylic acid containing copolymers as reactive compatibilizers for toughening nylon 6, *Polym. Eng. Sci.*, **34**, 33-41.
19. Mzrechal, P., Coppens, G., Legras, R. and Dekoninck, J.M. (1995) Amine/anhydride versus amide/anhydride reactions in polyamide/anhydride carriers, *J. Polym. Chem.: Part A: Polym. Chem.* **33**, 757-766.
20. Stewart, M.E., George, S.E., Miller, R.L. and Paul, D.R. (1993) Effect of catalyst on the reactive processing of polyesters with epoxy-functional polymers, *Polym. Eng. Sci.*, **33**, 675-685.
21. Hale, W., Keskkula, H. and Paul, D.R. (1999) Effect of crosslinking reactions and order of mixing on properties of compatibilized PBT/ABS blends," *Polymer* **40**, 3665-3676.
22. Martin, P., Devaux, J., Legras, R., van Furp, M., and van Duin, M. (2001) Competitive reactions during compatibilization of blends of poly(butylene terephthalate) with epoxide-containing rubber, *Polymer* **44**, 2463-2478.
23. Gan, P.P. and Paul, D.R. (1994) Interaction energies for blends based on glycidyl methacrylate copolymers, *Polymer* **35**, 3513-3524.
24. Majumdar, B., Keskkula, H. and Paul, D.R. (1994) Morphology development in toughened aliphatic polyamides," *Polymer* **35**, 1386-1398.
25. Paul, D.R. (1994) Effects of polymer-polymer interactions in multiphase blends or alloys, " *Makromol. Chem., Macromol. Symp.* **78**, 83-93.
26. Kitayama, N., Keskkula, H. and Paul, D.R. (2000) Reactive compatibilization of nylon 6/styrene-acrylonitrile copolymer blends: Part 2: Dispersed phase particle size, *Polymer* **41**, 8053-8060.
27. Oshinski, A.J., Keskkula, H. and Paul, D.R. (1992) Rubber toughening of polyamides with functionalized block copolymers: 1. Nylon-6," *Polymer* **33**, 268-283.
28. Oshinski, A.J., Keskkula, H. and Paul, D.R. (1992) Rubber toughening of polyamides with functionalized block copolymers: 2. Nylon-6,6, *Polymer* **33**, 284-293.

29. Takeda, Y., Keskkula, H. and Paul, D.R. (1992) Effect of polyamide functionality on the morphology and toughness of blends with a functionalized block copolymer, *Polymer* **33**, 3173-3181.
30. Majumdar, B., Keskkula, H. and Paul, D.R. (1994) Effect of extruder type on the properties and morphology of reactive blends based on polyamides, *J. Appl. Polym. Sci.*, **54**, 339-354.
31. Oshinski, A.J., Keskkula, H. and Paul, D.R. (1996) The effect of polyamide end-group configuration on morphology and toughness of blends with maleated elastomers, *J. Appl. Polym. Sci.,* **61**, 623-640.
32. Majumdar, B., Keskkula, H., Paul, D.R. and Harvey, N.G. (1994) Control of the morphology of polyamide/styrene-acrylonitrile copolymer blends via reactive compatibilizers, *Polymer* **35**, 4263-4279.
33. Majumdar, B., Keskkula, H. and Paul, D.R. (1994) Mechanical properties and morphology of nylon-6/acrylonitrile-butadiene-styrene blends compatibilized with imidized acrylic polymers, *Polymer* **35**, 5453-5467.
34. Majumdar, B., Keskkula, H. and Paul, D.R. (1994) Effects of the nature of the polyamide on the properties and morphology of compatibilized nylon/acrylonitrile-butadiene-styrene blends, *Polymer* **35**, 5468-5477.
35. Kudva, R.A., Keskkula, H. and Paul, D.R. (1998) Compatibilization of nylon 6/ABS blends using glycidyl methacrylate/methyl methacrylate copolymers, *Polymer* **39**, 2447-2460.
36. Kudva, R.A., Keskkula, H. and Paul, D.R. (2000) Properties of compatibilized nylon 6/ABS blends: Part 1: Effect of ABS type, *Polymer* **41**, 225-237.
37. Kudva, R.A., Keskkula, H. and Paul, D.R. (2000) Properties of compatibilized nylon 6/ABS blends: Part II: Effects of compatibilizer type and processing history, *Polymer* **41**, 239-258.
38. Kudva, R.A., Keskkula, H. and Paul, D.R. (2000) Fracture behavior of nylon 6/ABS blends compatibilized with an imidized acrylic polymer, *Polymer* **41**, 335-349.
39. Kitayama, N., Keskkula, H. and Paul, D.R. (2000) Reactive compatibilization of nylon 6/styrene-acrylonitrile copolymer blends: Part 1: Phase inversion behavior, *Polymer* **41**, 8041-8052.
40. Kitayama, N., Keskkula, H. and Paul, D.R. (2001) Reactive compatibilization of nylon 6/styrene-acrylonitrile copolymer blends: Part 3 Tensile stress-strain behavior, *Polymer* **42**, 3751-3759.
41. Majumdar, B., Keskkula, H. and Paul, D.R. (1994) Morphology of nylon 6/ABS blends compatibilized by a styrene/maleic anhydride copolymer, *Polymer* **35**, 3164 -3172.
42. Paul, D.R. (1994) Phase coupling and morphology generation in engineering polymer alloys," in C. L. Choy and F. G. Shin, (eds.), *Proceedings of the International Symposium on Polymer Alloys and Composites*, Hong Kong Polytechnic, pp. 21-37.
43. Majumdar, B., Keskkula, H. and Paul, D.R. (1994) Deformation mechanisms in nylon 6/ABS blends, *J. Polym. Sci.: Part B: Polym. Phys.* **32**, 2127-2133.

44. Keskkula, H. and Paul, D.R. (1995) Toughened nylons, Chapter 11.6 in Melvin I. Kohan, (ed.), *Nylon Plastics Handbook*, Hanser Publishers, Munich.
45. Oshinski, A.J., Keskkula, H. and Paul, D.R. (1996) The role of matrix molecular weight in rubber toughened nylon 6 blends: Part 1 - Morphology, *Polymer* **37**, 4891-4907.
46. Oshinski, A.J., Keskkula, H. and Paul, D.R. (1996) The role of matrix molecular weight in rubber toughened nylon 6 blends: Part 2 - Room temperature izod impact toughness, *Polymer* **37**, 4909-4918.
47. Oshinski, A.J., Keskkula, H. and Paul, D.R. (1996) The role of matrix molecular weight in rubber toughened nylon 6 blends: Part 3 - Ductile-brittle transition temperature, *Polymer* **37**, 4919-4928.
48. Kayano, Y., Keskkula, H. and Paul, D.R. (1997) Evaluation of the fracture behavior of nylon 6/SEBS-g-MA blends, *Polymer* **38**, 1885-1902.
49. Okada, O., Keskkula, H. and Paul, D.R. (2000) Fracture toughness of nylon 6 with maleated ethylene/propylene rubbers, *Polymer* **41**, 8061-8074.
50. Harada, T., Carone, E., Kudva, R., Keskkula, H. and Paul, D.R. (1999) Effect of adding an imidized acrylic polymer to super tough nylon 6 on stiffness and toughness, *Polymer* **40**, 3957-3969.
51. Bucknall, C.B. (2000) Deformation mechanisms in rubber-toughened polymers, Chapter 22 in D.R. Paul and C.B. Bucknall, (eds.), *Polymer Blends: Formulation and Performance*, Volumes 1 and 2, John Wiley, New York.
52. Scobbo, J.J. (2000) Thermomechanical performance of polymer blends, Chapter 29 in D.R. Paul and C.B. Bucknall, (eds.), *Polymer Blends: Formulation and Performance*, Volumes 1 and 2, John Wiley, New York.
53. Laura, D.M., Keskkula, H., Barlow, J.W. and Paul, D.R. (2002) Effect of glass fiber surface chemistry on the mechanical properties of glass fiber reinforced, rubber-toughened nylon 6, *Polymer* **43**, 4673-4687.
54. Laura, D.M., Keskkula, H., Barlow, J.W. and Paul, D.R. (2000) Effect of glass fiber and ethylene-propylene rubber content on tensile and impact properties of nylon 6, *Polymer* **41**, 7165-7174.
55. Laura, D.M., Keskkula, H., Barlow, J.W. and Paul, D.R. (2001) Effect of glass fiber and maleated ethylene-propylene rubber content on the impact fracture parameters of nylon 6, *Polymer* **42**, 6161-6172.
56. Laura, D.M., Keskkula, H., Barlow, J.W. and Paul, D.R. Effect of rubber particle size and rubber type on the mechanical properties of glass fiber reinforced, rubber-toughened nylon 6, *in press*.

CHARACTERIZATION OF COMPLEX POLYMER SYSTEMS BY FLUORESCENCE SPECTROSCOPY

STANISLAW SLOMKOWSKI
Center of Molecular and Macromolecular Studies, Polish Academy of Sciences, Sienkiewicza 112, 90-363 Lodz, Poland

Abstract: This chapter provides basic background of fluorescence spectroscopy with attention focused on methods useful for studies of macromolecules. Subsequently discussed are selected properties of linear, star and branched macromolecules (in particular dendrimers) determined by using these methods. Presented are results of studies of dynamics of linear chains, diffusion of small molecules inside of star and/or branched macromolecules and segmental mobility in dendrimers.

Key words: Fluorescence spectroscopy, end-to-end cyclization, diffusion into branched macromolecules, segmental mobility in dendrimers

1. INTRODUCTION

Return of an electron from excited state to the ground state in a molecule is often accompanied by emission of a photon.

Electron in the higher energy orbital may have either opposite or the same spin as second electron in the lower orbital. In the first case (opposite spins) the excited state is a singlet state in the second case (the same spins) the excited state is a triplet. In the ground state electrons occupying the same orbital have opposite spins. According to quantum mechanics transitions between singlet states are "allowed" whereas the triplet-singlet transitions are "forbidden". Emission of light accompanying singlet-singlet transition

F. Ciardelli and S. Penczek (eds.),
Modification and Blending of Synthetic and Natural Macromolecules, 317–335.

occurs usually within time periods ranging from ca 1 to 10 ns and is called fluorescence. However, there are known also compounds emitting light much longer, e.g. pyrene for which the fluorescence lifetime may exceed even 200 ns. Emission resulting from transition between triplet and singlet states requires much longer time (from miliseconds to seconds) and is called phosphorescence. Processes described above are conveniently illustrated by the diagram proposed by A. Jablonski[1]

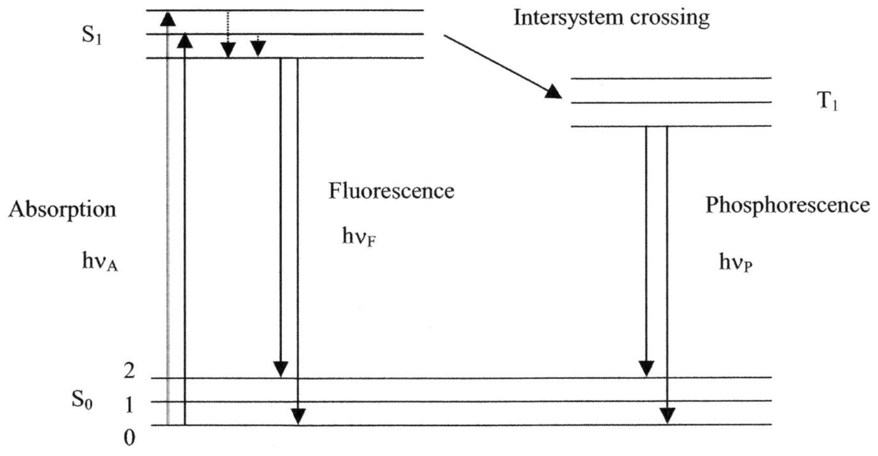

Figure 1. Jablonski diagram.

According to Jablonski diagram absorption of light occurs from the lowest (0) vibrational level. After excitation to various vibrational levels of the electronic excited S_1 state a rapid (within ca 1 ps) relaxation occurs to the lowest vibrational level of excited state. Thus, whereas absorption reflects distribution of vibrational levels in the excited state, fluorescence and phosphorescence characterize distribution of vibrational levels in the ground electronic state. Positions of electronic and vibrational levels depend not only on the structure of excited molecules but also on their interactions with environment. Thus, spectra of emitted light may be used for characterization of these interactions.

Absorption of exciting polarized light depends on mutual orientation of a chromophore and light polarization vector. A component of an electric vector of the light wave must be parallel to transition moment of the chromophore. Similarly, the direction of an electric vector of emitted light is determined by orientation of the light emitting chromophore. For experiments in which a sample is excited with a plane polarized light the following terms have been introduced:

a) Polarization P

$$P = \frac{I_{II} - I_{\perp}}{I_{II} + I_{\perp}} \qquad (1)$$

b) Emission anisotropy r

$$r = \frac{I_{II} - I_{\perp}}{I_{II} + 2I_{\perp}} \qquad (2)$$

In equations (1) and (2) I_{II} and $I_{\perp}$ denote intensities of emitted light polarized parallel and perpendicular to polarization vector of the exciting light. Any relative changes of polarization of absorbed and emitted light depend on rotation of a chromophore within a time period between light absorption and emission.

As it has been mentioned above characteristic times for fluorescence emission are in the nanosecond range. In pulse experiments a sequence of excitation followed by registration of emission signal could be repeated after ca few milisecond delay needed for the system (sample and electronics of the apparatus) to return to its original state. Obviously, each set of measurements could be repeated after minutes or even hours. Thus, structure of fluorescence spectra and fluorescence decay times could provide important information on changes in the environment of fluorophores in the extremely large time range.

Fluorescence emission is investigated by using spectrometers allowing monitoring emission spectra at the steady-state conditions and/or instruments with pulsed excitation allowing measurements of the fluorescence lifetime.

In the first type apparatus a sample is continuously illuminated with light. Emission spectra are obtained by illuminating samples with light with a constant selected wavelength and recording intensity of emitted light at various wavelengths. The so called excitation spectra are obtained by varying wavelength of exciting light and registering intensity of emitted light at a selected constant wavelength. Fluorescence lifetimes are usually determined using instruments in which sample is illuminated either with a lamp pulse (ca 2 ns) or with a laser pulse (ca 100 ps long). After pulse the intensity of emitted light is measured using photomultipliers linked to electronic set-ups designed for an analog or a single photon-counting detection. In the analog mode detection the incoming photons induces current averaged over a short time period characterizing the photomultiplier response time. In the single-photon counting mode measurements are based on determination of time between the excitation pulse and the first recording of emitted photon. Each experiment consists of thousands elementary measurement described above. Dependence of the total number of photons registered within the time period $\tau + \Delta\tau$ is plotted as a function of τ. This relation reflects the rate of deactivation of the excited state.

All processes resulting in deactivation of an excited molecule by bimolecular processes are caller quenching. Some of them require special attention because subsequent part of this paper will illustrate their application for studies of polymer systems. Quenching processes include:

a) Energy transfer – a process in which energy is transferred from the excited molecule (A^*) to another one (B).

$$A^* + B \rightarrow A + B^* \tag{3}$$

Energy of excited state of molecule B should be lower or equal to the energy of excited state of A. In a case when B is chemically identical with A the process is called energy migration.

b) Excimer formation – a process during which two chemically identical chromophores (A), that do not form complexes in the ground state, form a complex when one of them is in the excited state (A^*).

$$A^* + A \rightarrow (AA)^* \tag{4}$$

c) Exciplex formation – a process during which two chemically different chromophores (A and B), that do not form complexes in the ground state, form a complex when one of them is in the excited state (A^*).

$$A^* + B \rightarrow (AB)^* \tag{5}$$

The mentioned above processes may occur only when the excited chromophore is close enough to another one in the ground state. The maximum interaction distance depends on the nature of the process. For example, excimers are formed at a distance ca 4 Å, a distance for exciplex formation is within the range from 4 Å to 15 Å, and for energy transfer by dipole-dipole interactions the characteristic distance is in the range from

10 Å to 100 Å[2]. Thus, the occurrence of any of the mentioned above events provides information on the local mobility of chromophores (during lifetime of the chromophore in its excited state another chromophore has to approach the excited one at a distance equal or shorter than the interaction distance). For example, formation of excimers (exciplexes) between chromophores separated by few bonds along a polymer chain can be used for determination of the local chain dynamics, whereas fluorescence quenching (formation of excimers or exciplexes) by interaction between chromophores located far away along the chain provides information on global chain dynamics. In principle, various conformational changes of polymer chains affecting inter-chromophore distances could be investigated by energy transfer, excimer and/or exciplex formation.

There was published a very large number of papers on application of fluorescence methods for studies of polymers. This short review presents selected data on studies of conformations and dynamics of linear, star, hyperbranched and dendritic macromolecules.

2. CHAIN DYNAMICS OF LINEAR MACROMOLECULES

For many years polymer chemists and physicists concentrated their attention on properties of linear macromolecules. Important questions were related to conformations of linear chains in solution. Beginning from the early paper by Kuhn[3] and later studies by Jacobson and Stockmayer[4] and Flory and Semlyen[5] a theory allowing calculations of the probability that ends of linear macromolecule are in a close contact has been developed and used for explanation of the equilibrium distribution of cyclic oligomers formed in polymerization processes[6]. Subsequently Wilemski and Fixman[7], Perico and Cuniberti[8,9] and Doi[10,12] developed a theory of cyclization dynamics. In 1980-ties Winnik experimentally verified this theory in a series of papers on linear polystyrenes and dimethylsiloxanes labeled at chain ends with fluorescent groups (or with a fluorophore at one end and a quencher at another one)[13,17]. In his studies he used well characterized polymers with a narrow moleculer weight distribution labeled with pyrene (Py) – a fluorophore with long lifetime in the exited state (lifetime for isolated, excited pyrene substituted at 1 position is about 200 ns) able to form excimers. For polymers labeled with pyrene at both ends the end-to-end approach at a distance ca 4 Å leads to formation of an excimer. Kinetic scheme of this process is illustrated in Figure 2.

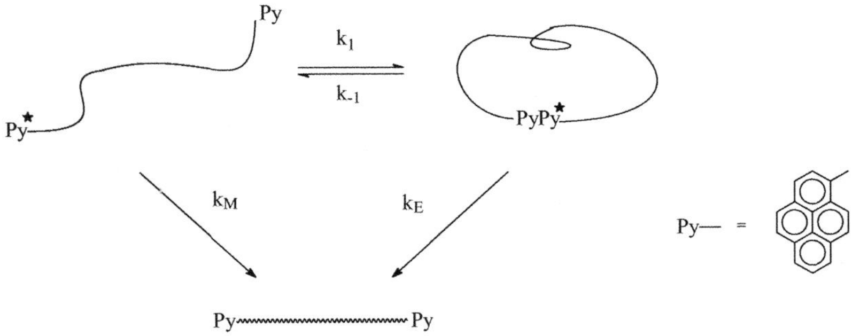

Figure 2. Scheme illustrating formation of $(PyPy)^*$ excimer as result of chain cyclization; k_1, k_{-1}, k_M, and k_E denote rate constants of cyclization, excimer dissociation, reciprocal lifetime of isolated Py in excited state and reciprocal lifetime of the excimer.

The typical emission spectra of polystyrene labeled with pyrene at both ends are shown in Figure 3. In these spectra the emission with fine vibronic structure and the the first intense band at 380 nm is due to the "isolated" (not interacting with other chrompohores) pyrene in the excited state (Py^*, often called the excited monomer) whereas the broad, structureless signal with a maximum at 480 nm correspond to emission by the $(PyPy)^*$ excimer. Studies of the kinetics of the decay of emission from the "isolated" pyrene and of the kinetics of emission build-up and decay from the pyrene excimer allowed determination of the cyclization rate constant. Measurement performed in good and θ solvents for polymer samples with various molecular weight (with very narrow moleculer weight distribution) allowed determination of relation between cyclization rate constant (k_1) and polymer chain length and its comparison with theoretical predictions. For polystyrenes in θ solvent k_1 was proportional to $n^{-1.52}$, in good solvent it was proportional to $n^{-1.7}$ and these relations were close to those found theoretically[16,18].

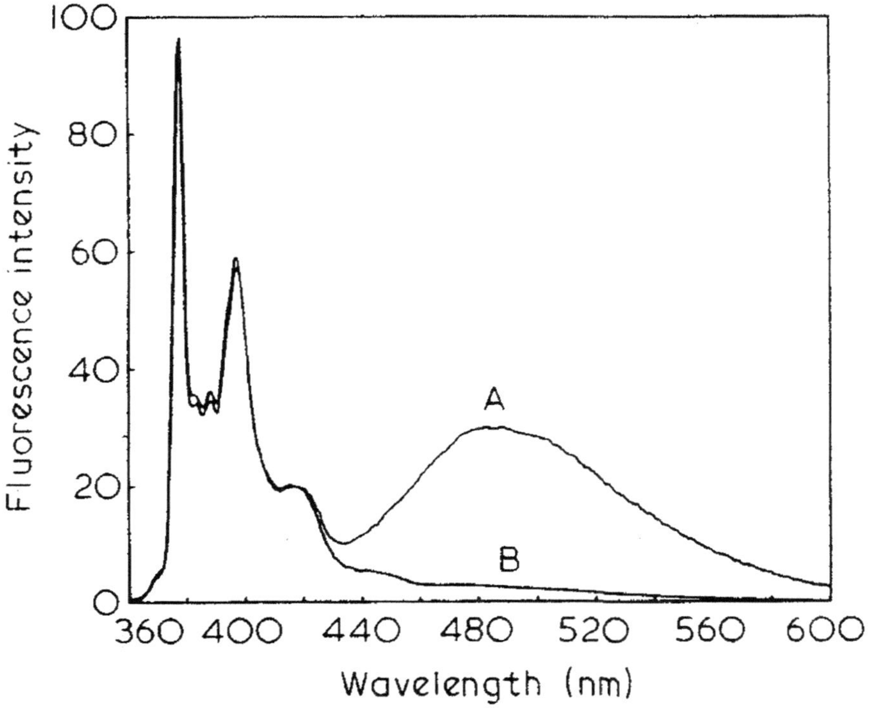

Figure 3. Fluorescence spectra of Py-polystyrene-Py ($2 \cdot 10^{-6}$ mol/L) in toluene at 22 °C; A $\overline{M}_n = 2{,}900$ and B $\overline{M}_n = 15{,}600$ (from Ref. 16).

3. DIFFUSION OF SMALL MOLECULES INTO HYPERBRANCHED MOLECULES

A large number of papers has been published on linear polymers labeled with fluorescent labels (often pyrene) along the chains (see e.g. the review by F.M.Winnik[19] and some later papers[20,21]). In these studies the changes in contribution of monomer and excimer to emission spectra resulting from variation of solvent, pH and ionic strength were used as a measure of polymer coil expansion (or collapse).

Recently, polymer chemists and physicists focused their attention on hyperbranched and dendritic molecules. There are hopes that macromolecules with this architecture will find important applications as carriers of catalysts, drugs, precursors used for formation of ordered nanoparticle arrays and others. Moreover, properties of molecules with the mentioned above geometry are much less understood than properties of the linear ones. In studies of branched macromolecules fluorescence spectroscopy has been used as a convenient tool. Due to the mentioned above possible applications of dendrimers and hyperbranched macromolecules the most important questions are related to segmental mobility inside of these structures and to diffusion of small molecules into and out of them.

Gauthier investigated properties of hyperbranched polystyrenes formed according to the strategy shown in Figure 4 by subsequent chloromethylation and reaction with living polystyrene chains[21].

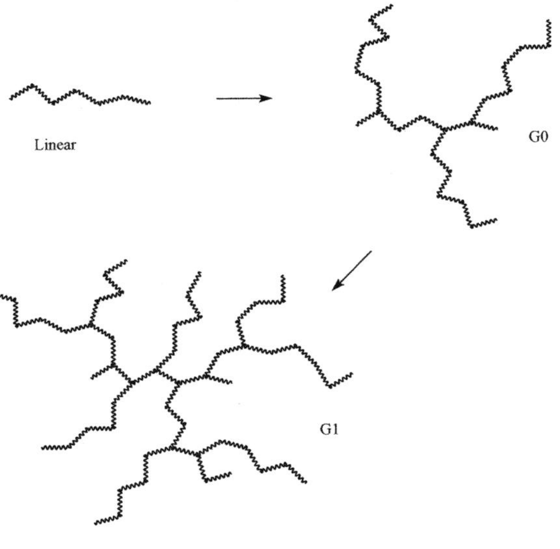

Figure 4. Strategy for generation of highly branched polystyrene macromolecules

These molecules were randomly labeled with pyrene (by chloromethylation and reaction with 1-pyrenemethnol)[22]. Quenching of pyrene emission with nitromethane revealed that diffusion of the quencher in a solution of hyperbranched polystyrene is ca four times slower than in the case of solution of linear chains. For example, in solution of linear polystyrene (polymer concentration ca 0.8 g/L) the diffusion constant of nitromethane was $5.2 \cdot 10^{-5}$ cm^2/s whereas in solution of hyperbranched polystyrene (G2, generation 2) the diffusion constant of nitromethane was equal $1.2 \cdot 10^{-5}$ cm^2/s. Differences in quenching were even more strongly pronounced when nitrated polystyrene was used as a quenchers.

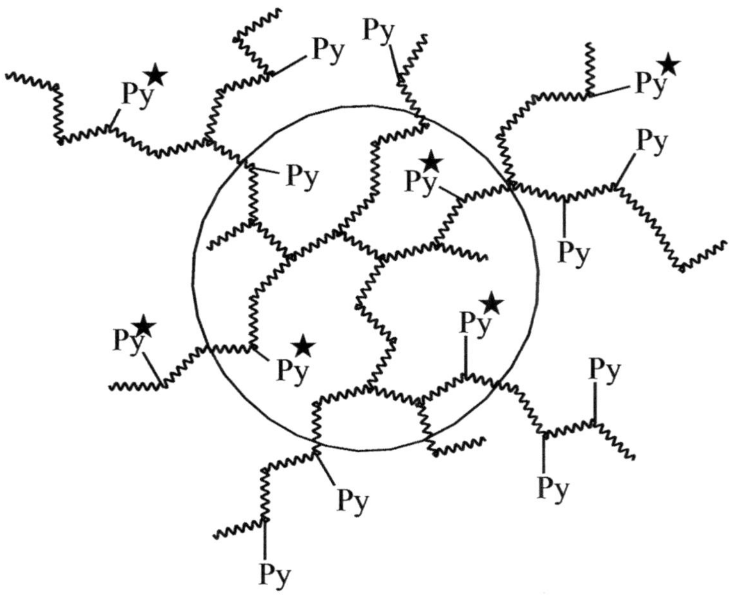

Figure 5. Schematic illustration of pyrene labeled hyperbranched polystyrene with central part not accessible for linear polystyrene chains bearing $-NO_2$ quenchers.

Accessibility of excited pyrene groups on linear and hyperbranched molecules for nitrated polystyrene macromolecules did vary as indicated in Table 1.

Table 1. Fraction of accessible pyrene groups on linear and branched polystyrenes for linear nitrated polystyrene ($\overline{M}_n = 1.2\cdot10^5$; 19 % of polystyrene monomeric units labeled with NO_2 groups) based on measurements of quenching of emission of pyrene moieties by NO_2 groups[22].

Sample	Fraction of accessible groups
Linear, 5.4 % Py, $\overline{M}_n = 4.4\cdot10^4$	0.73
G1, 1.3 % Py, $\overline{M}_n = 7.0\cdot10^5$	0.73
G2, 0.4 % Py, $\overline{M}_n = 1.3\cdot10^7$	0.57
G3, 0.9 % Py, $\overline{M}_n = 8.8\cdot10^7$	0.46

4. CONFORMATIONAL CHARACTERISTICS OF STAR POLYMERS BY FLUORESCENCE SPECTROSCOPY

Fluorescence spectroscopy has been successfully used not only for characterization of interactions of small molecules with cores of branched macromolecules but also for determination of interactions between the "peripheral" groups and the groups in the core of the same macromolecule. Interesting studies of such interactions were described recently by Frechet[24]. Authors investigated, among other, star macromolecules with porphyrin core and poly(ε-caprolactone) branches. Structures of these compounds are shown in Figures 6 and 7.

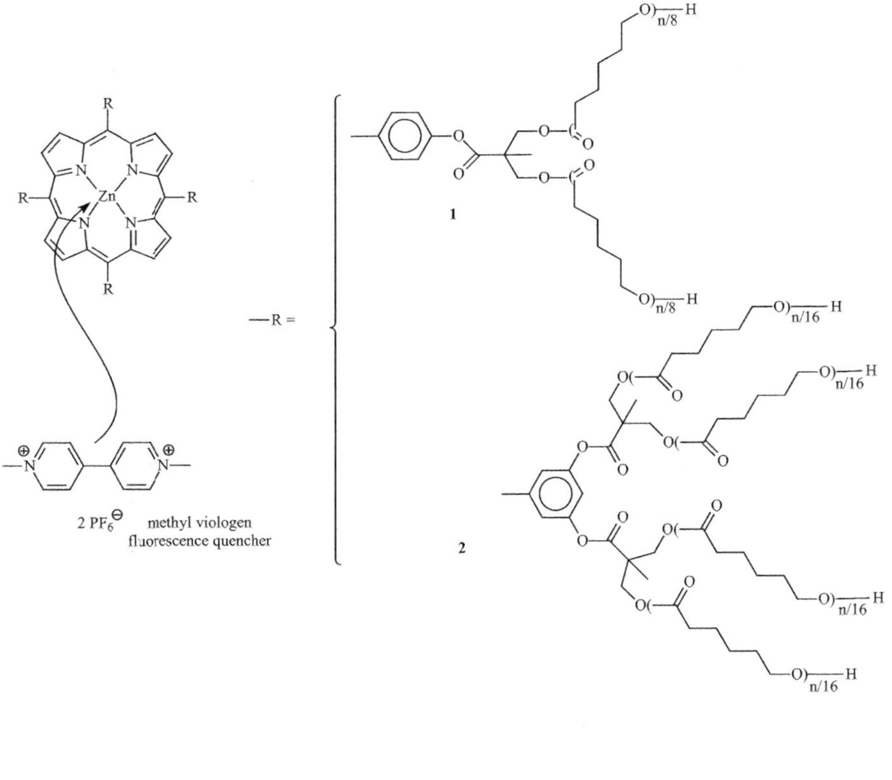

Figure 6. Star polymers with Zn porphyrin core and poly(ε-caprolactone) arms (8 or 16), n denotes the overall degree of polymerization of poly(ε-caprolactone) per one star macromolecule. Methyl viologen was used as quencher of Zn porphyrin emission.

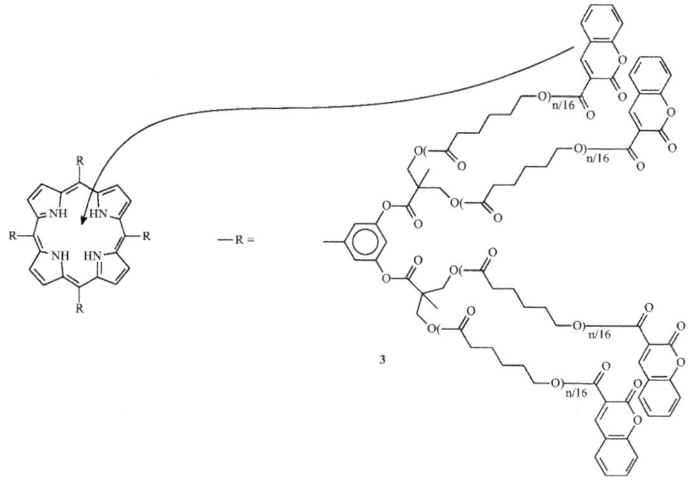

Figure 7. Star polymers with Zn porphyrin core and poly(ε-caprolactone) arms (16) with coumarin moieties at their ends, n denotes the overall degree of polymerization of poly(ε-caprolactone) per one star macromolecule.

In the case of compounds with structures 1 and 2 (see Figure 6) the intensity of porphyrin emission (λ_{max} = 605 and 659 nm) in presence and in absence of methyl viologen conformed to the standard Stern-Volmer equation,

$$\frac{F_0}{F} = 1 + k_q \tau_0 [MV] \tag{6}$$

In equation 6 F_0 and F denote intensity of fluorescence emission in absence and in presence of the quencher, τ_0 denotes the lifetime of fluorescence in the absence of the quencher, k_q the bimolecular collisional rate constant of quenching and [MV] concentration of methyl viologen. Slope in the plot of F_0/F as a function of [MV] allowed for determination of $k_q\tau_0$. The parameter $k_q\tau_0$ being proportional to k_q has been used for probing the diffusion of viologen to the core of investigated star macromolecules. Figure 8 shows the dependence of $k_q\tau_0$ on the degree of polymerization of poly(ε-caprolactone).

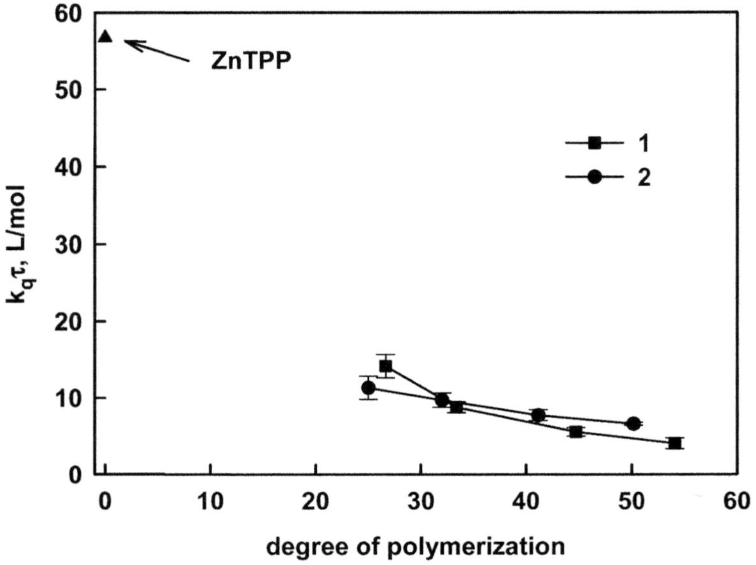

Figure 8. Florescence quenching for two series of zinc porphyrin star polymers with 8 (1) and 16 (2) arms (structures of compounds 1 and 2 are shown in Figure 8). Quenching of zinc tetraphorphyrin (ZnTPP) is used as the reference. Experiments were performed in acetonitrile using methyl viologen as a quencher. Ref.[23].

Comparison of quenching efficiency for zinc tetraphorphyrin and for star polymers revealed efficient shielding of the core by polymer chains strongly reducing penetration of methyl viologen into the macromolecule. Degree of polymerization has a significant influence on quenching efficiency. However, one has to note only a small difference between macromolecules with 8 and 16 arms. Apparently, in spite of the difference in number of arms segmental densities inside of investigated branched macromolecules are close.

In the case of star polymers with porphyrin core and poly(ε-caprolactone) arms labeled at their ends with coumarin the energy transfer from excited (at 350 nm) coumarin to porphyrin has been used for characterization of conformations of branched macromolecules in various solvents. Rate constant of energy transfer (k_T) depends on the distance between a donor (in investigated system coumarin) and acceptor (porphyrin) according to the following equation:

$$k_T = \frac{1}{\tau_d}\left(\frac{R_0}{r}\right)^6 \qquad (7)$$

in which τ_d denotes the rate of fluorescence decay in the absence of acceptor, r the distance between donor and acceptor and R_0 the so called Förster distance determined by spectral characteristics of the donor and acceptor (intensity of donor emission and extinction coefficient of acceptor in the range of wavelength at which emission and absorption do overlap) and on the refractive index of the medium. For investigated system (coumarin and porphyrin labels in $CHCl_3$ solvent) R_0 was found equal 8.6 nm.

Figure 9 shows the dependence of the energy transfer efficiency ($E = 1 - (F_{da}/F_d)$, where F_{da} and F_d denote the fluorescence yield in presence and in absence of an acceptor, respectively) on the degree of polymerization of poly(ε-caprolactone), for star polymers with porphyrin core and poly(ε-caprolactone) arms with coumarin moieties at their ends registered in good (toluene and chloroform) and bad (acetonitrile and DMSO) solvents.

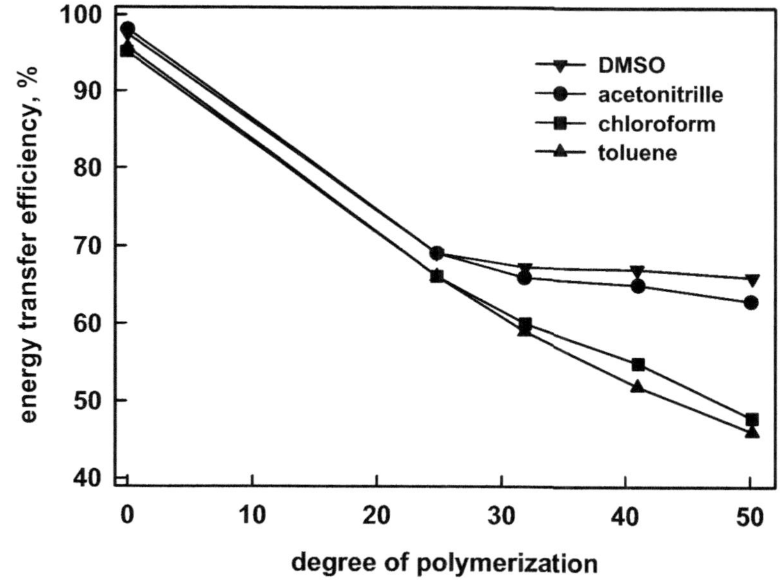

Figure 9. Resonance energy transfer from the terminal coumarin donor chromophores to the porphyrin core of star polymers with porphyrin core and poly(ε-caprolactone) arms (see Figure 7). Ref. [24].

According to Figure 9, in good solvents the energy transfer efficiency decreases gradually with increasing degree of polymerization indicating growing separation of the coumarin end groups and the porphyrin core. On the contrary, in bad solvents (at least for $\overline{DP}_n$ of poly(ε-caprolactone) in the range from 25 to 50) the energy transfer is almost independent from the degree of polymerization. This observation suggests collapsed conformation of branched macromolecules in which the frequency of encounters between the end groups and the core are almost constant.

5. SEGMENTAL MOBILITY IN DENDRIMERS

Fluorescence spectroscopy has been used for studies of relation between segmental mobility inside of dendrimer and solvent viscosity[25]. In these studies dendrimers of generation 3 labeled with pyrene containing labels at the generation 0 and a model compound with structure close to pyrene labels (see Figure 10) have been used (details of synthesis of these compounds are described in Ref. 25 and 26).

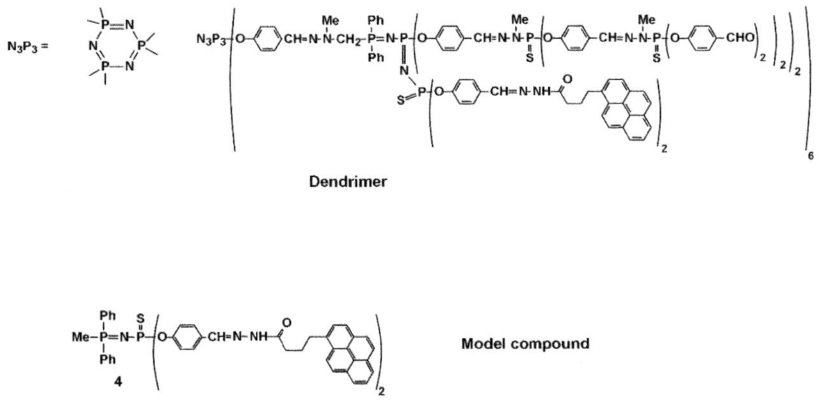

Figure 10. Dendrimer generation 3 labeled with pyrene containing label at generation 0 and a model compound.

Information on internal mobility in dendrimers could be obtained from studies of the rates of $(PyPy)^*$ excimer formation (see Figure 3 for examples of emission from systems with Py^* and $(PyPy)^*$). However, it is very well known that in some systems pyrene labels form dimers (PyPy) also in the ground state[19]. Formation of the ground state dimers results from geometrical constrictions and is facilitated by using solvents poorly solvating aromatic hydrocarbons. Ground state dimers are directly excited to $(PyPy)^*$ excimers and therefore, in the case of systems with ground state dimers, the excimmer emission cannot be used for probing segmental dynamics. Thus, before any further studies of dendrimers with pyrene labels were made it was necessary to check whether ground state dimers are formed in these structures. Formation of the ground state dimers (in dendrimers shown in Figure 10) has been excluded on the basis of the excitation spectra that were essentially identical when the intensities of monomer (at 376 nm) and excimer (at 480 nm) emissions were monitored. Identical excitation spectra indicated that isolated excited pyrenes and excimers were formed by excitation of the same species, the not dimerized pyrene groups.

The ratio of the intensities of excimer and monomer emissions (I_E/I_M; $\lambda_M = 378$ nm, $\lambda_E = 480$ nm) on solvent viscosity registered for the model and labeled dendrimer solutions is shown in Figure 11.

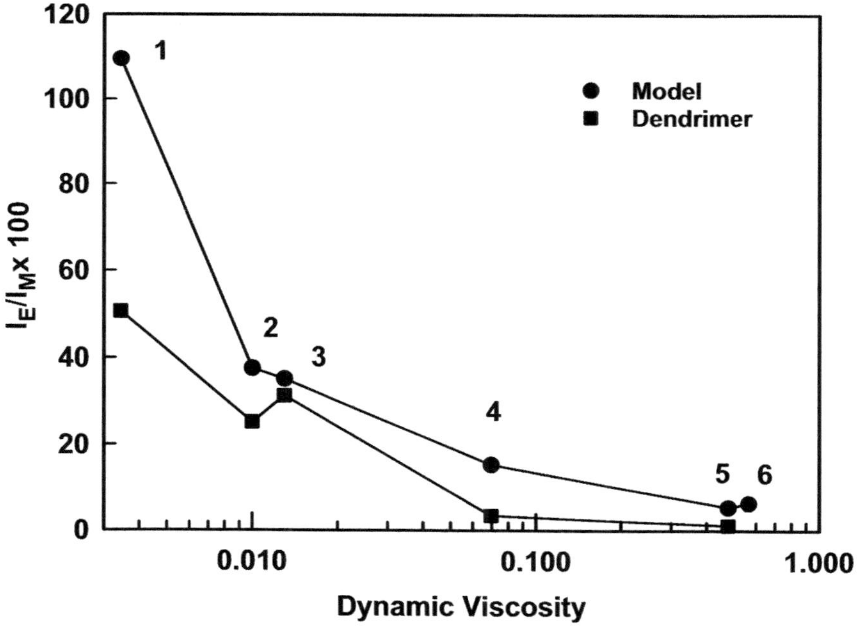

Figure 11. Dependence of the ratio of pyrene excimer and monomer emissions in model compound and in labeled dendrimers on solvent viscosity. Solvents: acetonitrile (1), diglyme (2), 1,4-dioxane (3), 1:1 v:v mixture of 1,4-dioxane and triethylene glycol (4), triethylene glycol (5), cyclohexanol (6). Ref. [26].

Ratio if I_E/I_M decreased significantly with an increasing solvent viscosity indicating that in more viscous media the excited pyrene chromophores more often emit photon before they are able to come close enough to another pyrene group in the ground state and form an excimer. It has to be mentioned also that with exception of acetonitrile for all other solvents values of I_E/I_M

for model compound and for dendrimer were very close. Close values of I_E/I_M for two systems suggested that the ratio of characteristic time constants for excimer and monomer decays was also close for these systems. Deeper insight into relations between mobility of pyrene labels in dendrimers dissolved in various solvents was obtained from decay curves.

Emission decays of pyrene monomer (I_M) and pyrene-pyrene excimers (I_E) can be described by following set of formulae derived by solving kinetic equations corresponding to schemes shown in Figure 2):

$$I_M = A_{M1}\exp(-t/\tau_1) + A_{M2}\exp(-t/\tau_2) \tag{8}$$

$$I_E = -A_{E1}\exp(-t/\tau_1) + A_{E2}\exp(-t/\tau_2) \tag{9}$$

$$\frac{1}{\tau_{1,2}} = \frac{k_1 + k_{-1} + k_M + k_E \pm \sqrt{(k_1 + k_M - k_{-1} - k_E)^2 - 4k_1k_{-1}}}{2} \tag{10}$$

In which k_1, k_{-1}, k_M, and k_E denote rate constants of excimer formation, excimer dissociation, reciprocal lifetime of isolated Py in excited state and reciprocal lifetime of excimer

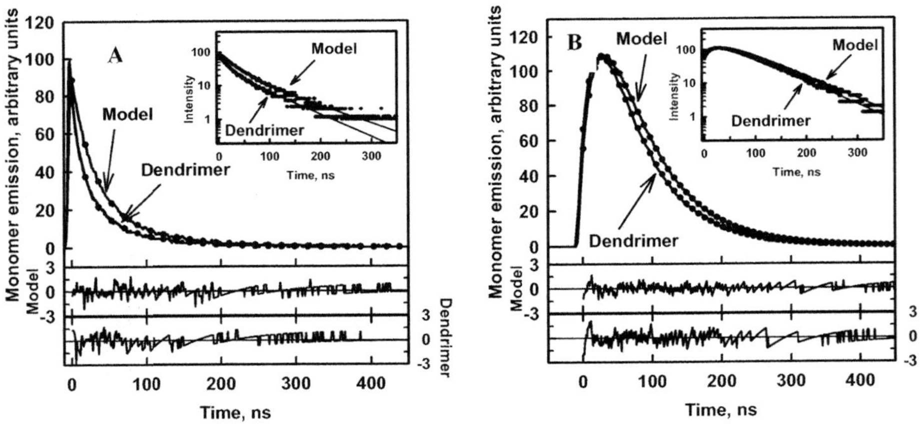

Figure 12. Decay curves of: A - pyrene monomer emission (at 378 nm), B - excimer emission (at 480 nm) in model compound and in labeled dendrimers. Solvent: 1,4-dioxane. Lines represent experimental traces. Points were calculated on the basis of fits to biexponential equation (for clarity only every tenth point is shown). (In inserts points are experimental)[26].

Values of τ_1, τ_2, and values of the ratio of pre-exponential factors for excimer decays (A_{E1}/A_{E2}) in various solvents are listed in Table 2.

Table 2. Values of the fluorescence emission decay time constants for the model and for the dendrimer in various solvents[25].

Solvent	Model			Dendrimer			$1/\tau_1$(dendrimer)
	A_{E2}/A_{E1}	τ_1 ns	τ_2 ns	A_{E2}/A_{E1}	τ_1 ns	τ_2 ns	$-1/\tau_1$(model) s^{-1}
Acetonitrile	0.998	7.1	74.1	0.984	6.2	61.3	$2.5\cdot10^7$
Diglyme	0.876	34.1	68.3	0.852	31.4	56.1	$2.5\cdot10^6$
1,4-Dioxane	0.998	25.3	68.8	1.07	21.8	66.4	$6.3\cdot10^6$
Triethyleneglycol	0.883	40.1	110.2	0.976	36.4	117.4	$2.5\cdot10^6$
Cyclohexanol	0.995	34.7	124.3	-	-	-	-

It has been found that for pyrene substituted at position 1 with propyl linker the rate constant of excimer dissociation to excited and nonexcited pyrene groups (k_{-1}) is very low at room temperature. For example, in the case of polystyrene molecules substituted at both ends with pyrene groups, in cyclohexane and in toluene, at room temperature, $1/\tau_1$ and $1/\tau_2$ could be approximated by putting in formula (10) k_{-1} equal zero. At this approximation $1/\tau_1 = k_1 + k_M$ and $1/\tau_2 = k_E$. Relation between time constant of excimer emission build-up (τ_1) and solvent viscosity is shown in Figure 13. For both, model and dendrimers, the increased solvent viscosity resulted in lower values of $1/\tau_1$, i.e. in lower values of $k_1 + k_M$.

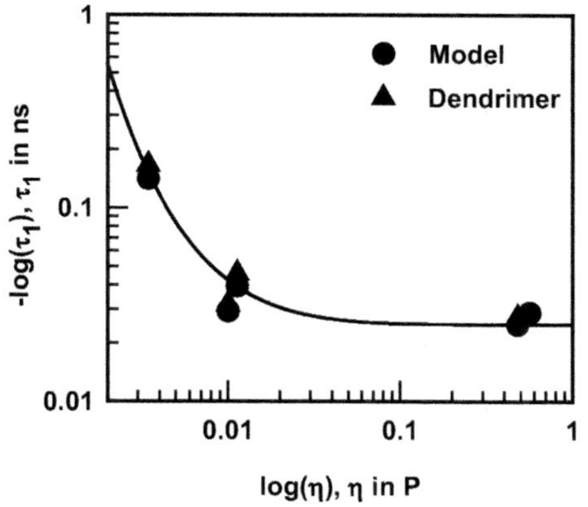

Figure 13. Dependence of $-\log(\tau_1)$ on solvent viscosity[26]

In solvents used for investigations the values of $k_E = 1/\tau_2$ (calculated from data in Table 1) are equal $(1.3 \pm 0.1)10^7$ 1/s and $(1.4 \pm 0.2)10^7$ 1/s, for model compound and for labeled dendrimers, respectively. This means that for the investigated systems k_E can be considered as a constant, for both the model compound and the dendrimer, regardless of the solvent. It is reasonable to assume that for the analyzed systems also k_M does not vary significantly. This assumption is supported by the relatively small differences $\Delta(1/\tau_1) = 1/\tau_1(\text{dendrimer}) - 1/\tau_1(\text{model}) = \Delta k_1(\text{dendrimer,model}) + \Delta k_M(\text{dendrimer,model})$ in each solvent constituting not more than 20% of $1/\tau_1$. Thus, for the large changes of $1/\tau_1$ in different solvents (ca 6-fold for acetonitrile and triethylene glycol) responsible are changes of k_1 reflecting changes in the mobility of pyrene labels.

Interesting conclusions were made on the basis of the studies of end-group dynamics of small (G(0), G(1) and G(2)) dendrimers labeled with pyrene at their exterior. Structures of these dendrimers are shown in Figure 14.

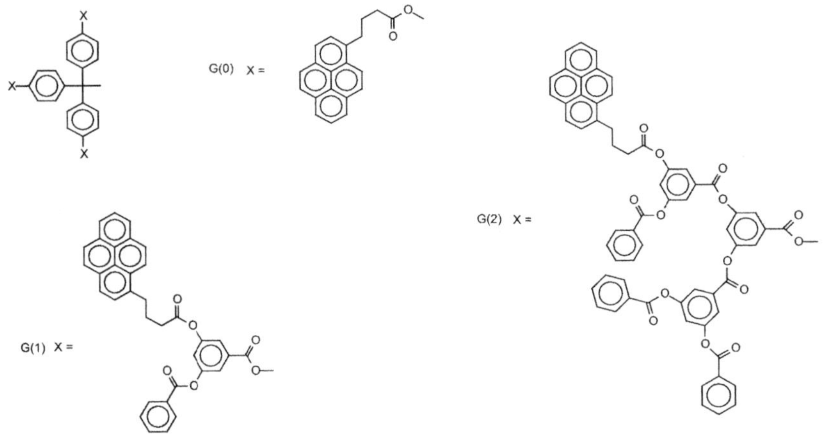

Figure 14. Structures of dendrimers labeled with pyrene at their exterior, one pyrene label per dendron.

In THF the rate constants of excimer formation increased for these dendrimers in the following order: $5.7 \cdot 10^6\ s^{-1}$ (G(0), $1.87 \cdot 10^7\ s^{-1}$ (G(1), $1.33 \cdot 10^8\ s^{-1}$ (G(2)[27]. This dependence has been explained by a conformational effect. Small molecules are rigid and number of conformations allowing formation of excimer in them is low. With increased length of arms the number of conformations suitable for excimer formation increases. However, one has to remember (see subsection on cyclization of linear polymers) that when a chain separating two pyrene chromophores is long enough and could be treated as randomly coiled the rates of excimer formation decrease with the increased chain lenght.

6. CONCLUSIONS

Fluorescence spectroscopy is an extremely useful method for characterization of conformations and mobility of linear, star and branched macromolecules, including dendrimers. Depending on the nature and location of fluorescent labels fluorescence spectroscopy provides information on the dynamisc of end-to-end cyclization of linear macromolecules, on diffusion of small and large molecules into branched structures and on internal mobility in dendrimers.

REFERENCES

1. Jablonski,A., Phys,Z., 94, 38 (1935).
2. Tazuke,S., Winnik,M.A., in Photophysical and Photochemical Tools in Polymer Science – Conformation, Dynamics, Morphology, M.A.Winnik (ed), NATO ASI Series, D.Reidel Publishing Company, Dordrecht, Holland, 1986, 15.
3. Kuhn,W.H., Kolloid Z. 68, 2 (1934).
4. Jacobson,H., Stockmayer,W.H., Chem.J., Phys. 16, 1600 (1950).
5. .Flory,P.J.,.Semlyen,J.A., J. Am. Chem. Soc. 88, 3209 (1966).
6. .Brown,J.F., Slusarczuk,G.M., J. Am. Chem. Soc., 87, 931 (1965).
7. .Wilemski,G., .Fixman,M., J. Chem. Phys. 60, 866 (1974).
8. .Cuniberti,C., .Perico,A., Prog. Polym. Sci. 10, 271 (1984).
9. .Perico,A., Cuniberti,C., J. Polym. Sci. Polym. Phys. Ed. 15, 1435 (1977).
10. .Doi,M., Chem. Phys. 9, 455 (1975).
11. .Sunagawa,S., .Doi,M., Polym. J. 7, 604 (1975).
12. .Sunagawa,S., .Doi,M., Polym. J. 8, 239 (1976).
13. .Winnik,M.A.,.Redpath,T., Richards,D.H. Macromolecules, 13, 328 (1980).
14. .Swirskaya,P., Danhelka,J., .Redpath,A.E.C., Winnik,M.A., Polymer, 24, 319 (1983).
15. .Redpath,A.E.C., Winnik,M.A., Polymer, 24, 1286 (1983).
16. Winnik,M.A., Redpath,A.E.C., Paton,K., .Danhelka,J., Polymer, 25, 91 (1984).
17. .Martinho,J M.G., Winnik,M.A., Macromolecules, 19, 2281 (1986).
18. Winnik,M.A., in Photophysical and Photochemical Tools in Polymer Science – Conformation, Dynamics, Morphology, M.A.Winnik (ed), NATO ASI Series, D.Reidel Publishing Company, Dordrecht, Holland, 1986, 293.
19. Winnik,F.M., Chem. Rev. 93, 587 (1993).
20. .Spafford,M., .Polozova,A., Winnik,F.M., Macromolecules, 31, 7099 (1998).
21. .Principi,T., Goh,C.C.E., .Liu,R.C.W.,.Winnik,F.M., Macromolecules, 33, 2958 (2000).
22. Gauthier,M., .Möller,M., Macromolecules, 24 4548 (1991).
23. Frank,R.S.,.Merkle,G.,.Gauthier,M., Macromolecules, 30, 5397 (1997).
24. .Hecht,S., Vladimitov,N.,.Frechet,J.M.J, J. Am. Chem. Soc. 123, 18 (2001).
25. Galliot,Ch., Larré,Ch., Caminade,A.M., .Majoral,J.P., Science, 227, 1981 (1997)
26. Brauge,L., Caminade,A.M., .Majoral,J.P.,Slomkowski,S.,Wolszczak,M., Macromolecules, 34, 5599 (2001).
27. Wilken,R., Adams,J., Macromol. ERapid Commun. 18, 659 (1997).

Index

GPSR Compliance

The European Union's (EU) General Product Safety Regulation (GPSR) is a set of rules that requires consumer products to be safe and our obligations to ensure this.

If you have any concerns about our products, you can contact us on ProductSafety@springernature.com

In case Publisher is established outside the EU, the EU authorized representative is:

Springer Nature Customer Service Center GmbH
Europaplatz 3
69115 Heidelberg, Germany

Batch number: 10755284

Printed by Printforce, the Netherlands